INTEGRATED MATRIX ANALYSIS OF STRUCTURES

Theory and Computation

INTEGRATED MATRIX ANALYSIS OF STRUCTURES

Theory and Computation

by

Mario Paz
Speed Scientific School
University of Louisville
Louisville, Kentucky

and

William Leigh
University of Central Florida
Orlando, Florida

SPRINGER SCIENCE+BUSINESS MEDIA, LLC

ISBN 978-1-4613-5640-0 ISBN 978-1-4615-1611-8 (eBook)
DOI 10.1007/978-1-4615-1611-8

Library of Congress Cataloging-in-Publication Data
Additional material to this book can be downloaded from http://extras.springer.com.
A C.I.P. Catalogue record for this book is available
from the Library of Congress.

Printed on acid-free paper.

To Annis

Contents

Preface

This is the first volume of a series of integrated textbooks for the analysis and design of structures. The series is projected to include a first volume in Matrix Structural Analysis to be followed by volumes in Structural Dynamics and Earthquake Engineering as well as other volumes dealing with specialized or advanced topics in the analysis and design of structures. An important objective in the preparation of these volumes is to integrate and unify the presentation using common notation, symbols and general format. Furthermore, all of these volumes will be using the same structural computer program, SAP2000, developed and maintained by Computers and Structures, Inc., Berkeley, California. SAP2000 is a commercially available program with a most powerful capability for the analysis and computer aided design of structures under static or dynamic loads that may result in elastic or inelastic behavior of the structure. The CD-ROM included in this volume contains the student version of SAP2000 with limitations in both the size and behavior of the structure. Even with these limitations, the student version of SAP2000 is extremely useful in introducing the student to a powerful structural program used by professional engineers world-wide. In particular, for this first volume on Matrix Structural Analysis, the only perceptible limitation of the student version is the size of the structure. The CD-ROM accompanying this book contains, in addition to the student version of SAP2000, various tutorials, example problems and a complete set of user manuals for SAP2000.

In order to maintain the continuity of the text, we present in the various chapters the development of mathematical formulas in a section on Analytical Problems, which is followed by another section containing Practice Problems. Throughout the book numerous Illustrative Examples are given with detailed numerical hand solutions as well as solutions obtained with SAP2000. In using SAP2000, the student is guided (as he or she becomes acquainted with the program) with detailed documentation of the commands implemented as well as input/output tables and graphs.

This book is organized into nine chapters. The first two chapters deal with the analysis of beams, serving to introduce and acquaint the reader with Matrix Analysis of Structures. The first chapter presents the analytical material for the analysis of beams using the matrix stiffness method and the second chapter provides the software documentation and commands in SAP2000 for the solution of a large number of Illustrative Examples. The next five chapters, Chapters 3 through 7, present the analysis of frame type structures; Plane Frames, Grid Frames, Space Frames, Plane Trusses and Space Trusses. The following chapter, Chapter 8, presents the special topic of substructuring which is a method by which the structure is separated into component parts to be analyzed separately, thus facilitating the analysis and design of a large or complex structure. The various component parts of the structure are separated by condensation of internal degrees of freedom. The final chapter of this book, Chapter 9, introduces the reader to the Finite Element Method of Analysis as a natural extension of matrix structural analysis. Both the analysis of Plane Elasticity Problems for plates with forces applied on their plane, as well as the analysis of plates with forces normal to the plane of the plates are presented.

Appendix I of the book provides the formulae of the Equivalent Nodal Forces for fixed end beams under some common loading. Appendix II contains the analytical expressions of the lateral displacements at fixed end beams with some common loading conditions. These analytical expressions are needed to evaluate the total lateral displacements in beams resulting from nodal displacements and from loads applied on the spans of a beam. Finally, please note that throughout the book certain key phrases are underlined. This denotes that these analytical expressions and analytical expressions of Matrix Structural Analysis are included in the Glossary that is provided for consultation and review.

Acknowledgements

In the preparation of this book and as the senior author, I am indebted to many people from whom I received valuable assistance. First, I wish to express my appreciation for the discussions and comments offered by my colleague Dr. Michael A. Cassaro, Civil Engineering Department of the University of Louisville, who helped me in refining the presentation of material in this book. I also would like to thank Dr. Michael Day, Department of Mechanical Engineering of the University of Louisville, for his assistance in comparing various commercial computer applications for structural analysis. I am also particularly grateful to Dr. Robert Sennett, professor at California Polytechnic State University, who most generously exchanged information with me on the presentation of several topics in the book. I wish to thank my friend Jack Bension for his professional help in editing selected sections of this book. I also wish to thank Professor Robert Matthew and his assistant David Haag of the Center for Professional and Continuing Education at the University of Louisville for their patient help in assisting me to become acquainted with the intricacies of various computer applications to properly incorporate figures with the text material of the book.

I am especially grateful to Dr. Ashraf Habibullah, president of Computers and Structures, Inc., who most kindly authorized me to include in this volume the student evaluation version of SAP2000. In addition, Dr. Habibullah provided me with the full version of SAP2000 so I could solve problems beyond the capability of the student version. I am also most grateful to Syed Hasanain and G. Robert Morris from Computers and Structures, Inc., who most patiently tutored me and clarified for me many of the intricacies of using SAP2000.

The preparation of this book would not have been possible without the competent and dedicated assistance of Annis Kay Young, who patiently transcribed the many drafts until it reached its final form. She also most diligently applied her expertise in AutoCAD to prepare and refine many of the figures in the book. In recognition of her indispensable help, this book is duly dedicated to her.

Mario Paz

1 Beams: Analytical Methodology

1.1 Introduction

Beams are defined as structures subjected to loads normal to their longitudinal direction, thereby producing lateral displacements and flexural stresses. We restrict consideration to straight beams with symmetric cross-sectional areas; the loads are applied along the plane of symmetry of the beam. Figure 1.1 shows an example of a continuous beam, with two fixed supports at the ends (fixed for translation and for rotation) and two simple support rollers (fixed for lateral translation)

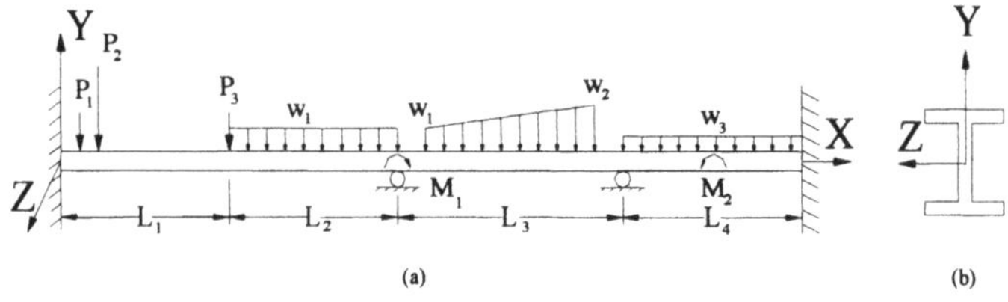

Fig. 1.1 (a) Example of a continuous beam; (b) Cross-sectional area.

The beam in Fig 1.1 carries several distributed loads (w_1) that are expressed in units of force per unit of length (kip/in), for example, and several concentrated forces (P_i) and moments (M_i) applied at different locations along the beam. As shown in Fig.1.1(b), the beam has an "I" cross-sectional area with a vertical axis of symmetry forming a plane of symmetry along the length of the beam on which the external loads are applied. Coordinate axes *X, Y,* and *Z* are also shown in the figure. These coordinate axes are designated as the global or the system coordinate axes, to distinguish this system of axes from the member or element coordinate axes as defined in Section 1.3.

1.2 Elements, Nodes and Nodal Coordinates

In preparation for matrix analysis, the structure is divided into beam elements, or simply elements, connected at nodes or joints. The beam of Fig.1.1 has been divided into four elements numbered consecutively as 1, 2, 3 and 4 connected at nodes, also numbered consecutively as ①, ②, ③, ④ and ⑤ as shown in Fig 1.2. Both the elements and the nodes may be numbered in any order. However, it is customary to number these items consecutively from the left to the right end of the beam.

The division of the beam into elements is generally arbitrary; however, this division aims to locate the nodes or connections between elements at points of support or at locations along the beam where there is a change of the cross-section properties.

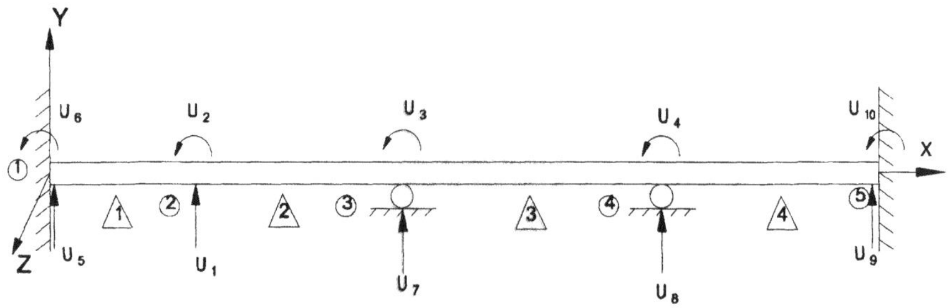

Fig. 1.2. Example of a beam showing the elements 1 through 4, nodes ① through ⑤, and nodal coordinates u_1 through u_{10}.

Structures modeled as beams are allowed two possible displacements at each node; translatory displacement normal to the longitudinal direction of the beam and a rotational displacement around a perpendicular axis to the plane (X, Y) of the beam. These displacements are designated by nodal coordinates which in Fig. 1.2 are labeled u_1 through u_{10}.

Free nodal coordinates are those coordinates which are not restricted for displacement, while the fixed nodal coordinates are constrained to have no displacement. For convenience, the free nodal coordinates u_1 to u_4 were numbered first, followed by the fixed nodal coordinates u_5 through u_{10}. The nodal coordinates in Fig.1.2 are shown in the positive direction to agree with the direction of the Y-axis for nodal translations and with the Z-axis for nodal rotations according to the right-hand rule.

1.3 Shape Functions and Stiffness Coefficients

In Fig. 1.3, we have isolated one element of the beam shown in Fig. 1.2. The nodal coordinates at the two ends of this element are now designated δ_1, δ_2, δ_3, and δ_4 (linear or angular displacement). This figure also shows the forces (or moments) P_1, P_2, P_3 and P_4 corresponding to the element nodal displacements δ_i. The <u>local or element system of coordinate axes</u> x, y and z which are fixed on the element with the origin at its left end are also shown in Fig. 1.3.

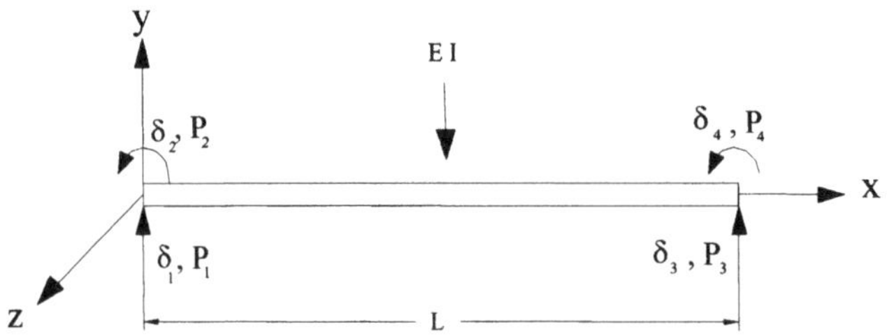

Fig. 1.3 Beam element showing the nodal coordinates δ_1 through δ_4, nodal forces P_1 through P_4 and the element coordinate axes x., y and z.

The <u>stiffness coefficient</u> k_{ij} is defined as the force at the nodal coordinate i when a unit displacement is given at the nodal coordinate j (all other nodal coordinates are kept fixed). Figure 1.4(a) shows the stiffness coefficients for a beam element corresponding to a unit displacement given at nodal coordinate 1, that is, for $\delta_1 = 1.0$. The shape of the curve due to this unit displacement is labeled in Fig.1.4 (a) as $N_1(x)$. Analogously, the other diagrams in Fig. 1.4(b), (c) and (d), show the stiffness coefficients and the shape functions for a unit displacement $\delta_2 = 1$, $\delta_3 = 1$ and $\delta_4 = 1$, respectively.

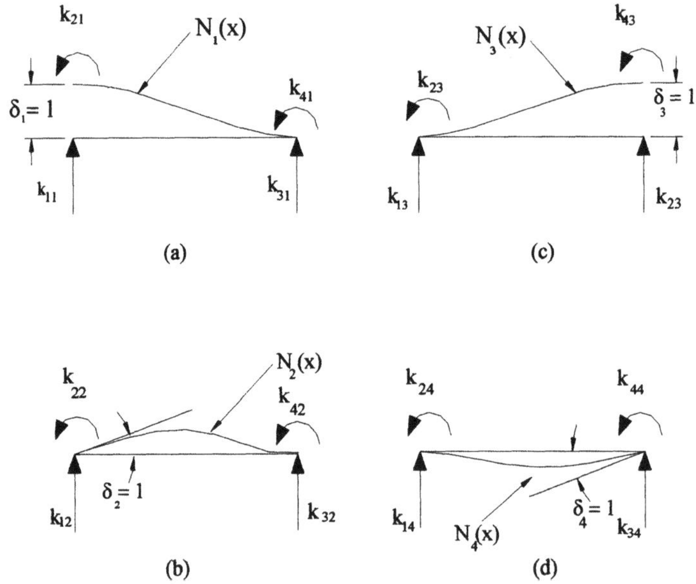

Fig. 1.4 Beam element showing [in (a), (b), (c), and (d)], the stiffness coefficients due to unit displacement at the nodal coordinates δ_1, δ_2, δ_3, and δ_4, respectively.

The evaluation of the stiffness coefficients k_{ij}, and of the shape functions, $N_1(x)$, $N_2(x)$, $N_3(x)$ and $N_4(x)$ for the curves shown in Fig 1.4 corresponding to a unit displacement at the nodal coordinates δ_1 , δ_2 , δ_3 , or δ_4, respectively, may be obtained by integrating the differential equation of the beam which for a uniform beam is given by

$$\frac{d^4y}{dx^4} = \frac{w(x)}{EI} \tag{1.1}$$

in which

$w(x)$ = external force per unit length
E = modulus of elasticity
I = cross-sectional moment of inertia

For the beams shown in Fig. 1.4 with no load on the span, that is, for $w(x) = 0$, eq.(1.1) reduces to

$$\frac{d^4y}{dx^4} = 0 \tag{1.2}$$

Integrating eq.(1.2) successively four times yields the equation for displacement of a beam in terms of the four constants of integration; C_1, C_2, C_3, C_4:

$$y = \frac{1}{6}C_1x^3 + \frac{1}{2}C_2x^2 + C_3x + C_4 \tag{1.3}$$

The constants of integration C_1 through C_4 in eq.(1.3) are evaluated from the corresponding boundary conditions. For example, for the beam in Fig.1.4(a) the boundary conditions are:

$$\begin{aligned} &\text{at } x = 0 \quad y(0) = 1 \quad \text{and} \quad \frac{dy(0)}{dx} = 0 \\ &\text{at } x = L \quad y(L) = 0 \quad \text{and} \quad \frac{dy(L)}{dx} = 0 \end{aligned} \tag{1.4}$$

The use of these boundary conditions in eq. (1.3) results in an algebraic system of four equations to determine C_1, C_2, C_3, and C_4 which subsequently substituted into eq.(1.3) results in the equation of the shape function for the beam in Fig. 1.4(a):

$$N_1(x) = 1 - 3\left(\frac{x}{L}\right)^2 + 2\left(\frac{x}{L}\right)^3 \tag{1.5a}$$

in which $N_1(x)$ is used instead of $y(x)$ to correspond to the condition $\delta_1 = 1$ imposed on this beam. Proceeding in analogous fashion, we obtain the equations of the other shape functions depicted in Fig. 1.4 as:

$$N_2(x) = x\left(1 - \frac{x}{L}\right)^2 \tag{1.5b}$$

$$N_3(x) = 3\left(\frac{x}{L}\right)^2 - 2\left(\frac{x}{L}\right)^3 \tag{1.5c}$$

$$N_4(x) = \frac{x^2}{L}\left(\frac{x}{L} - 1\right) \tag{1.5d}$$

The total displacement $y(x)$ at coordinate x due to arbitrary displacements δ_1, δ_2, δ_3 and δ_4 at the nodal coordinates of a beam element is then given by superposition as

$$y(x) = N_1(x)\,\delta_1 + N_2(x)\,\delta_2 + N_3(x)\,\delta_3 + N_4(x)\,\delta_4 \tag{1.6}$$

1.4 Element Stiffness Matrix

The general expression to calculate the stiffness coefficients for a beam element (Problem 1.1) is given by

$$k_{ij} = \int_0^L EI\, N_i''(x)\, N_j''(x)\,dx \tag{1.7}$$

in which $N_i''(x)$ and $N_j''(x)$ are the second derivatives of the shape functions [eqs.(1.5)] with respect to x, E is the modulus of elasticity, and I is the cross-sectional moment of inertia of the beam. For example, to calculate the coefficient k_{12}, we substitute into eq. (1.7) the second derivatives of the shape functions $N_1(x)$ and $N_2(x)$ given, respectively, by eqs.(1.5a) and (1.5b), to obtain

$$k_{12} = \int_0^L EI\left(\frac{-6}{L^2} + \frac{12x}{L^3}\right)\left(\frac{-4}{L} + \frac{6x}{L^2}\right)dx$$

which upon integration results in

$$k_{12} = \frac{6EI}{L^2} \tag{1.8}$$

The stiffness coefficient k_{ij} has been defined as a force at the nodal coordinate i due to a unit displacement at the nodal coordinate j. Consequently, the forces at nodal coordinate ① due to successive displacements δ_1, δ_2, δ_3, and δ_4 at the four nodal coordinates of the beam element are $k_{11}\,\delta_1$, $k_{12}\,\delta_2$, $k_{13}\,\delta_3$, and $k_{14}\,\delta_4$, respectively. Therefore, the total force P_1 at the nodal coordinate ① resulting from these nodal displacements is obtained by the superposition of these four forces, that is,

$$P_1 = k_{11}\delta_1 + k_{12}\delta_2 + k_{13}\delta_3 + k_{14}\delta_4$$

Analogously, the forces at the other nodal coordinates are:

$$\begin{aligned} P_2 &= k_{21}\delta_1 + k_{22}\delta_2 + k_{23}\delta_3 + k_{24}\delta_4 \\ P_3 &= k_{31}\delta_1 + k_{32}\delta_2 + k_{33}\delta_3 + k_{34}\delta_4 \\ P_4 &= k_{41}\delta_1 + k_{42}\delta_2 + k_{43}\delta_3 + k_{44}\delta_4 \end{aligned} \tag{1.9}$$

The above equations may be conveniently written in matrix notation as

$$\begin{Bmatrix} P_1 \\ P_2 \\ P_3 \\ P_4 \end{Bmatrix} = \begin{bmatrix} k_{11} & k_{12} & k_{13} & k_{14} \\ k_{21} & k_{22} & k_{23} & k_{24} \\ k_{31} & k_{32} & k_{33} & k_{34} \\ k_{41} & k_{42} & k_{43} & k_{44} \end{bmatrix} \begin{Bmatrix} \delta_1 \\ \delta_2 \\ \delta_3 \\ \delta_4 \end{Bmatrix} \tag{1.10}$$

The use of eq.(1.7) in the manner shown to determine the coefficient k_{12} in eq.(1.8) will result in the evaluation of all the coefficients of the element stiffness matrix in eq.(1.10). The result for a uniform beam element is

$$\begin{Bmatrix} P_1 \\ P_2 \\ P_3 \\ P_4 \end{Bmatrix} = \frac{EI}{L^3} \begin{bmatrix} 12 & 6L & -12 & 6L \\ 6L & 4L^2 & -6L & 2L^2 \\ -12 & -6L & 12 & -6L \\ 6L & 2L^2 & -6L & 4L^2 \end{bmatrix} \begin{Bmatrix} \delta_1 \\ \delta_2 \\ \delta_3 \\ \delta_4 \end{Bmatrix} \tag{1.11}$$

or in condensed notation as

$$\{P\} = [k]\{\delta\} \tag{1.12}$$

in which $[k]$ is the element stiffness matrix.

1.5 Assembly of the System Stiffness Matrix

Thus far, we have established in eq.(1.11) or eq.(1.12) the stiffness equation for a uniform beam element, that is, we have obtained for a beam element the relationship between nodal displacements $\{\delta\}$ (linear and angular) and nodal forces $\{P\}$ (forces and moments). Our next objective is to obtain the same type of relationship for the entire structure between the nodal displacements $\{u\}$ and the nodal forces $\{F\}$. The relationship establishing the system stiffness equation for the entire structure may be expressed in condensed notation as

$$\{F\} = [K]\{u\} \tag{1.13}$$

in which $[K]$ is the system stiffness matrix. Furthermore, our aim is to obtain the system stiffness matrix from the stiffness matrices of the elements of the system. The procedure is perhaps better explained through a specific example such as the beam shown with its load in Fig. 1.1 and with the nodal coordinates numbered as shown in Fig. 1.2. This beam has been divided into four beam elements, labeled 1 through 4. The nodal coordinates are indicated in this figure as u_1 through u_{10} with the four free coordinates conveniently numbered first, as already stated.

The procedure of assembling the system stiffness matrix consists of transferring and adding appropriately the coefficients in each element stiffness matrix to obtain the system stiffness matrix. This method of assembling the system stiffness matrix is called the direct method. In effect, any coefficient k_{ij} of the system may be obtained by adding together the corresponding element stiffness coefficients associated with those nodal coordinates. For example, to obtain the stiffness coefficient k_{11} it is necessary to add the stiffness coefficients of the beam elements 1 and 2 that correspond to the nodal coordinates, u_I. These coefficients are designated, respectively, as $k_{33}^{(1)}$ and $k_{11}^{(2)}$ for these two beam elements. The upper indices (1) and (2) serve to identify the beam elements and the lower indices to locate the appropriate stiffness coefficients in the corresponding element stiffness matrix.

Illustrative Example 1.1

For the beam shown in Fig 1.5, which has been divided into four beam elements, obtain: (a) the stiffness matrices for each element and (b) the system stiffness matrix for the beam. The material modulus of elasticity is E = 29,000 ksi and the cross-

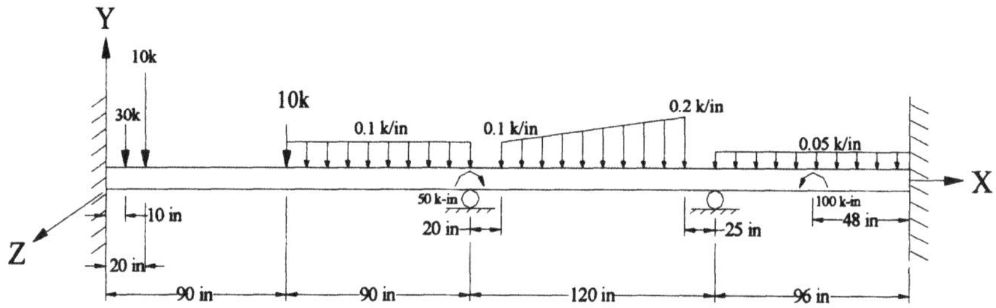

sectional moment of inertia for the entire beam is 882 in^4.

Fig. 1.5 Beam for Illustrative Examples 1.1 and 1.2

Solution:

1. Element Stiffness Matrices

The substitution of numerical values into the element stiffness matrix in eq.(1.11) for each of the four beam elements of this beam gives the following element stiffness matrices:

ELEMENT 1

$$[k]_1 = \frac{29\text{x}10^3\text{x}882}{90^3}\begin{bmatrix} 12 & 6\text{x}90 & -12 & 6\text{x}90 \\ 6\text{x}90 & 4\text{x}90^2 & -6\text{x}90 & 2\text{x}90^2 \\ -12 & -6\text{x}90 & 12 & -6\text{x}90 \\ 6\text{x}90 & 2\text{x}90^2 & -6\text{x}90 & 4\text{x}90^2 \end{bmatrix}$$

or

$$
\begin{array}{cccc} 5 & 6 & 1 & 2 \end{array}
$$

$$
[k]_1 \begin{bmatrix} 420 & 18900 & -420 & 18900 \\ 18900 & 1134200 & -18900 & 567110 \\ -420 & -18900 & 420 & -18900 \\ 18900 & 567110 & -18900 & 1134200 \end{bmatrix} \begin{array}{c} 5 \\ 6 \\ 1 \\ 2 \end{array} \qquad \text{(a)}
$$

ELEMENT 2

$$
\begin{array}{cccc} 1 & 2 & 7 & 3 \end{array}
$$

$$
[k]_2 = \begin{bmatrix} 420 & 18900 & -420 & 18900 \\ 18900 & 1134200 & -18900 & 567110 \\ -420 & -18900 & 420 & -18900 \\ 18900 & 567100 & -18900 & 1134200 \end{bmatrix} \begin{array}{c} 1 \\ 2 \\ 7 \\ 3 \end{array} \qquad \text{(b)}
$$

ELEMENT 3

$$
\begin{array}{cccc} 7 & 3 & 8 & 4 \end{array}
$$

$$
[k]_3 = \begin{bmatrix} 177 & 10633 & -177 & 10633 \\ 10633 & 850667 & -10633 & 425333 \\ -177 & -10633 & 177 & -10633 \\ 10633 & 425333 & -10633 & 850667 \end{bmatrix} \begin{array}{c} 7 \\ 3 \\ 8 \\ 4 \end{array} \qquad \text{(c)}
$$

ELEMENT 4

$$
\begin{array}{cccc} 8 & 4 & 9 & 10 \end{array}
$$

$$
[k]_4 = \begin{bmatrix} 346 & 16615 & -346 & 16615 \\ 16615 & 1063333 & -16615 & 531667 \\ -346 & -16615 & 346 & -16615 \\ 16615 & 531667 & -16615 & 1063333 \end{bmatrix} \begin{array}{c} 8 \\ 4 \\ 9 \\ 10 \end{array} \qquad \text{(d)}
$$

2. System Stiffness Matrix

As previously stated, the system stiffness matrix is obtained by transferring and adding appropriately the coefficients of the element stiffness matrices. To perform this transfer one should realize that the nodal coordinates for the beam elements should be labeled according to the system coordinates assigned to the two ends of a specific beam element. For example, for element 1 this assignment is u_5, u_6, u_1 and u_2, as shown in Fig. 1.2, and not δ_1, δ_2, δ_3, and δ_4, as indicated in Fig. 1.3 for an isolated beam element. A simple way to indicate this allocation of nodal coordinates when working by hand is to write at the top and on the right of the element stiffness matrix the coordinate numbers corresponding to the system nodal coordinates for the elements, as it is shown in the matrices (a), (b), (c) and (d) for the four beam elements of this structure.

Proceeding systematically to assemble the system stiffness matrix, we transfer each entry in matrices (a), (b), (c) and (d) to the row and column indicated respectively on the right and at the top of these matrices. Each transferred coefficient is added to the total accumulated at the transferred location of the system stiffness matrix. For example, the coefficient 420 in the third row and third column of matrix (a) is transferred to the location row 1 and column 1, of the system stiffness matrix. Also, the coefficient 420 in the first row and first column of matrix (b) is transferred to the same location; row 1 and column 1 of the system stiffness matrix to give a total of 840 at this location. The final complete system stiffness matrix $[K]$, obtained after transferring all the coefficients of element stiffness matrices (a), (b), (c) and (d), is given by eq. (e).

$$[K]=\left[\begin{array}{cccc|cccccc}
840 & 0 & 18900 & 0 & -420 & -18900 & -420 & 0 & 0 & 0 \\
0 & 2268400 & 567110 & 0 & 18900 & 567110 & -18900 & 0 & 0 & 0 \\
18900 & 567110 & 1984867 & 425333 & 0 & 0 & -8267 & -10633 & 0 & 0 \\
0 & 0 & 425333 & 1914000 & 0 & 0 & 10633 & 5982 & -16615 & 531667 \\
\hline
-420 & 18900 & 0 & 0 & 420 & 18900 & 0 & 0 & 0 & 0 \\
-18900 & 567110 & 0 & 0 & 18900 & 1134200 & 0 & 0 & 0 & 0 \\
-420 & -18900 & -8267 & 10633 & 0 & 0 & 597 & -177 & 0 & 0 \\
0 & 0 & -10633 & 5982 & 0 & 0 & -177 & 523 & -346 & 16615 \\
0 & 0 & 0 & -16615 & 0 & 0 & 0 & -346 & 346 & -16615 \\
0 & 0 & 0 & 531667 & 0 & 0 & 0 & 16615 & -16615 & 106333
\end{array}\right] \qquad (e)$$

The system stiffness matrix in eq.(e) has been partitioned to separate the first four free nodal coordinates from the last six fixed nodal coordinates. The separation of free and fixed coordinates will be needed to calculate the four unknown displacements at the free nodal coordinates.

1.6 System Force Vector

The force vector $\{F\}$ in the system stiffness matrix [eq.(1.13)] consists of forces directly applied to nodal coordinates and forces denoted as equivalent nodal forces, (for those forces applied along the spans of the beam elements). The forces applied at the nodal coordinates of the structure are directly assigned to the corresponding entry in the force vector $\{F\}$ of eq.(1.13). On the other hand, for forces applied on the elements, it is necessary to determine the equivalent forces at the nodal coordinates of the element and then allocate these equivalent nodal forces to the corresponding entry of the system force vector $\{F\}$ in eq.(1.13).

The equivalent nodal force Q_i for a distributed load $w(x)$ applied on the span of a beam element is given, in general, by (see Problem 1.2):

$$Q_i = \int_0^L N_i(x)\; w(x)\, dx \qquad (1.14a)$$

and for a distributed moment $m(x)$ by (see Problem 1.3):

$$Q_i = \int_0^L N_i'(x)\, m(x)\, dx \tag{1.14b}$$

in which

Q_i = equivalent force at the element nodal coordinate i
$w(x)$ = distributed load on the beam element
($w(x)$ is positive for an upward load)
$m(x)$ = distributed moment on the beam element
$N_i(x)$ = shape function for nodal displacement $\delta_i=1$ given by eqs.(1.5)
$N_i'(x)$ = derivative of the shape function given by eqs. (1.5)

For example, for an element carrying a uniformly distributed load w as shown in Fig.1.6, the application of eq.(1.14a) to determine Q_2 yields

$$Q_2 = \int_0^L w\, N_2(x) dx$$

and substituting $N_2(x)$ from eq.(1.5b)

$$Q_2 = \int_0^L w\, x \left(1 - \frac{x}{L}\right)^2 dx$$

which upon integration yields

$$Q_2 = \frac{wL^2}{12}$$

Analogously, substituting into eq.(1.14a) the expression for the shape functions $N_i(x)$ [given by eqs.(1.5)], we obtain all the equivalent nodal forces:

$$Q_1 = \frac{wL}{2} \qquad Q_2 = \frac{wL^2}{12} \qquad Q_3 = \frac{wL}{2} \qquad Q_4 = -\frac{wL^2}{12}$$

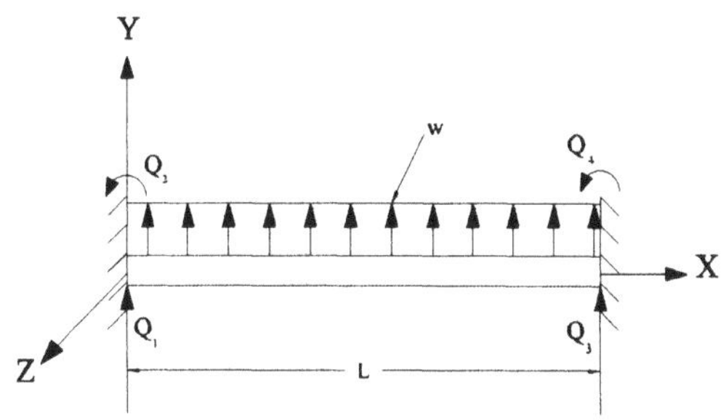

Fig. 1.6 Equivalent nodal forces Q_i for a beam element supporting a uniformly distributed loal.

Alternatively, the equivalent nodal forces $\{Q\}$ may be evaluated by determining the Fixed End Reactions $\{FER\}$ for the loaded beam by assuming that the beam element is completely fixed for translation or rotation at its two ends and then reversing the sense of these reactions. To demonstrate the validity of this method for determining the equivalent nodal forces, we consider in Fig.1.7(a) a simply supported beam divided into three elements carrying a general distributed load $w(x)$. Figure 17(b) shows element 2 of this beam equated to the element fixed at its two ends (loaded with the external force $w(x)$ and the fixed end reactions) and the same element supporting the equivalent end-forces equal in magnitude to the fixed end reactions but acting in the opposite sense. It may be seen from Fig.1.7(b) that since the fixed-end beam does not develop end displacements, the fixed end reactions applied in the opposite sense produce in the beam element the actual nodal displacements. Therefore, these fixed-end reactions applied in opposite directions are precisely the equivalent nodal forces.

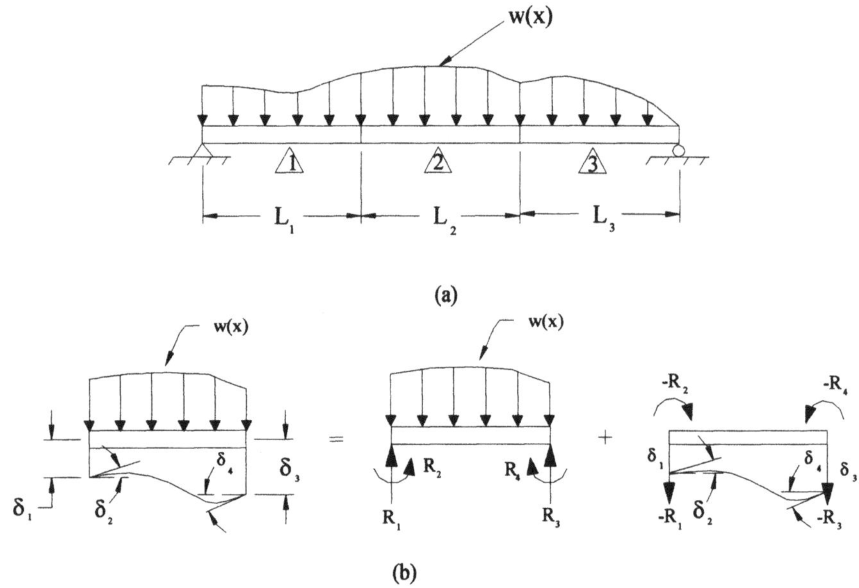

Fig. 1.7 (a) Simply supported beam divided into three elements carrying a distributed load *w(x)*,
(b) Element 2 equated to the superposition of the element fixed-end reactions and the element loaded with these reactions acting in opposite sense.

Appendix I provides the expressions for the element nodal equivalent forces for some common loads. The expressions for the nodal equivalent forces listed in Appendix I may be obtained by either using eq.(1.14), or alternatively, by calculating the element fixed-end reactions and reversing their directions. (See Problems 1.3 and 1.4).

Illustrative Example 1.2

For the loaded beam shown in Fig.1.5 determine:
(a) The equivalent nodal forces for each of the four elements of the beam.
(b) The assembled system force vector.

Solution:

1. Equivalent Nodal Forces

ELEMENT 1

From Appendix I, Case (a), with the concentrated loads W_1 = -30(kip) and W_2 = -10(kip) applied from the left end of element 1 at 10 in. and 20 in., respectively, we obtain

$$Q_1 = -\frac{30x80^2(3x10+80)}{90^3} - \frac{10x70^2(3x20+70)}{90^3} = -37.71(\text{kip})$$

$$Q_2 = -\frac{30x10x80^2}{90^2} - \frac{10x20x70^2}{90^2} = -358.02 \ (\text{kip}\cdot\text{in})$$

Analogously, we obtain

$$Q_3 = -2.29(\text{kip})$$
$$Q_4 = +64.20(\text{kip}\cdot\text{in})$$

ELEMENT 2

From Appendix I, Case (c), for a uniformly distributed load w = -0.10 (kip/in):

$$Q_1 = \frac{wL}{2} = -\frac{0.10x90}{2} = -4.50\,(\text{kip})$$

$$Q_2 = \frac{wL^2}{12} = -\frac{0.10x90^2}{12} = -67.50(\text{kip}\cdot\text{in})$$

$$Q_3 = \frac{wL}{2} = -4.50\,(\text{kip})$$

$$Q_4 = -\frac{wL^2}{12} = +67.50(\text{kip}\cdot\text{in})$$

ELEMENT 3

From Appendix I, case (d), with w_1 = -0.10 (kip/in), w_2 = -0.20 (kip/in), L_1 = 20 in, L_2 = 25 in, and L = 120 in:

$$Q_1 = -5.39 \text{ (kip)}$$
$$Q_2 = -142.82 \text{ (kip.in)}$$
$$Q_3 = -5.86 \text{ (kip)}$$
$$Q_4 = 152.10 \text{ (kip.in)}$$

ELEMENT 4

From Appendix I, case (c), for a uniformly distributed load w= -0.05 (kip/in) and case (b) for a concentrated moment M = 100 (kip.in) at a distance of 48 in. from the left end of the element, we have

$$Q_1 = \frac{wL}{2} - \frac{6ML_1L_2}{L^3} = -\frac{0.05\text{x}96}{2} - \frac{6\text{x}100\text{x}48\text{x}48}{96^3} = -3.96 \text{ (kip)}$$

$$Q_2 = \frac{wL^2}{12} + \frac{ML_2(L_2 - 2L_1)}{L^2} = -\frac{0.05\text{x}96^2}{12} + \frac{100\text{x}48(48-96)}{96^2} = -63.40 \text{ (kip·in)}$$

Analogously, we obtain

$$Q_3 = -0.84 \text{ (kip)}$$
$$Q_4 = 13.40 \text{ (kip·in)}$$

2. Assemble System Force Vector

The equivalent nodal forces for the four beam elements as calculated above are given by the following element force vectors:

$$\{Q\}_1 = \begin{Bmatrix} -37.71 \\ -358.02 \\ -2.29 \\ +64.20 \end{Bmatrix} \begin{matrix} 5 \\ 6 \\ 1 \\ 2 \end{matrix} \quad \{Q\}_2 = \begin{Bmatrix} -4.50 \\ -67.50 \\ -4.50 \\ +67.50 \end{Bmatrix} \begin{matrix} 1 \\ 2 \\ 7 \\ 3 \end{matrix} \quad \{Q\}_3 = \begin{Bmatrix} -5.39 \\ -142.82 \\ -5.86 \\ +152.10 \end{Bmatrix} \begin{matrix} 7 \\ 3 \\ 8 \\ 4 \end{matrix} \quad \{Q\}_4 = \begin{Bmatrix} -3.96 \\ -63.40 \\ -0.84 \\ +13.40 \end{Bmatrix} \begin{matrix} 8 \\ 4 \\ 9 \\ 10 \end{matrix} \qquad \text{(a)}$$

The coefficients of these element force vectors are transferred to the system force vector according to the system nodal coordinates assigned to each beam element. For example, for element 1 the assignment of nodal coordinates is 5, 6, 1 and 2 have been indicated in Fig. 1.2. For convenience, the assignment of nodal coordinates for each element force vector is indicated in eq.(a) on the right side of the equivalent nodal force vector $\{Q\}_1$ and corresponding assignments for the nodal vectors $\{Q\}_2$, $\{Q\}_3$, and $\{Q\}_4$. We proceed to transfer each coefficient in these vectors to the corresponding location in the system force vector {F} as shown in the first force vector of eq.(b):

$$\{F\} = \begin{Bmatrix} -2.29-4.50 \\ +64.20-67.50 \\ +67.50-142.82 \\ +152.10-63.40 \\ \hline -37.71 \\ -358.02 \\ -4.50-5.39 \\ -5.86-3.96 \\ -0.84 \\ +13.40 \end{Bmatrix} + \begin{Bmatrix} -10 \\ 0 \\ -50 \\ 0 \\ \hline R_5 \\ R_6 \\ R_7 \\ R_8 \\ R_9 \\ R_{10} \end{Bmatrix} = \begin{Bmatrix} -16.79 \\ -3.30 \\ -125.32 \\ +88.70 \\ \hline R_5-37.71 \\ R_6-358.02 \\ R_7-9.89 \\ R_8-9.82 \\ R_9-0.84 \\ R_{10}+13.40 \end{Bmatrix} \tag{b}$$

The second force vector in eq.(b) contains the forces applied directly to the free nodal coordinate and the reactions R_i at the fixed nodal coordinates labeled u_5 through u_{10} as shown for this beam in Fig. 1.2.

1.7 System Nodal Displacements and Support Reactions

The unknown nodal displacements and support reactions are calculated by partitioning the system stiffness equation [eq.(1.13)], solving first for the unknown displacements, and then calculating the unknown reactions. Therefore, eq.(1.13) is conveniently partitioned to separate the unknown nodal displacements from the fixed displacements at the supports. Thus, by eq.(1.13)

$$\{F\} = [K]\{u\} \tag{1.15}$$

or in partitioned form

$$\begin{Bmatrix} \{F\}_1 \\ \hline \{F\}_2 \end{Bmatrix} = \begin{Bmatrix} \{K\}_{11} & \{K\}_{12} \\ \hline \{K\}_{21} & \{K\}_{22} \end{Bmatrix} \begin{Bmatrix} \{u\}_1 \\ \hline \{u\}_2 \end{Bmatrix} \tag{1.16}$$

in which

$\{F\}_1$ = external force vector at the free nodal coordinates
$\{u\}_1$ = unknown displacements at the free nodal coordinates
$\{F\}_2$ = external force vector at the fixed nodal coordinates
$\{u\}_2$ = $\{0\}$ = displacements at the fixed nodal coordinates

After setting $\{u\}_2 = \{0\}$ and expanding the two equations in eq.(1.16), we obtain

$$\{F\}_1 = [K]_{11}\{u\}_1 \tag{1.17}$$

$$\{F\}_2 = [K]_{21}\{u\}_1 \tag{1.18}$$

Equation (1.17) is then solved for the unknown nodal displacements $\{u\}_1$, which are subsequently substituted into eq.(1.18) to yield the nodal forces at the fixed nodal coordinates; thus allowing the determination of the reactions at the supports. The following Illustrative Example uses eqs.(1.17) and (1.18) to determine the displacements at the free nodal coordinates and the reactions at the supports.

Illustrative Example 1.3

Use the results in Illustrative Examples 1.1 and 1.2 to establish eqs.(1.17) and (1.18). Then solve for the unknown displacements at the free nodal coordinates and calculate the reactions at the fixed nodal coordinates.

Solution:

Substituting into eq.(1.17) the first four forces in the partitioned force vector, eq.(b) of Illustrative Example 1.2, and the top-left partitioned stiffness matrix [eq.(e)] of Illustrative Example 1.1, we obtain

$$\begin{Bmatrix} -16.79 \\ -3.30 \\ -125.32 \\ +88.70 \end{Bmatrix} = \begin{bmatrix} 840 & 0 & 18900 & 0 \\ 0 & 2268400 & 567110 & 0 \\ 18900 & 567110 & 1984867 & 425333 \\ 0 & 0 & 425333 & 1914000 \end{bmatrix} \begin{Bmatrix} u_1 \\ u_2 \\ u_3 \\ u_4 \end{Bmatrix} \tag{a}$$

Solving eq.(a) for the unknown nodal displacements u_1 through u_4 results in

$$\begin{aligned} u_1 &= -2.40\text{E-}2 \text{ (in)} \\ u_2 &= -4.56\text{E-}5 \text{ (rad)} \\ u_3 &= 1.77\text{E-}4 \text{ (rad)} \\ u_4 &= 7.12\text{E-}6 \text{ (rad)} \end{aligned} \tag{b}$$

To solve for the reactions at the fixed supports, we substitute into eq.(1.18) the lower partition vector of eq.(b) in Illustrative Example 1.2 and the left lower partition of the stiffness matrix from eq.(e) of Illustrative Example 1.1:

$$\begin{Bmatrix} R_5 - 37.71 \\ R_6 - 358.02 \\ R_7 - 9.89 \\ R_8 - 9.82 \\ R_9 - 0.84 \\ R_{10} + 13.40 \end{Bmatrix} = \begin{bmatrix} -420 & -18900 & 0 & 0 \\ -18900 & 567110 & 0 & 0 \\ -420 & 18900 & -8267 & 10633 \\ 0 & 0 & 10633 & 5982 \\ 0 & 0 & 0 & -16615 \\ 0 & 0 & 0 & 531667 \end{bmatrix} \begin{Bmatrix} -2.40\text{E}-2 \\ -4.56\text{E}-5 \\ 1.77\text{E}-4 \\ 7.12\text{E}-6 \end{Bmatrix} = \begin{Bmatrix} 9.20 \\ 427.09 \\ 9.54 \\ -1.83 \\ -0.12 \\ 3.78 \end{Bmatrix} \tag{c}$$

Thus, the reactions $\{R\}$ at the supports of the beam are then calculated from eq.(c) as

$$\begin{Bmatrix} R_5 \\ R_6 \\ R_7 \\ R_8 \\ R_9 \\ R_{10} \end{Bmatrix} = \begin{Bmatrix} 9.20 \\ +427.09 \\ +9.54 \\ -1.83 \\ -0.12 \\ 3.78 \end{Bmatrix} + \begin{Bmatrix} +37.71 \\ +358.02 \\ +9.89 \\ +9.82 \\ +0.84 \\ -13.40 \end{Bmatrix} = \begin{Bmatrix} +46.91 \\ +785.11 \\ +19.43 \\ +7.99 \\ +0.72 \\ -9.62 \end{Bmatrix} \qquad \text{(d)}$$

Alternatively, the reactions at the fixed nodal coordinates could also be determined by calculating the element end forces as presented in the following Section 1.8.

1.8 Element End Forces

The element end forces $\{P\}$, at the nodal coordinates of a beam element, are given by the superposition of the forces $[k]\{\delta\}$ due to the element nodal displacements and the fixed end reactions $\{FER\}$ at the nodal coordinates of the element due to the applied loads on the span of the element; that is,

$$\{P\} = \{k\}\{\delta\} + \{FER\} \qquad (1.19)$$

or

$$\{P\} = [k]\{\delta\} - \{Q\} \qquad (1.20)$$

because as shown in Section 1.6, the fixed end reactions $\{FER\}$ are equal to the equivalent nodal forces $\{Q\}$, but with opposite sign.

In eqs.(1.19) and (1.20),

$[k]$ = element stiffness matrix
$\{\delta\}$ = displacement vector at the element nodal coordinates
$\{FER\}$ = Fixed End Reactions at the element nodal coordinates due to applied loads
$\{Q\}$ = equivalent forces at the nodal coordinates of a loaded beam element

Illustrative Example 1.4

For the beam in Illustrative Examples 1.1 and 1.2 determine:

1. Displacements at the nodal coordinates for each element.
2. End forces at the nodal coordinates for each element of the beam.
3. Support reactions.

Solution:

1. Element Nodal Displacements

The nodal displacement vectors for each element of the beam are identified from Fig. 1.2 as follows:

$$\{\delta\}_1=\begin{Bmatrix}0\\0\\u_1\\u_2\end{Bmatrix}\quad \{\delta\}_2=\begin{Bmatrix}u_1\\u_2\\0\\u_3\end{Bmatrix}\quad \{\delta\}_3=\begin{Bmatrix}0\\u_3\\0\\u_4\end{Bmatrix}\quad \{\delta\}_4=\begin{Bmatrix}0\\u_4\\0\\0\end{Bmatrix} \tag{a}$$

or substituting numerical values for u_1, u_2, u_3 *and* u_4 calculated in Illustrative Example 1.3:

$$\{\delta\}_1=\begin{Bmatrix}0\\0\\-2.40\text{E}-2\\-4.56\text{ E}-5\end{Bmatrix}\quad \{\delta\}_2=\begin{Bmatrix}-2.40\text{ E}-2\\-4.56\text{ E}-5\\0\\-1.77\text{ E}-4\end{Bmatrix}\quad \{\delta\}_3=\begin{Bmatrix}0\\1.77\text{ E}-4\\0\\7.12\text{ E}-6\end{Bmatrix}\quad \{\delta\}_4=\begin{Bmatrix}0\\7.12\text{E}-6\\0\\0\end{Bmatrix} \tag{b}$$

2. Element End Forces

Substituting into eq.(1.20): (a) the numerical values for the element stiffness matrix $[k]_1$ of element 1 (Illustrative Example 1.1), (b) the displacement vector $\{\delta\}_1$, (Illustrative Example 1.4), and (c) the equivalent nodal force $\{Q\}_1$ (Illustrative Example 1.2), we obtain

$$\begin{Bmatrix}P_1\\P_2\\P_3\\P_4\end{Bmatrix}_1=\begin{bmatrix}420 & 18900 & -420 & 18900\\ 18900 & 1134200 & -18900 & 567100\\ -420 & -18900 & 420 & -18900\\ 18900 & 567110 & -18900 & 1134200\end{bmatrix}\begin{Bmatrix}0\\0\\-2.40\text{ E}-2\\-4.56\text{ E}-5\end{Bmatrix}-\begin{Bmatrix}-37.71\\-358.02\\-2.29\\+64.20\end{Bmatrix}=\begin{Bmatrix}46.91\\785.04\\-6.91\\336.97\end{Bmatrix} \tag{c}$$

or

$$\begin{aligned}P_{11}&=46.91\text{ (kip)}\\ P_{21}&=785.04\text{ (kip}\cdot\text{in)}\\ P_{31}&=-6.91\text{ (kip)}\\ P_{41}&=336.9\text{ (kip}\cdot\text{in)}\end{aligned} \tag{d}$$

Analogously, for elements 2, 3 and 4, we obtain:

$$P_{12} = -3.08(\text{kip}) \qquad P_{13} = 7.34(\text{kip}) \qquad P_{14} = 4.08(\text{kip})$$

$$P_{22} = -336.97(\text{kip}\cdot\text{in}) \quad P_{23} = 296.01(\text{kip}\cdot\text{in}) \quad P_{24} = 70.97(\text{kip}\cdot\text{in}) \tag{e}$$

$$P_{32} = 12.09(\text{kip}) \qquad P_{33} = 3.91(\text{kip}) \qquad P_{34} = 0.72(\text{kip})$$

$$P_{42} = -346.01(\text{kip}\cdot\text{in}) \quad P_{43} = -70.97(\text{kip}\cdot\text{in}) \quad P_{44} = -9.62(\text{kip}\cdot\text{in})$$

In eqs. (e) the second sub-index of P_{ij} serves to identify the beam element number. The end-forces for each of the four elements of this beam are shown in their proper direction in Fig. 1.8.

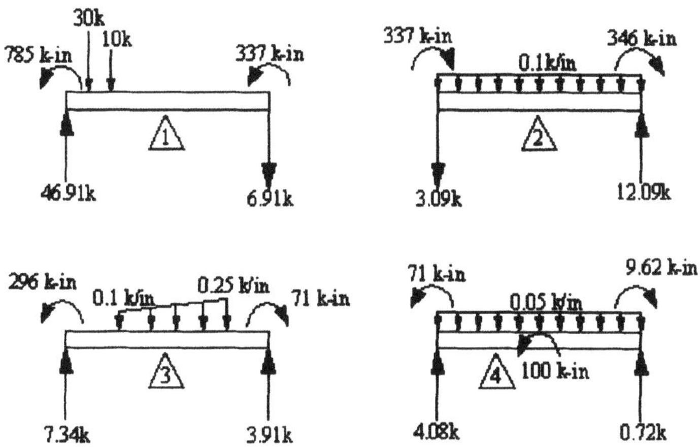

Fig. 1.8 Beam elements of the beam in Fig. 1.5 showing the applied loads and the calculated end forces.

3. Reactions at the Supports

As stated in the preceding Section (1.7), the reactions at the fixed nodal coordinates could also be determined from the calculated element end forces. For the beam in Illustrative Examples 1.1 and 1.2, the reactions R_5 and R_6 at the fixed support of joint ① are identified as the forces at the nodal coordinates 1 and 2 of element 1. The reactions R_9 and R_{10} at the fixed support of joint ⑤ are identified as the forces at the nodal coordinate 3 and 4 of element 4. That is, from eq.(d) we have:

$$R_5 = P_{11} = 46.91 \ \ (\text{kip}) \qquad R_6 = P_{21} = 785.04 \ \ (\text{kip.in})$$

and from eq.(e)

$$R_9 = P_{34} = 0.72 \text{ (kip)} \qquad R_{10} = P_{44} = -9.62 \text{ (kip.in)}$$

At the interior support at joint ③, the reaction R_7 is equal to the addition of the end forces at this joint for elements 2 and 3. Analogously, the reaction R_8 at the interior support at joint ④ is equal to the sum of the end forces at this joint for elements 3 and 4. That is,

$$R_7 = P_{32} + P_{13} = 12.09 + 7.34 = 19.43 \text{ (kip)}$$

and

$$R_8 = P_{33} + P_{14} = 3.91 + 4.08 = 7.99 \text{ (kip)}$$

The values for these reactions obtained from end element forces, are certainly equal to those calculated in eq.(d) of Illustrative Example 1.3.

1.9 End Releases in Beam Elements

The stiffness matrix method of analysis for beams presented in the preceding sections is based on the condition that each beam element is rigidly connected to the nodes or joints at both ends. This method of analysis has to be modified for cases in which elements of the beam are connected to the nodes through hinges. When an element of a beam is connected to the adjacent node by a hinge, the moment at the hinge end must be zero. Such connections through hinges are often referred to as member or element releases.

The effect of end releases in a beam element can be conveniently considered in the stiffness method of analysis by modifying both the element stiffness matrix and the equivalent force vector. Only released moments are considered in this section; these are the most common releases in structural systems such as beams and frames. Other types of releases could also be considered in the development of the element stiffness matrix.

The substitution of the stiffness matrix for a beam element from eq.(1.11) into eq.(1.20) followed by the matrix multiplication $[k]\{\delta\}$ yields

$$P_1 = \frac{EI}{L^3}\left[12\delta_1 + 6L\delta_2 - 12\delta_3 + 6L\delta_4\right] - Q_1 \qquad (1.21a)$$

$$P_2 = \frac{EI}{L^3}\left[6L\delta_1 + 4L^2\delta_2 - 6L\delta_3 + 2L^2\delta_4\right] - Q_2 \qquad (1.21b)$$

$$P_3 = \frac{EI}{L^3}\left[-12\delta_1 - 6L\delta_2 + 12\delta_3 - 6L\delta_4\right] - Q_3 \qquad (1.21c)$$

$$P_4 = \frac{EI}{L^3}\left[6L\delta_1 + 2L^2\delta_2 - 6L\delta_3 + 4L^2\delta_4\right] - Q_4 \qquad (1.21d)$$

We distinguish three cases of element end releases as shown in Fig. 1.9:

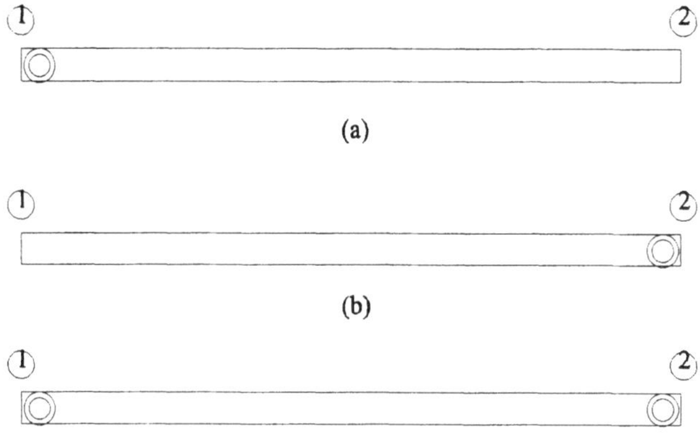

Fig.1.9. Element end releases: (a) Case 1 (hinge at node①); (b) Case 2 (hinge at node ②) ; and (c) Case 3 (hinges at the two ends of the beam element).

Case 1: Beam element with a hinge at the first node ①

When a beam element has a hinge at node ① as shown in Fig.1.8(a), the bending moment at this end as expressed by eq. (1.21b) is equal to zero. Thus, after equating eq.(1.21b) to zero and solving for the rotational displacement δ_2, we obtain

$$\delta_2 = -\frac{3}{2L}\delta_1 + \frac{3}{2L}\delta_3 - \frac{1}{2}\delta_4 + \frac{L}{4EI}Q_2 \tag{1.22}$$

Equation (1.22) indicates that in this case (a hinge at the first node of a beam element) the angular displacement δ_2 at this node is not an independent nodal coordinate, but a function of the nodal displacements δ_1, δ_3, and δ_4. Therefore, the number of independent nodal displacements for the hinged beam element is reduced to three. We can then eliminate the rotational displacement δ_2 from the element stiffness equation by substituting δ_2 from eq.(1.22) into eqs.(1.21) to obtain

$$\begin{aligned}
P_1 &= \frac{EI}{L^3}(3\delta_1 - 3\delta_3 + 3L\delta_4) + \left(-Q_1 + \frac{3}{2L}Q_2\right) \\
P_2 &= 0 \\
P_3 &= \frac{EI}{L^3}(-3\delta_1 + 3\delta_3 - 3L\delta_4) + \left(-Q_3 - \frac{3}{2L}Q_2\right) \\
P_4 &= \frac{EI}{L}\left(3L\delta_1 - 3L\delta_3 + 3L^2\delta_4\right) + \left(-Q_4 + \frac{1}{2}Q_2\right)
\end{aligned} \tag{1.23}$$

Equations (1.23), which establish the modified relationship between nodal forces and nodal displacement for a beam element released at node ①, can be expressed in matrix form as

$$\begin{Bmatrix} P_1 \\ P_2 \\ P_3 \\ P_4 \end{Bmatrix} = \frac{EI}{L^3} \begin{bmatrix} 3 & 0 & -3 & 3L \\ 0 & 0 & 0 & 0 \\ -3 & 0 & 3 & -3L \\ 3L & 0 & -3L & 3L^2 \end{bmatrix} \begin{Bmatrix} \delta_1 \\ \delta_2 \\ \delta_3 \\ \delta_4 \end{Bmatrix} + \begin{Bmatrix} -Q_1 + \frac{3}{2L} Q_2 \\ 0 \\ -Q_3 - \frac{3}{2L} Q_2 \\ -Q_4 + \frac{1}{2} Q_2 \end{Bmatrix} \tag{1.24}$$

or in condensed notation

$$\{P\} = \{k\}_m \{\delta\} + \{Q\}_m$$

in which the modified element stiffness matrix $[k]_m$ is given by

$$[k]_m = \frac{EI}{L^3} \begin{bmatrix} 3 & 0 & -3 & 3L \\ 0 & 0 & 0 & 0 \\ -3 & 0 & 3 & -3L \\ 3L & 0 & -3L & 3L^2 \end{bmatrix} \tag{1.25}$$

and the modified equivalent force vector $\{Q\}_m$ by

$$\{Q\}_m = \begin{Bmatrix} -Q_1 + \frac{3}{2L} Q_2 \\ 0 \\ -Q_3 - \frac{3}{2L} Q_2 \\ -Q_4 + \frac{1}{2} Q_2 \end{Bmatrix} \tag{1.26}$$

Case 2: Beam element with a hinge at node ②

Proceeding as in Case 1, we set the moment P_4 in eq.1.21(d) equal to zero followed by the calculation of the angular displacement δ_4 and its subsequent substitution into eqs.(1.21) to obtain for this case the modified element stiffness matrix $[k]_m$ and the modified equivalent force vector $\{Q\}_m$ as:

$$[k]_m = \frac{EI}{L^3} \begin{bmatrix} 3 & 3L & -3 & 0 \\ 3L & 3L^2 & -3L & 0 \\ -3 & -3L & 3 & 0 \\ 0 & 0 & 0 & 0 \end{bmatrix} \tag{1.27}$$

and

$$\{Q\}_m = \begin{Bmatrix} -Q_1 + \dfrac{3}{2L} Q_4 \\ -Q_2 + \dfrac{1}{2} Q_4 \\ -Q_3 - \dfrac{3}{2L} Q_4 \\ 0 \end{Bmatrix} \tag{1.28}$$

Case 3: Beam element with hinges at both ends

Finally, for the case in which both ends of the beam element are released (hinges at both ends), the modified stiffness matrix and the modified equivalent force vector calculated as in Cases 1 and 2 are

$$\{k\}_m = \begin{bmatrix} 0 & 0 & 0 & 0 \\ 0 & 0 & 0 & 0 \\ 0 & 0 & 0 & 0 \\ 0 & 0 & 0 & 0 \end{bmatrix} \tag{1.29}$$

and

$$\{Q\}_m = \begin{Bmatrix} -Q_1 + \dfrac{1}{L}(Q_2 + Q_4) \\ 0 \\ -Q_3 - \dfrac{1}{L}(Q_2 + Q_4) \\ 0 \end{Bmatrix} \tag{1.30}$$

Illustrative Example 1.5

Determine the system stiffness matrix and the system force vector for the beam of Illustrative Examples 1.1 and 1.2 after releasing the left end of element 4 by introducing a hinge at this location as shown in Fig. 1.10.

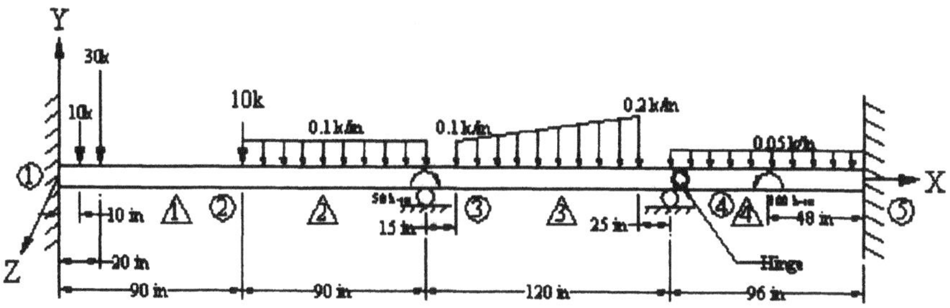

Fig. 1.10 Beam of Illustrative Example 1.5 with a hinge (release) at the left end of element 4.

Solution:

Using the results calculated in Illustrative Example 1.1 for the stiffness matrices for elements 1, 2 and 3, only the modified stiffness matrix and the modified equivalent force vector for element 4 must be determined. In this case (hinge at the first node), the modified element stiffness matrix is obtained from eq.(1.25) as

$$[K]_m = \frac{29\text{x}10^3\text{x}882}{96^3}\begin{bmatrix} 3 & 0 & -3 & 3\text{x}96 \\ 0 & 0 & 0 & 0 \\ -3 & 0 & 3 & -3\text{x}96 \\ 3\text{x}96 & 0 & -3\text{x}96 & 3\text{x}96^2 \end{bmatrix}$$

or

$$[k]_m = \begin{array}{c} \begin{matrix} 8 & \quad 4 & \quad 9 & \quad 10 \end{matrix} \\ \begin{bmatrix} 86.53 & 0 & -86.53 & 8307 \\ 0 & 0 & 0 & 0 \\ -86.53 & 0 & 86.53 & -8307 \\ 8307 & 0 & -8307 & 797500 \end{bmatrix} \end{array} \begin{matrix} 8 \\ 4 \\ 9 \\ 10 \end{matrix} \qquad (1.31)$$

The modified equivalent force vector for element 4 is then obtained by substituting into eq.(1.26) the values calculated in Illustrative Example 1.2 for Q_1, Q_2, Q_3, and Q_4:

$$\{Q\}_m = \begin{Bmatrix} 3.96 - \dfrac{3}{2\text{x}96}(63.40) \\ 0 \\ 0.84 + \dfrac{3}{2\text{x}96}(63.40) \\ -13.40 - \dfrac{1}{2}(63.40) \end{Bmatrix} = \begin{Bmatrix} 2.97 \\ 0 \\ 1.83 \\ -45.10 \end{Bmatrix} \begin{matrix} 8 \\ 4 \\ 9 \\ 10 \end{matrix} \qquad (1.32)$$

The system stiffness matrix is then assembled from the stiffness matrices developed in Illustrative Example 1.1 for elements 1, 2, and 3, and the modified stiffness matrix determined in eq.(1.31) for element 4. It should be realized that only the sub-matrix 4 x 4 corresponding to the first four nodal coordinates is needed for further calculation. Thus the reduced system stiffness matrix $[K]_R$ assembled from stiffness matrices for elements 1, 2, 3 (calculated in Illustrative Example1.1) and the modified stiffness matrix (given by eq.(1.31) for element 4), results in

$$[K]_R = \begin{Bmatrix} 840 & 0 & 18900 & 0 \\ 0 & 2268445 & 967111 & 0 \\ 18900 & 567111 & 1984889 & 425333 \\ 0 & 0 & 425333 & 850667 \end{Bmatrix} \quad (1.33)$$

The system force vector is obtained by substituting in eq.(b) of Illustrative Example (1.2) the new values calculated in eq.(1.32) for the modified equivalent force vector for the released element 4 of this beam. The reduced system force vector is then obtained as

$$\{F\}_R = \begin{Bmatrix} -2.29-4.50 \\ +64.20-67.50 \\ +67.50-142.82 \\ +152.10 \end{Bmatrix} + \begin{Bmatrix} -10 \\ 0 \\ -50 \\ 0 \end{Bmatrix} = \begin{Bmatrix} -16.79 \\ -3.30 \\ -125.32 \\ +152.10 \end{Bmatrix} \quad (1.34)$$

1.10 Support Displacements

When displacements at supports do occur due to yielding at the foundation or to imposed displacements at the nodal coordinates, the values of these displacements may be introduced in the corresponding entries of the vector $\{u\}$ of the system of nodal displacements. Alternatively, the effects of small imposed or support displacements are introduced in the analysis as equivalent forces at the nodal coordinates.

Consider a beam element having both displacements δ_1, δ_2, δ_3, and δ_4, due to the applied loads and imposed displacements Δ_1, Δ_2, Δ_3, and Δ_4, as shown in Fig.1.11(a) and (b), respectively. In this case the element nodal forces $\{P\}$ resulting from the total nodal displacement of $[\{\delta\}+\{\Delta\}]$ are given by eq.(1.19) as

$$\{P\} = [k]\left[\{\delta\} + \{\Delta\}\right] + \{FER\} \quad (1.35)$$

in which

$[k]$ = element stiffness matrix
$\{\delta\}$ = element nodal displacements resulting from external loads
$\{\Delta\}$ = imposed element nodal displacements
$\{FER\}$= fixed end reaction vector for loads applied on the span of the element

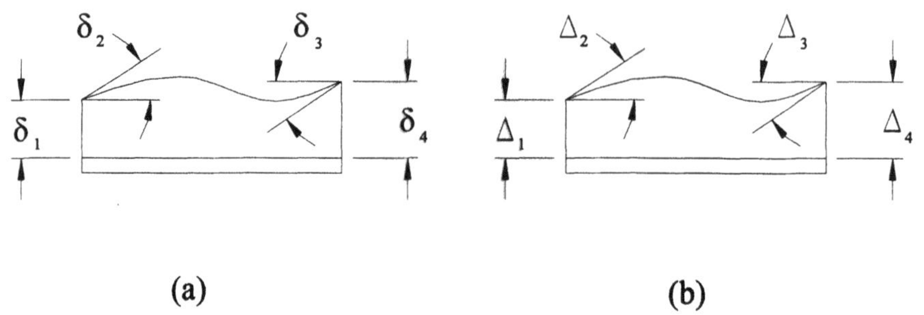

Fig. 1.11 Beam element showing: (a) nodal displacement $\delta_1, \delta_2, \delta_3$ and δ_4, resulting from applied loads; and (b) imposed nodal displacements Δ_1, Δ_2, Δ_3, and Δ_4.

From eq.(1.35) we have

$$\{P\} = [k]\{\delta\} + [k]\{\Delta\} + \{FER\}$$

or

$$\{P\} - [k]\{\Delta\} = [k]\{\delta\} + \{FER\}$$

or

$$\{P\} + \{Q\}_\Delta = [k]\{\delta\} + \{FER\} \qquad (1.36)$$

in which the equivalent nodal force vector $\{Q\}_\Delta$ resulting from the imposed displacements $\{\Delta\}$ is given by

$$\{Q\}_\Delta = -[k]\{\Delta\} \qquad (1.37)$$

It is seen from eq.(1.36) that the effect of the displacements $\{\Delta\}$ imposed at the nodal coordinates may be considered in the analysis by introducing the equivalent nodal forces $\{Q\}_\Delta$ given in eq.(1.37) for each element of the beam.

As presented in Section 1.5, the system stiffness matrix is assembled by transferring the coefficients of the element stiffness matrices $[k]$ to the proper locations in the system stiffness matrix $[K]$. The elements of the equivalent nodal forces $\{Q\}_\Delta$ due to imposed or support nodal displacements $\{\Delta\}$ are also transferred to the proper locations in the reduced system force vector $\{F\}_R$. The following example illustrates the necessary calculations.

Illustrative Example 1.6

Assume that the beam analyzed in Illustrative Examples 1.1 and 1.2 is subjected to a displacement of 1.0 (in) downward at joint ③ and of 2.0 (in) downward at joint ④. Determine:

1. Displacements at the free nodal coordinates.
2. Element end forces.
3. Reactions at the fixed nodal coordinates.

Solution:

1. Displacements at the Free Nodal Coordinates.

The reduced system stiffness matrix and the reduced system force vector resulting from the applied loads are given by the reduced stiffness matrix $[K]_R$ and the reduced force vector $\{F\}_R$ in eq.(a) of Illustrative Example 1.3 as

$$[K]_R = \begin{bmatrix} 840 & 0 & 18900 & 0 \\ 0 & 2668445 & 567110 & 0 \\ 18900 & 567110 & 1984867 & 425333 \\ 0 & 0 & 425333 & 1914000 \end{bmatrix} \tag{a}$$

and

$$\{F\}_R = \begin{Bmatrix} -16.79 \\ -3.30 \\ -125.32 \\ +88.70 \end{Bmatrix} \begin{matrix} 1 \\ 2 \\ 3 \\ 4 \end{matrix} \tag{b}$$

To include the effect of the imposed nodal displacements it is necessary to recalculate the equivalent force vectors for the elements 2, 3, and 4, since these elements are affected by the displacements and supports ③ and ④. The application of eq.(1.37) to elements 2, 3 and 4 for which the stiffness matrices are given in, respectively, eqs.(b), (c) and (d) of Illustrative Example 1.1 results in

ELEMENT 2

$$\begin{Bmatrix} Q_1 \\ Q_2 \\ Q_3 \\ Q_4 \end{Bmatrix}_{(2)} = -\begin{bmatrix} 420 & 18900 & -420 & 18900 \\ 18900 & 1134200 & -18900 & 567110 \\ -420 & -18900 & 420 & -18900 \\ 18900 & 567110 & -18900 & 1134200 \end{bmatrix} \begin{Bmatrix} 0 \\ 0 \\ -1 \\ 0 \end{Bmatrix} = \begin{Bmatrix} -420 \\ -18900 \\ 420 \\ -18900 \end{Bmatrix}_{(2)} \begin{matrix} 1 \\ 2 \\ 7 \\ 3 \end{matrix} \tag{c}$$

ELEMENT 3

$$\begin{Bmatrix} Q_1 \\ Q_2 \\ Q_3 \\ Q_4 \end{Bmatrix}_{(3)} = -\begin{bmatrix} 177 & 10633 & -177 & 10633 \\ 10633 & 850667 & -10633 & 425333 \\ -177 & -10633 & 177 & -10633 \\ 10633 & 425333 & -10633 & 850667 \end{bmatrix} \begin{Bmatrix} -1 \\ 0 \\ -2 \\ 0 \end{Bmatrix} = \begin{Bmatrix} -177 \\ -10633 \\ 177 \\ -10633 \end{Bmatrix}_{(3)} \begin{matrix} 7 \\ 3 \\ 8 \\ 4 \end{matrix} \tag{d}$$

ELEMENT 4

$$\begin{Bmatrix} Q_1 \\ Q_2 \\ Q_3 \\ Q_4 \end{Bmatrix}_{(4)} = -\begin{bmatrix} 346 & 16615 & -346 & 16615 \\ 16615 & 1063333 & -16615 & 531667 \\ -346 & -16615 & 346 & -16615 \\ 16615 & 531667 & -16615 & 1063333 \end{bmatrix} \begin{Bmatrix} -2 \\ 0 \\ 0 \\ 0 \end{Bmatrix} = \begin{Bmatrix} 692 \\ 33230 \\ -692 \\ 33230 \end{Bmatrix}_{(4)} \begin{matrix} 8 \\ 4 \\ 9 \\ 10 \end{matrix} \tag{e}$$

The reduced system force vector, including the imposed support displacements, is then obtained by transferring into the force vector $\{F\}_R$ in eq.(b) the equivalent element nodal forces calculated in eqs. (c) through (e). The equivalent nodal forces calculated for elements 2, 3, and 4 are transferred to the reduced system force vector at locations corresponding to the system nodal coordinates as indicated on the right of eqs.(c), (d), and (e):

$$\begin{Bmatrix} F_1 \\ F_2 \\ F_3 \\ F_4 \end{Bmatrix}_R = \begin{bmatrix} -16.79-420 \\ -3.30-18900 \\ -125.32-18900-10633 \\ +88.70-10633+33230 \end{bmatrix} = \begin{Bmatrix} -436.79 \\ -18903.30 \\ -29658.32 \\ +22685.70 \end{Bmatrix} \tag{f}$$

Substituting the system force vector [eq.(f)] for the force vector in the system stiffness equation [eq.(a)] of Illustrative Example 1.3, gives

$$\begin{Bmatrix} -436.79 \\ -18903.30 \\ -29658.32 \\ +22685.70 \end{Bmatrix} = \begin{Bmatrix} 840 & 0 & 18900 & 0 \\ 0 & 2268445 & 567110 & 0 \\ 18900 & 567110 & 1984867 & 425333 \\ 0 & 0 & 425333 & 1914000 \end{Bmatrix} \begin{bmatrix} u_1 \\ u_2 \\ u_3 \\ u_4 \end{bmatrix} \tag{g}$$

The solution of eq.(g) results in

$$u_1 = -0.177\,in \qquad u_2 = -4.53(10^{-3})\,rad$$
$$u_3 = -1.52(10^{-2})\,rad \qquad u_4 = 1.52(10^{-2})\,rad$$

2. Element End Forces.

The element end-forces are calculated using eq.(1.20), which for convenience is repeated here:

$$\{P\} = [k]\{\delta\} - \{Q\} \qquad \text{(1.20) repeated}$$

in which

$[k]$ = element stiffness matrix
$\{\delta\}$ = element nodal displacements
$\{Q\}$ = element equivalent nodal forces including the equivalent nodal forces due to imposed support displacements

The element nodal displacements, excluding imposed displacements (the effect of these imposed displacements are considered as equivalent nodal forces) are identified from the solution of eq.(g) as

$$\{\delta\}_1 = \begin{Bmatrix} 0 \\ 0 \\ -0.177 \\ -4.53\text{x}10^{-3} \end{Bmatrix} \quad \{\delta\}_2 = \begin{Bmatrix} -0.177 \\ -4.53\text{x}10^{-3} \\ 0 \\ -1.52\text{x}10^{-2} \end{Bmatrix}$$
$$\{\delta\}_3 = \begin{Bmatrix} 0 \\ 1.52\text{x}10^{-2} \\ 0 \\ 1.52\text{x}10^{-2} \end{Bmatrix} \quad \{\delta\}_4 = \begin{Bmatrix} 0 \\ 1.52\text{x}10^{-2} \\ 0 \\ 0 \end{Bmatrix} \qquad \text{(h)}$$

The total element equivalent nodal forces are then obtained by adding the equivalent forces due to imposed nodal displacements as given by eq.(c) through (e) to the nodal forces calculated in Illustrative Example 1.2, namely,

$$\{Q\}_1 = \begin{Bmatrix} -37.71 \\ -358.02 \\ -2.29 \\ 64.20 \end{Bmatrix} \quad \{Q\}_2 = \begin{Bmatrix} -4.50-4.20 \\ -67.50-18900 \\ -4.50+4.20 \\ +67.50-18900 \end{Bmatrix}$$
$$\{Q\}_3 = \begin{Bmatrix} -5.39-177 \\ -142.82-106.33 \\ -5.86+177 \\ 152.10-10633 \end{Bmatrix} \quad \{Q\}_4 = \begin{Bmatrix} -3.96+692 \\ -63.40+33230 \\ -0.84-692 \\ +13.40+33230 \end{Bmatrix} \qquad \text{(i)}$$

Finally, substituting the nodal displacement vector $\{\delta\}_1$ from eqs.(h), the equivalent nodal force vector $\{Q\}_1$ from eqs.(i) and the element stiffness matrix $[k]^{(1)}$ calculated in Illustrative Example 1.1 into eq.(1.20), for element 1 results in

$$\{P\}_1 = \begin{bmatrix} -420 & 18900 & -420 & 18900 \\ 18900 & 1134200 & -18900 & 567110 \\ -420 & -18900 & 420 & -18900 \\ 18900 & 567110 & -18900 & 1134200 \end{bmatrix} \begin{Bmatrix} 0 \\ 0 \\ -0.177 \\ -4.53(10^{-3}) \end{Bmatrix} - \begin{Bmatrix} +37.71 \\ -358.02 \\ -2.29 \\ -64.20 \end{Bmatrix} = \begin{Bmatrix} 26.63 \\ 1143.71 \\ 13.37 \\ -1846.66 \end{Bmatrix}_1 \quad \text{(j)}$$

Analogously, we obtain from elements 2 , 3 and 4 the end forces

$$\{P\}_2 = \begin{Bmatrix} -23.36 \\ 1846.66 \\ 32.36 \\ -4354.61 \end{Bmatrix}_2 \quad \{P\}_3 = \begin{Bmatrix} 182.72 \\ 4304.61 \\ -171.47 \\ 16965.57 \end{Bmatrix}_3 \quad \{P\}_4 = \begin{Bmatrix} -435.18 \\ -16965.57 \\ 439.98 \\ -25142.47 \end{Bmatrix}_4 \quad \text{(k)}$$

3. Reactions at the Fixed Nodal Coordinates.

The reactions R_5 and R_6 at the fixed support of joint ① are identified as the forces P_{11} and P_{21} at the nodal coordinates of joint ① of element 1. Analogously, the reactions R_9 and R_{10} at the fixed support of joint ⑤ are identified as the forces P_{34} and P_{44} of element 4. Thus, from eqs.(j) and (k), we have:

$$R_5 = P_{11} = 26.63 \text{ (kip)} \qquad R_6 = P_{21} = 1143.71 \text{ (kip.in)}$$

and

$$R_9 = P_{34} = 439.98 \text{ (kip)} \qquad R_{10} = P_{44} = -25142.47 \text{ (kip.in)}$$

At the interior support joint ③ the reaction R_7 is equal to the sum of the end forces at this joint for elements 2 and 3. Analogously, the reactions R_8 at the interior support at joint ④ is equal to the sum of the end forces at this joint for elements 3 and 4. Thus, from eqs.(j) and (k) we have:

$$R_7 = P_{32} + P_{13} = 32.36 + 182.72 = 215.08 \text{ (kip)}$$
$$R_8 = P_{33} + P_{14} = -171.47 - 435.18 = -606.65 \text{ (kip)}$$

1.11 Element Displacement Functions

The displacement function $y(x)$ due to the nodal displacements δ_1 , δ_2, δ_3 , and δ_4 for an element of the beam is given by eq.(1.6) in which the coefficients $N_1(x)$, $N_2(x)$, $N_3(x)$ and $N_4(x)$ are given by eqs.(1.5). The total displacement $y_T(x)$ at any section x of a beam element is then calculated as the superposition of $y(x)$ due to the element nodal displacements and the displacement $y_L(x)$ resulting from the loads applied to the beam element assumed to be fixed at its two nodes; that is,

$$y_T(x) = y(x) + y_L(x) \tag{1.38}$$

The determination of the displacement function $y_L(x)$ due to an applied load on a beam element may be obtained by integrating twice the differential equation for a beam:

$$\frac{d^2y(x)}{dx^2} = \frac{M(x)}{EI} \tag{1.39}$$

in which

$M(x)$ = bending moment
E = modulus of elasticity
I = cross-sectional moment of inertia

The determination of the displacement function $y_L(x)$ of a fixed-end beam supporting a uniformly distributed force is presented in the following Illustrative Example 1.7.

Illustrative Example 1.7

Determine the displacement function for a fixed-end beam element with an applied uniformly distributed force w as shown in Fig. 1.9(a).

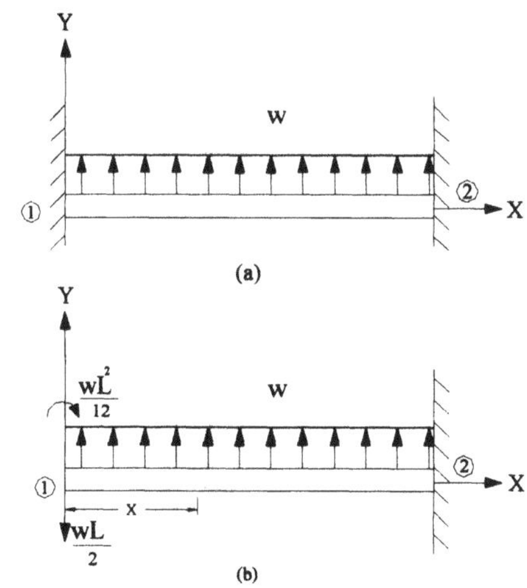

Fig. 1.12 (a) Fixed-end beam element supporting a uniformly distributed force
(b) Beam element showing the fixed-end reactions at node ①.

Solution:

Consider in Fig. 1.12(a) a fixed-end beam element supporting a uniformly distributed force w applied in the positive direction of axis y and in Fig. 1.12(b) the support reactions at node ① for this beam element. These support reactions are numerically equal to the equivalent nodal forces but with opposite signs as given in Appendix I, Case c. The bending moment at any section x is then given from Fig. 1.12(b) by

$$M(x) = \frac{wL^2}{12} - \frac{wL}{2}x + \frac{w}{2}x^2 \tag{1.40}$$

The substitution of eq.(1.40) into eq.(1.39) followed by two successive integrations results in

$$EI\frac{d^2y_L(x)}{dx^2} = \frac{wL^2}{12} - \frac{wL}{2}x + \frac{w}{2}x^2$$

$$EI\frac{dy_L(x)}{dx} = \frac{wL^2}{12}x - \frac{wL}{4}x^2 + \frac{w}{6}x^3 + C_1 \tag{1.41}$$

$$EIy_L(x) = \frac{wL^2}{24}x^2 - \frac{wL}{12}x^3 + \frac{w}{24}x^4 + C_1x + C_2$$

The constants of integration C_1 and C_2 are evaluated from the boundary conditions of zero displacement and zero slope at $x = 0$, giving $C_1 = C_2 = 0$. Thus from eq.(1.41), the displacement function y_L for a fixed-end beam element with a uniformly distributed force is:

$$y_L(x) = \frac{1}{EI}\left[\frac{wL^2}{24}x^2 - \frac{wL}{12}x^3 + \frac{w}{24}x^4\right] \tag{1.42}$$

Analogously, the displacement functions for other common loads applied to a fixed-end beam element may be determined by integrating eq.(1.39) for each specific load, as it has been presented for a uniformly distributed force in Illustrative Example 1.7.

1.12 Shear Force and Bending Moment Functions

The shear force and bending moment at a section of beam element may readily be found from static equations of equilibrium. Consider in Fig. 1.13(a) a loaded beam element showing the end element forces and in Fig.1.13(b) the free body diagram of a segment of this beam element.

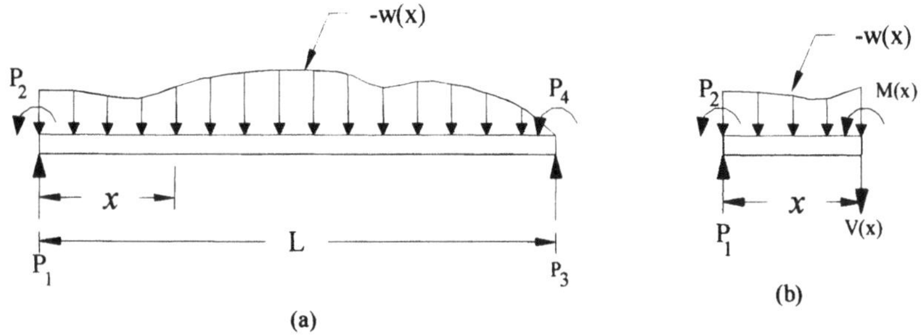

Fig. 1.13 (a) Beam element showing end element forces and the applied external forces, and (b) Free body diagram up to section *x* of the beam

By equating to zero the sum of the forces and also by equating to zero the sum of the moments of forces in the free body diagram shown in Fig 1.13(b), we obtain:

$$-V(x) + P_1 + \int_0^x w(x)dx = 0$$

and

$$M(x) + P_2 - P_1 x - \int_0^x x\, w(x)dx = 0$$

or

$$V(x) = P_1 + \int_0^x w(x)dx \tag{1.43}$$

and

$$M(x) = -P_2 + P_1 x + \int_0^x x\, w(x)dx \tag{1.44}$$

Equations (1.43) and (1.44) provide, respectively, the functions to calculate the shear force $V(x)$ and the bending moment $M(x)$ for a beam element in terms of the element end forces P_1 and P_2 and of the force $w(x)$ applied to the element.

1.13 The Temperature Effect

Temperature changes, like support displacements, can cause large stresses in statically indeterminate structures, but not in statically determined structures in which members are free to expand or contract due to changes in temperature. In the case of beams, a uniform change in temperature in the cross-section does not induce stresses, since the analysis of beams does not consider axial deformations along the longitudinal axis of the beam. However, stresses can develop in beams that are subjected to a temperature variation across the depth of the beam.

As in the case of loads applied on the beam elements, the effect of a temperature change in the cross section of the beam is considered in the analysis by introducing equivalent forces at the nodes of the beam elements subjected to temperature changes. Consider in Fig. 1.14(a) a beam element that is subjected to a linear temperature change over the depth of the cross sectional area with a temperature T_1 at the bottom and T_2 at the top of the beam element. Assuming that $T_2 > T_1$, an isolated beam element will curve as shown in Fig.1.14(a). For a symmetric cross sectional area with respect to the centroidal axis, the expression for the temperature function *T(y)* in the cross section of the beam is depicted in Fig. 1.14(b).

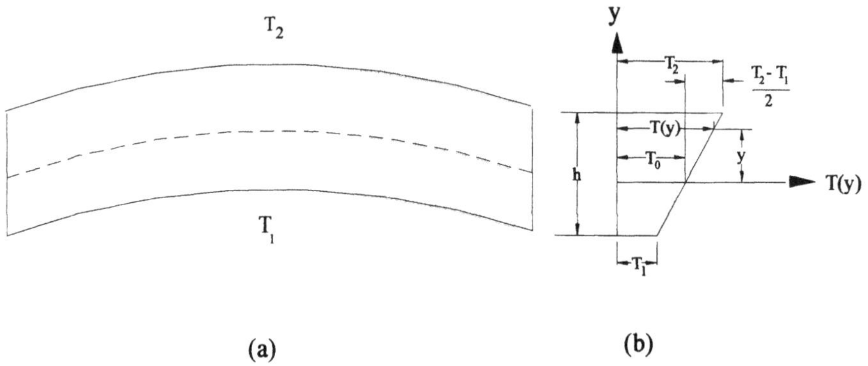

Fig. 1.14 (a) Beam element subjected to a linear temperature change in the cross section (b) Temperature function *T(y)* along the depth of the beam

The temperature T_0 at the center of the cross sectional area of the beam is given by the average of the temperatures at the two faces of the beam:

$$T_0 = \frac{T_1 + T_2}{2} \tag{1.45}$$

Then, from Fig. 1.14(b)

$$T(y) = T_0 + (T_2 - T_1)\frac{y}{h}$$

or substituting T_0 from eq.(1.45)

$$T(y)=\frac{T_1+T_2}{2}+(T_2-T_1)\frac{y}{h} \tag{1.46}$$

The strain, $\epsilon(y)$ in the cross section of any point at a distance y from the centroid is $\epsilon(y) = \alpha T(y)$ where α is the coefficient of thermal expansion. The corresponding stress for fixed-end conditions is $\sigma(y) = E\alpha T(y)$. The thermal axial force P_T in the axially constrained beam element will be

$$P_T = \int_{(A)} \sigma(y)\, dA = \int_{(A)} E\alpha\ T(y)\, dA \tag{1.47}$$

and the thermal bending moment

$$M_T = \int_A y\,\sigma(x)\, dA = \int_A E\alpha\ y T(y)\, dA \tag{1.48}$$

in which A is the cross-sectional area of the beam.

For a beam having a rectangular cross sectional area of width b, eqs. (1.47) and (1.48) become

$$P_T = \int_{-h/2}^{h/2} E\alpha\ T(y)\, b\, dy = E\alpha A T_0$$

and

$$M_T = \int_{-h/2}^{h/2} \alpha E T(y)\, b y\, dy = \alpha \frac{EI}{h}(T_2 - T_1) \tag{1.49}$$

The equivalent nodal forces for a beam element subjected to a linear temperature change in its cross section are then given by

$$Q_1=0 \qquad Q_2=-\frac{\alpha EI}{h}(T_2-T_1) \quad Q_3=0 \qquad Q_4=\frac{\alpha EI}{h}(T_2-T_1) \tag{1.50}$$

or in the vector notation

$$\{Q\}_T = \begin{Bmatrix} 0 \\ -\frac{\alpha EI}{h}(T_2 - T_1) \\ 0 \\ +\frac{\alpha EI}{h}(T_2 - T_1) \end{Bmatrix} \tag{1.51}$$

These equivalent nodal forces for a beam element subjected to a linear temperature variation along the height of the cross-section are shown in Fig. 1.15.

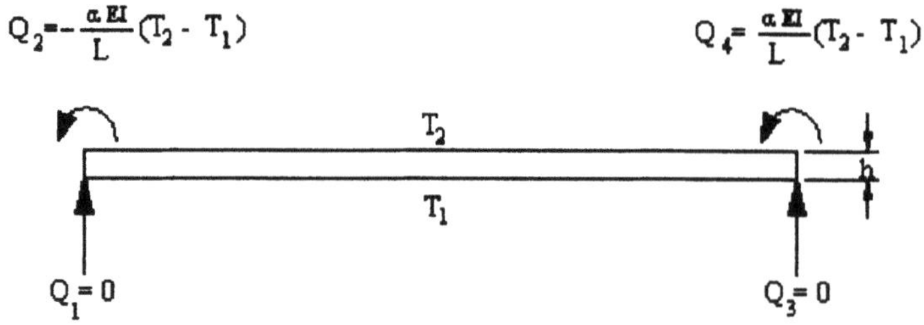

Fig. 1.15 Beam element showing the equivalent nodal forces due to a linear temperature variation in the cross-section.

Illustrative Example 1.8

Element 3 of the beam shown in Fig. 1.5 is now subjected to a linear temperature change with $T_1 = 70°F$ at the bottom face and $T_2 = 120°F$ at the top face. The coefficient of thermal expansion is $\alpha = 6.5 \times 10^{-6}/°F$ and the height of the cross sectional area of the beam is h = 10(in). Determine the nodal displacements resulting from this temperature variation.

Solution:

The element stiffness matrices [eq.(a) through eq.(d)] and the system stiffness matrix [eq.(e)] for this beam have been determined in Illustrative Example 1.1. The equivalent nodal force vector for element 3 subjected to a linear variation of temperature is given by eq.(1.51), which after substituting numerical values results in

$$\{Q\}_T = \begin{Bmatrix} 0 \\ -829.4 \\ 0 \\ 829.4 \end{Bmatrix} \tag{a}$$

Substituting $(Q\}_T$ from eq.(a) for the force vector in the left hand side of eq.(a) of Illustrative Example 1.3 yields

$$\begin{Bmatrix} 0 \\ -829.4 \\ 0 \\ 829.4 \end{Bmatrix} = \begin{bmatrix} 840 & 0 & 18900 & 0 \\ 0 & 2268400 & 567110 & 0 \\ 18900 & 567110 & 1984867 & 425333 \\ 0 & 0 & 425333 & 1914000 \end{bmatrix} \begin{Bmatrix} u_1 \\ u_2 \\ u_3 \\ u_4 \end{Bmatrix} \tag{b}$$

The solution of eq.(b) for the unknown displacements results in

$$\begin{aligned} u_1 &= 1.724\ 10^{-2}\ \text{(in)} \\ u_2 &= 1.915\ 10^{-4}\ \text{(rad)} \\ u_3 &= -7.661\ 10^{-4}\ \text{(rad)} \\ u_4 &= 6.036\ 10^{-4}\ \text{(rad)} \end{aligned}$$

1.14 Elastic Supports

Linear elastic supports on a beam, whether resulting from axial or torsional springs may be considered in the analysis by simply adding the value of the spring constant (stiffness coefficient) to the corresponding coefficient in the diagonal of the system stiffness matrix. In the following Illustrative Example, the stiffness method of analysis is applied to a beam having an elastic support.

Illustrative Example 1.9

Consider in Fig. 1.16 a fixed-end beam supported by a spring at its center having an axial stiffness k_a = 100 kip/in and a torsional stiffness k_t = 1000 kip-in/rad. For this beam determine: (a) the vertical and rotational displacements at the central section, (b) the reactions at the supports, and (c) the end forces on the two elements of the beam.

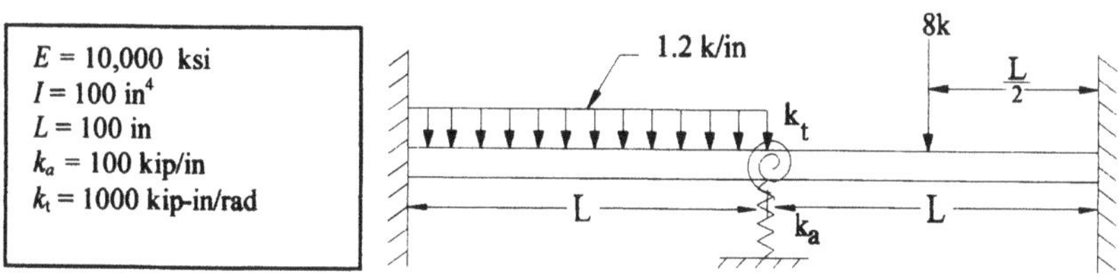

Fig. 1.16 Beam of Illustrative Example 1.9.

Solution:

1. Analytical model.

The analytical model for this beam is shown in Fig. 1.17. It has 2 beam elements, 3 nodes, and 2 free nodal coordinates labeled as u_1, and u_2.

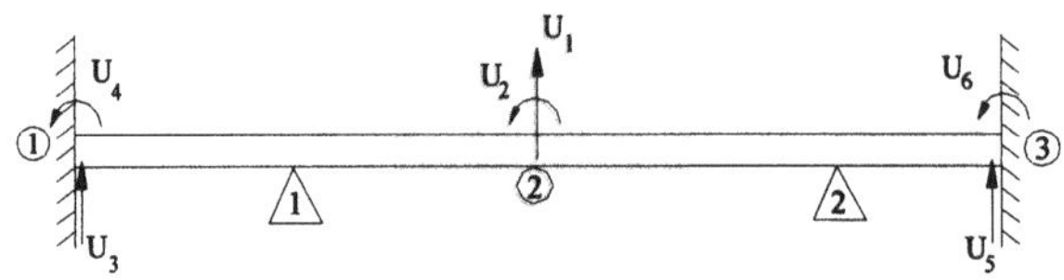

Fig. 1.17 Analytical model for the beam of Illustrative Example 1.9

2. Element stiffness matrices.

Substituting numerical values into the element stiffness matrix eq.(1.11) results in

$$
\begin{array}{cccc} 1 & 2 & 5 & 6 \\ 3 & 4 & 1 & 2 \end{array}
$$

$$
[k]_1 = [k]_2 = \frac{10^4 \text{ x } 10^2}{(100)^3} \begin{bmatrix} 12 & 600 & -12 & 600 \\ 600 & 40000 & -600 & 20000 \\ -12 & -600 & 12 & -600 \\ 600 & 20000 & -600 & 40000 \end{bmatrix} \begin{array}{cc} 3 & 1 \\ 4 & 2 \\ 1 & 5 \\ 2 & 6 \end{array} \tag{a}
$$

3. Reduced system stiffness matrix.

The reduced system stiffness matrix is assembled by transferring to this matrix the coefficients of the matrix in eq.(a) to locations indicated for the two elements of this beam at the top and on the right side of this matrix. The axial stiffness k_a = 100 kip/in and the torsional stiffness k_t = 1000 kip-in/rad are then added, respectively, to the diagonal coefficients on rows 1 and 2 of the reduced system stiffness matrix, namely,

$$
[K]_R = \begin{bmatrix} 12+12+100 & -600+600 \\ -600+600 & 40000+40000+1000 \end{bmatrix}
$$

or

$$
[K]_R = \begin{bmatrix} 124 & 0 \\ 0 & 81000 \end{bmatrix} \tag{b}
$$

4. Element equivalent force vectors.

ELEMENT 1:

From Case (c) in Appendix I (with w = -1.2 kip/in):

$$Q_1 = \frac{wL}{2} = -\frac{1.2\text{x}100}{2} = -60\,\text{kip}$$

$$Q_2 = \frac{wL^2}{12} = -\frac{1.2\text{x}100^2}{12} = -1000\,\text{kip.in/rad}$$

$$Q_3 = \frac{wL}{2} = -\frac{1.2\text{x}100}{2} = -60\,\text{kip}$$

$$Q_4 = -\frac{wL^2}{12} = \frac{1.2\text{x}100^2}{12} = 1000\,\text{kip.in/rad}$$

or in vector form

$$\{Q\}_1 = \begin{Bmatrix} -60 \\ -1000 \\ -60 \\ 1000 \end{Bmatrix} \begin{matrix} 3 \\ 4 \\ 1 \\ 2 \end{matrix} \qquad \text{(c)}$$

ELEMENT 2:

From Case (a) in Appendix I (with W = -8 kip):

$$Q_1 = \frac{W}{2} = -\frac{8}{2} = -4\,\text{kip}$$

$$Q_2 = \frac{WL}{8} = -\frac{8\text{x}100}{8} = -100\,\text{kip-in/rad}$$

$$Q_3 = \frac{W}{2} = -\frac{8}{2} = -4\,\text{kip}$$

$$Q_4 = -\frac{WL}{8} = \frac{8\text{x}100}{8} = 100\,\text{kip-in/rad}$$

or in vector notation:

$$\{Q\}_2 = \begin{Bmatrix} -4 \\ -100 \\ -4 \\ 100 \end{Bmatrix} \begin{matrix} 1 \\ 2 \\ 5 \\ 6 \end{matrix} \qquad \text{(d)}$$

5. Reduced system force vector.

The coefficients of equivalent force vectors in eqs. (c) and (d) are transferred to reduced system force vector in the location indicated on the right of these vectors, namely,

$$\{F\}_R = \begin{Bmatrix} -60-4 \\ 1000-100 \end{Bmatrix} = \begin{Bmatrix} -64 \\ 900 \end{Bmatrix} \tag{e}$$

6. Reduced system stiffness equation.

The reduced system stiffness matrix is given by

$$\{F\}_R = [K]_R\{u\}$$

Substitution of $\{F\}_R$ from eq. (e) and $[K]_R$ from eq. (b) results in

$$\begin{Bmatrix} -64 \\ 900 \end{Bmatrix} = \begin{bmatrix} 124 & 0 \\ 0 & 81000 \end{bmatrix} \begin{bmatrix} u_1 \\ u_2 \end{bmatrix} \tag{f}$$

7. Solution of eq. (f).

$$u_1 = -0.516 \text{ (in)} \quad \text{and} \quad u_2 = 0.0111 \text{ (rad)}$$

8. Element end forces.

Element end forces are given by eq.(1.20) as

$$\{P\} = [k]\{\delta\} - \{Q\} \tag{g}$$

ELEMENT 1:

The nodal displacement for element 1 are identified from Fig. 1.17 as

$$\{\delta\}_1 = \begin{Bmatrix} u_3 \\ u_4 \\ u_1 \\ u_2 \end{Bmatrix} = \begin{Bmatrix} 0 \\ 0 \\ -0.516 \\ 0.0111 \end{Bmatrix} \tag{h}$$

The substitution of eqs.(a) and (h) into eq.(g) results in

$$\begin{Bmatrix} P_1 \\ P_2 \\ P_3 \\ P_4 \end{Bmatrix}_1 = \begin{bmatrix} 12 & 600 & -12 & 600 \\ 600 & 40000 & -600 & 20000 \\ -12 & -600 & 12 & -600 \\ 600 & 20000 & -600 & 40000 \end{bmatrix} \begin{Bmatrix} 0 \\ 0 \\ -0.516 \\ 0.0111 \end{Bmatrix} - \begin{Bmatrix} -60 \\ -1000 \\ -60 \\ 1000 \end{Bmatrix}$$

ELEMENT 2:

Analogously, for element 2 , we obtain

$$\begin{Bmatrix} P_1 \\ P_2 \\ P_3 \\ P_4 \end{Bmatrix}_2 = \begin{bmatrix} 12 & 600 & -12 & 600 \\ 600 & 40000 & -600 & 20000 \\ -12 & -600 & 12 & -600 \\ 600 & 20000 & -600 & 40000 \end{bmatrix} \begin{Bmatrix} -0.516 \\ 0.0111 \\ 0 \\ 0 \end{Bmatrix} - \begin{Bmatrix} -4 \\ -100 \\ -4 \\ 100 \end{Bmatrix}$$

or

$$\begin{Bmatrix} P_1 \\ P_2 \\ P_3 \\ P_4 \end{Bmatrix}_1 = \begin{Bmatrix} 72.85 \\ 1531.6 \\ 47.15 \\ -245.4 \end{Bmatrix}_1 \quad \text{and} \quad \begin{Bmatrix} P_1 \\ P_2 \\ P_3 \\ P_4 \end{Bmatrix}_2 = \begin{Bmatrix} 4.47 \\ 234.40 \\ 3.53 \\ -187.6 \end{Bmatrix}_2$$

9. Reactions.

$$R_3 = P_{11} = 72.85\,\text{kip}$$
$$R_4 = P_{21} = 1531.6\,\text{kip-in}$$
$$R_5 = P_{32} = 3.53\ \text{kip}$$
$$R_6 = P_{42} = -187.6\,\text{kip-in}$$

10. Forces at the elastic support.

Spring axial force:

$$\begin{aligned} S_a &= P_{31} + P_{12} \\ &= 47.15 + 4.47 \\ &= 51.62\ \text{kip (Compression)} \end{aligned}$$

Spring torsional moment S_t:

$$\begin{aligned} S_t &= P_{41} + P_{22} \\ &= -245.4 + 234.4 \\ &= 11.0\ \text{kip·in} \end{aligned}$$

11. Notation: P_{ik} = Force at nodal coordinate *i* of beam element *k*.

Alternatively, the forces S_a and S_t may be calculated as the product of the respective spring constants and the corresponding displacements. Namely,

$$S_a = (100\text{kip/in})(-0..516\,\text{in}) = -51.6\,\text{kip} \quad \text{(Compression)}$$

and

$$S_t = (100\,\text{kip·in/rad})(0.0111\,\text{rad}) = 11.1 \;\; \text{kip·in}$$

1.15 Analytical Problems

Problem 1.1

Derive the general expression [eq.(1.7)] to calculate the stiffness coefficient for a beam element.

$$k_{ij} = \int_0^L EI\; N_i''(x)\; N_j''(x)\, dx \qquad \text{(1.7) repeated}$$

in which $N_i''(x)$ and $N_j''(x)$ are the second derivatives of the shape functions given by eqs.(1.5), I is the cross-sectional moment of inertia, and E the modulus of elasticity.

Solution:

Consider in Fig.P1.1(a), a beam element and in (b) and (c) the plots, respectively, of the shape function corresponding to unit displacements $\delta_1 = 1.0$ and $\delta_2 = 1.0$.

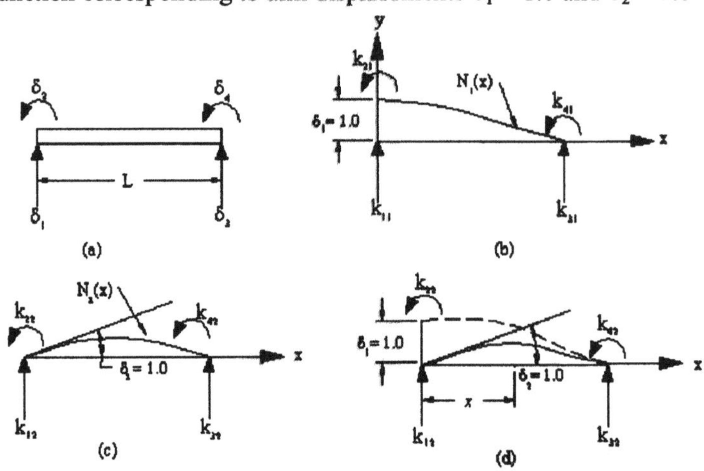

Fig. P1.1 Beam element: (a) nodal displacements; (b) stiffness coefficients for a unit displacement $\delta_1 = 1.0$, (c) stiffness coefficients for $\delta_2 = 1.0$., and (d) virtual displacement $\delta_1 = 1$ given to beam at (c).

The displacement functions $N_1(x)$ and $N_2(x)$ corresponding to these unit displacements are given by eqs.(1.5a) and (1.5b). These functions are repeated here for convenience:

$$N_1(x) = 1 - 3\left(\frac{x}{L}\right)^2 + 2\left(\frac{x}{L}\right)^3 \qquad \text{(1.5a) repeated}$$

$$N_2(x) = x\left(1 - \frac{x}{L}\right)^2 \qquad \text{(1.5b) repeated}$$

We assume, as it is shown in Fig. P1.1(d), that the beam element in Fig. P1.1(c) undergoes an additional displacement equal to the deflected curve shown in Fig. P1.1(b). We then apply the principle of virtual work which states that for an elastic system in equilibrium, the work done by the external forces is equal to the work of the internal forces during the virtual displacement. In order to apply this principle, we note that the external work W_E is equal to the product of the force k_{12} displaced by $\delta_1 = 1.0$, that is

$$W_E = k_{12}\delta_1 = k_{12} \qquad \text{(a)}$$

This work, as stated, is equal to the work performed by the elastic forces during the virtual displacement. Considering the work performed by the bending moment, we obtain for the internal work

$$W_i = \int_0^L M(x)\; d\theta \qquad \text{(b)}$$

in which $M(x)$ is the bending moment at section x of the beam and $d\theta$ the relative angular displacement at this section. For the virtual displacement under consideration, the transverse deflection of the beam is given by eq.(1.5b) which is related to the bending moment through the differential equation [eq.(1.43)]

$$\frac{d^2y}{dx^2} = \frac{M(x)}{EI} \qquad \text{(1.39) repeated}$$

Substitution of the second derivative $N_2''(x)$ from eq.(1.5b) into eq.(1.39) results in

$$EI\, N_2''(x) = M(x)$$

The angular deflection $d\theta$ produced during this virtual displacement is related to the transverse deflection $N_1(x)$ by

$$\frac{d\theta}{dx} = \frac{d^2N_1(x)}{dx^2} = N_1''(x)$$

or

$$d\theta = N_1''(x)\,dx \tag{d}$$

Equating the virtual external work W_E from eq.(a) with the internal virtual work W_i [(eq.(b)] after substituting $M(x)$ and $d\theta$ from eqs.(c) and (d), respectively, results in the stiffness coefficient k_{12} as

$$k_{12} = \int_0^L EI\ N_1''(x)\ N_2''(x)\ dx$$

Therefore, in general, any stiffness coefficient associated with beam flexure may be expressed as

$$k_{ij} = \int_0^L EI\ N_i''(x)\ N_j''(x)\ dx \qquad Q.E.D.$$

Problem 1.2

Demonstrate that the equivalent nodal force Q_i for a distributed force $w(x)$ applied to a beam element is given by

$$Q_i = \int_0^L N_i(x)\,w(x)\ dx \qquad \text{(1.14a) repeated}$$

Solution:

Consider in Fig. P1.2(a) a beam element showing a general applied force $w(x)$ and in Fig. P1.2(b) the beam element showing the nodal equivalent forces Q_1 through Q_4. We give to both beam elements a unit virtual displacement at nodal coordinate 1 resulting in the shape function $N_1(x)$ given by eq.(1.5a) which is repeated here for convenience:

$$N_1(x) = 1 - 3\left(\frac{x}{L}\right)^2 + 2\left(\frac{x}{L}\right)^3 \qquad \text{(1.5a) repeated}$$

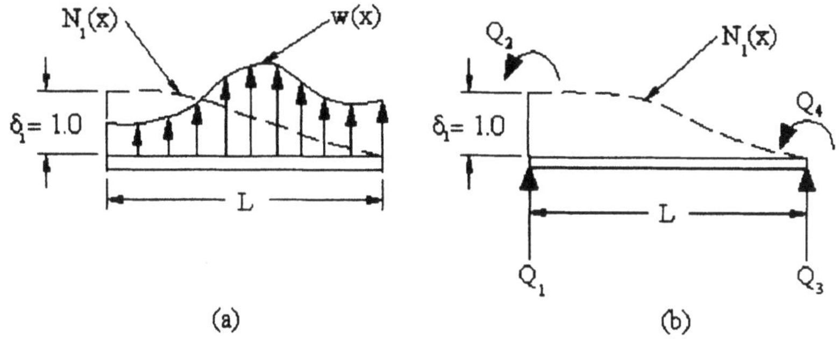

Fig. P1.2 (a) Beam element acted on by a distributed force $w(x)$ undergoing a virtual displacement $\delta_1 = 1.0$ resulting in the shape function $N_1(x)$, and (b) The beam element acted upon by the equivalent nodal forces Q_i undergoing the same virtual displacement, $\delta_1 = 1.0$.

Now we require that the work done as a result of this virtual displacement by the equivalent forces Q_i be equal to the work performed by the externally applied forces $w(x)$. The work $W_E^{'}$ of the equivalent forces Q_i is simply

$$W_E^{'} = Q_1\delta_1 = Q_1 \tag{a}$$

since Q_1 is the only equivalent force undergoing displacement and $\delta_1 = 1.0$.

The work performed by the applied force $w(x)dx$ during the virtual displacement on the differential beam segment of length dx is $N_1(x)\ w(x)\ dx$. The total work is then

$$W_E^{''} = \int_0^L N_1(x)\ w(x)\ dx \tag{b}$$

Equating these two calculations [eqs.(a) and (b)] of the virtual work results in the following expression for the equivalent nodal force Q_1:

$$Q_1 = \int_0^L N_1(x)\ w(x)\ dx$$

In general, the expression for the equivalent virtual nodal force Q_i is given by

$$Q_i = \int_0^L N_i(x)\ w(x)\ dx \qquad Q.E.D.$$

Problem 1.3

Demonstrate that the equivalent nodal forces Q_i at the nodal coordinates of a beam element acted upon by a distributed moment $m(x)$ [Fig. P1.3] is given by

$$Q_i = \int_0^L N_i^{'}(x)\ m(x)\ dx \qquad \text{(1.14b) repeated}$$

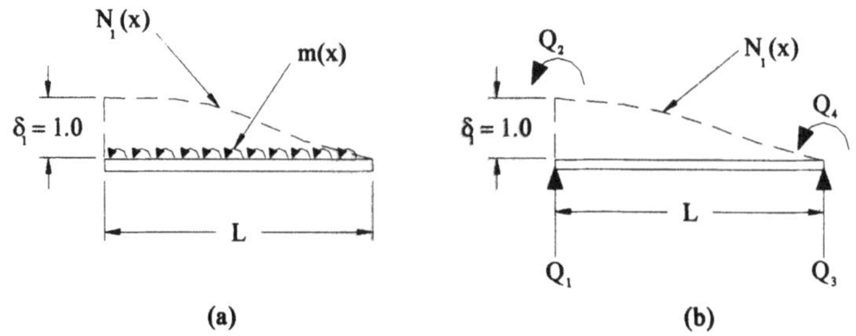

Fig. P1.3 Beam element undergoing a virtual displacement δ_1 = 1.0
(a) Supporting a distributed moment $m(x)$, and
(b) Acted upon by nodal equivalent force Q_i.

Analogous to Problem 1.2, to calculate the equivalent nodal force Q_1 we give a unit displacement $\delta_1 = 1.0$ to the beam element shown in Fig. P1.3(a) and (b) resulting in the shape function $N_1(x)$ which is given by eq.(1.5a). For this virtual displacement the external work performed by the equivalent nodal forces in Fig. P1.3(b) is

$$W_E = Q_1\delta_1 = Q_1 \qquad \text{(a)}$$

and the work done by the distributed moment $m(x)$ on a differential element beam segment of length dx is $N_1^{'}(x)\,m(x)\,dx$ where $N_1^{'}(x)$ is the derivative of the shape function $N_1(x)$. Thus $N_1^{'}(x)$ is the slope of this function. Therefore, the total work is given by

$$W_E = \int_0^L N_1^{'}(x)\ m(x)\ dx \qquad \text{(b)}$$

Finally, equating the expressions in (a) and (b) results in

$$Q_1 = \int_0^L N_1^{'}(x)\ m(x)\ dx \qquad \text{(c)}$$

The generalization of eq.(c) gives the expression to calculate the equivalent nodal force Q_i as

$$Q_i = \int_0^L N_i^{'}(x)\ m(x)\ \text{dx} \qquad Q.E.D.$$

Problem 1.4

Determine the equivalent nodal forces for a beam element supporting a concentrated force W as shown in Fig. P1.4.

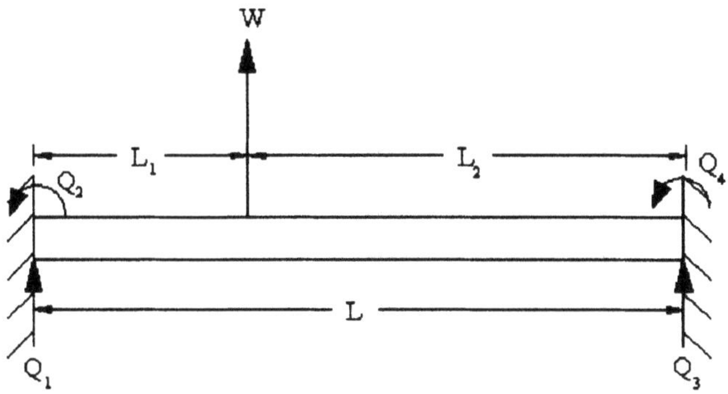

Fig. P1.4 Beam element showing a concentrated force W and the equivalent nodal forces Q_1, Q_2, Q_3 and Q_4.

Solution:

The use of eq.(1.14a) provides the general expression to determine the equivalent nodal force Q_i as

$$Q_i = \int_0^L N_i(x)\ w(x)\ dx \qquad \text{(1.14a) repeated}$$

where $w(x)$ is the force applied to the beam and $N_i(x)$ is the shape function given by eq.(1.5). To evaluate Q_1, we substitute into eq.(1.14a) $N_1(x)$ given by eq.(1.5a) and $w(x)$ for W applied at location $x = L_1$. We then obtain:

$$Q_1 = W\left[1 - 3\left(\frac{L_1}{L}\right)^2 + 2\left(\frac{L_1}{L}\right)^3\right]$$

Analogously, the successive substitution of eqs.(1.5b), (1.5c) and (1.5d) into eq.(1.14a) gives, respectively,

$$Q_2 = \frac{W\, L_1\, L_2^2}{L^3}$$

$$Q_3 = \frac{W\, L_1^2}{L^3}(L_1 + 3L_2)$$

and

$$Q_4 = -\frac{W L_1^2 L_2}{L^2}$$

Problem 1.5

Consider a beam element in Fig. P1.5 supporting a flexural moment M at the distance L_1 from the left end of the beam. Determine the expression for the equivalent nodal forces Q_i.

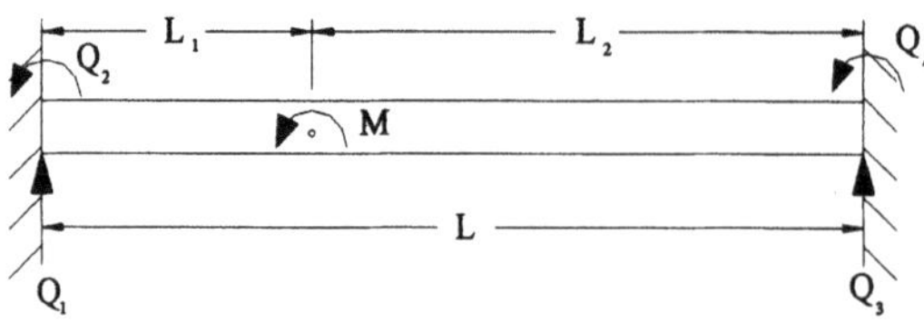

Fig. P1.5 Fixed end beam supporting moment *M*, showing equivalent nodal forces Q_i.

Solution:

The use of eq.(1.14b), repeated here for convenience, provides the expression to calculate the equivalent nodal forces:

$$Q_i = \int_0^L N_i'(x)\ m(x)\ dx \qquad \text{(1.14b) repeated}$$

where $N_i'(x)$ is the derivative of the shape function, eq.(1.5) and $m(x)$ is the applied moment along the beam. In this case, to calculate Q_1 we substitute into eq.(1.14b) the derivative of the shape function $N_1(x)$ given by eq.(1.5a) to obtain

$$Q_1 = \int_0^L \left(\frac{-6}{L^2} x + \frac{6}{L^3} x^2 \right) m(x)\ dx$$

Since the external moment M exists only at $x = L_1$ we obtain

$$Q_1 = \left(-\frac{6}{L^2} L_1 + \frac{6}{L^3} L_1^2 \right) M$$

or

$$Q_1 = -\frac{6M\, L_1\, L_2}{L^3}$$

Analogously, the equivalent nodal forces Q_2, Q_3 and Q_4 are calculated by substituting successively into eq.(1.14b) the derivatives $N_2'(x)$, $N_3'(x)$ and $N_4'(x)$ which are obtained, respectively, from eqs.(1.5b), (1.5c) and (1.5d):

$$Q_2 = \frac{M L_2}{L^2}(L_2 - 2L_1)$$

$$Q_3 = \frac{6M L_1 L_2}{L^3}$$

and

$$Q_4 = \frac{M L_1}{L^2}(L_1 - 2L_2)$$

1.16 Practice Problems

Problem 1.6

For the beam shown in Fig. P1.6 determine:

(a) Displacements at nodes ① and ②

(b) Reactions at the support

$E = 30{,}000$ ksi
$I = 100$ in^4

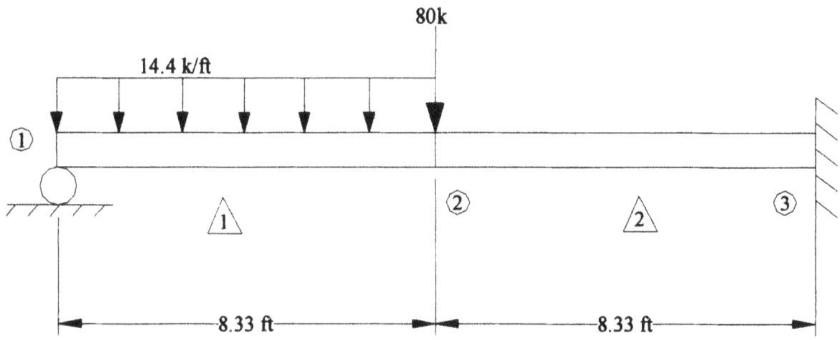

Fig. P1.6

Problem 1.7

For the beam shown in Fig. P1.7 determine

a) Displacements at node ②
b) End forces on the beam elements
c) Reactions at the supports

$E = 30{,}000$ ksi
$I = 200$ in^4

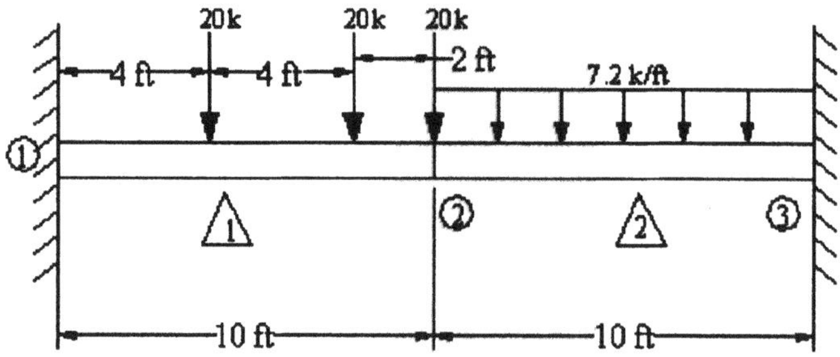

Fig P1.7

Problem 1.8

For the beam shown in Fig. P1.8 determine:

(a) Displacement at the nodes ② and ③,
(b) End-forces on the beam elements
(c) Reactions at the supports

$E = 200\ GP_a$
$I = 40{,}000$ cm^4

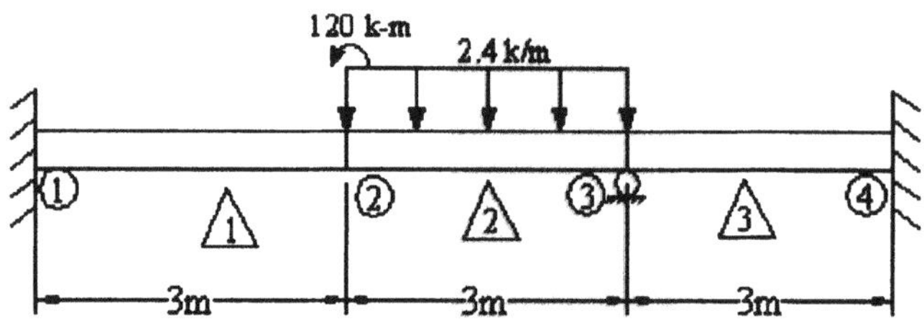

Fig. P1.8

Problem 1.9

For the beam shown in Fig. P1.9 determine:

(a) Displacement at node ②
(b) End-forces on the beam elements
(c) Reactions at the supports

$E = 4500$ ksi
$I = 1200\ \text{in}^4$

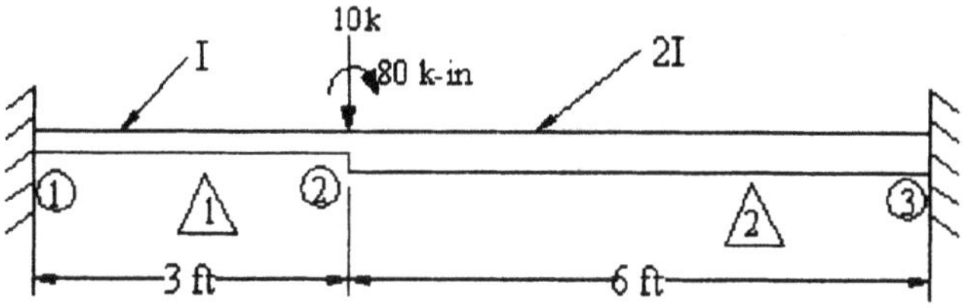

Fig. P1.9

Problem 1.10

For the beam shown in Fig. P1.10 determine :

(a) displacements at nodes ② and ③,
(b) End-forces on the beam elements, and
(c) Reactions at the supports.

$E = 200\ GP_a$
$I = 40{,}000\ \text{cm}^4$

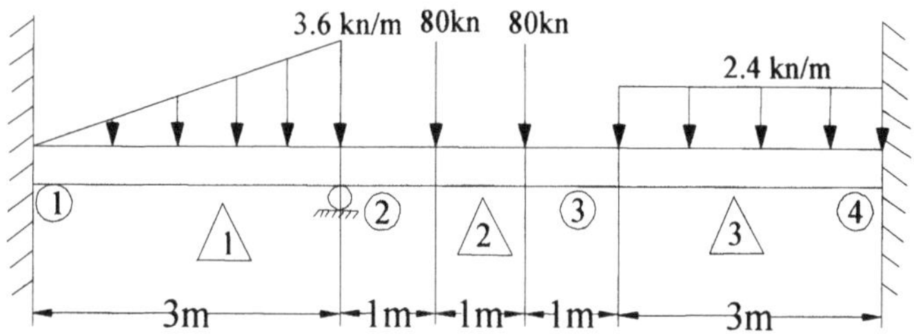

Fig. P1.10

Problem 1.11

For the beam shown in Fig. P1.11 determine:

(a) Displacements at nodes ① ② and ③,
(b) End-forces on the beam elements, and
(c) Reactions at the supports.

$E = 30{,}000$ ksi
$I = 120$ in^4

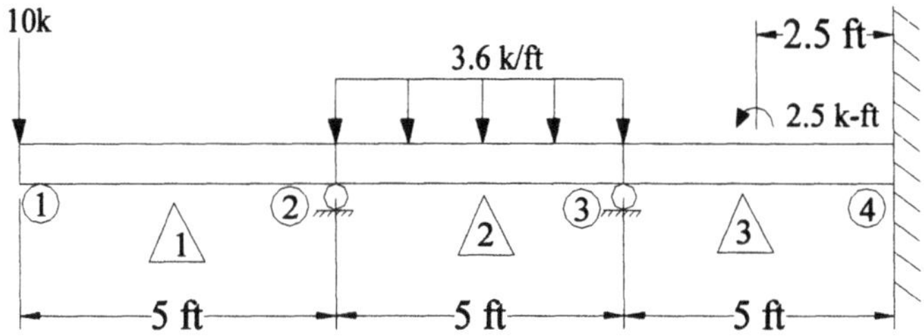

Fig. P1.11

Problem 1.12

Solve Problem 1.6 assuming that a hinge exists at the right end of beam element 1.

Problem 1.13

Solve Problem 1.7 assuming a hinge at the left end of beam element 2.

Problem 1.14

Solve Problem 1.8 assuming a hinge at the right end of beam element 2.

Problem 1.15

Solve Problem 1.9 assuming a hinge at node ② (right end of beam element 1 and left end of element 2).

Problem 1.16

Solve Problem 1.10 assuming a hinge at the left end of beam element 2.

Problem 1.17

Solve Problem 1.11 assuming hinges at the two ends of beam element 2.

Problem 1.18

Solve Problem 1.6 knowing that the support at node ① underwent a downward displacement of 1.0 in and rotational displacement of 5 degrees in the clockwise direction.

Problem 1.19

Solve Problem 1.8 knowing that the support at node ③ underwent a downward displacement of 2.0 in.

Problem 1.20

Solve Problem 1.9 knowing that the support at node ① underwent a downward displacement of 0.5 in and clockwise rotation of 10 degrees.

Problem 1.21

Solve Problem 1.6 knowing that the temperature at the top side of beam element 2 is 30°F and at the bottom side is 80°F.

Problem 1.22

Solve Problem 1.7 knowing that the temperature along the top of the beam is 40°F and at its bottom side is 100°F.

Problem 1.23

For the beam shown in Fig. P1.23 determine:

(a) Displacements at nodes ① and ②.

(b) The force and the moment of the spring at node ②.

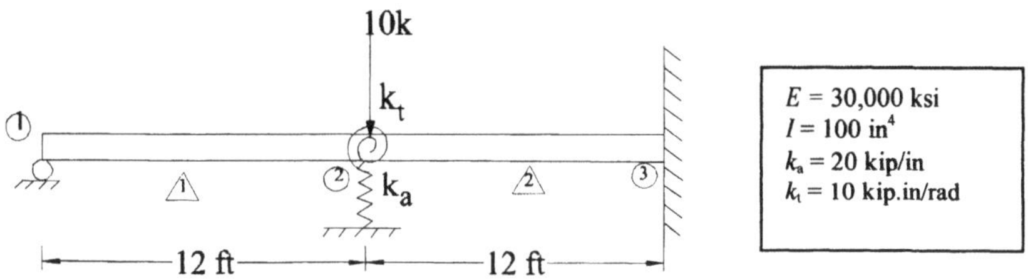

Fig. P1.23

Problem 1.24

For the beam shown in Fig. P1.24 determine:

(a) Displacements at nodes ② and ③.

(b) Force in the spring at node ②.

E = 30,000 ksi
I = 400 in^4
k_a = 10 kip/in
L = 100 in

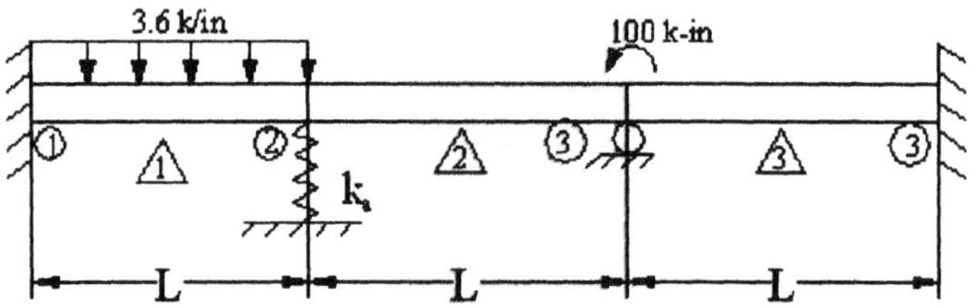

Fig. P1.24

2 Beams: Computer Applications

2.1 Computer Program

The computer program used throughout this book is the educational version of the powerful structural analysis program SAP2000. In its large version, the program includes static and dynamic analysis of structures (frames, plates, shells etc.) as well as the analysis of structures with non-linear behavior.

The CD-ROM accompanying this volume includes the educational version of the program SAP2000, user manuals and a tutorial. The reader may use or install the educational version of SAP2000 in his or her computer. The following examples describe in detail the use of SAP2000 to analyze the beam of the Illustrative Examples presented in Chapter 1.

Illustrative Example 2.1

Use SAP2000 to analyze the beam in Illustrative Examples 1.1 through 1.4 of Chapter 1. For convenience, the beam used in this example is reproduced in Fig 2.1 with the axis Z in the upward direction. The edited input data for this example is given in Table 2.1.

Table 2.1 Edited Input Data for Illustrative Example 2.1 (Units: kips and inches)

JOINT DATA

JOINT	GLOBAL-X	GLOBAL-Y	GLOBAL-Z	RESTRAINTS *
1	0.00000	0.00000	0.00000	1 1 1 1 1 1
2	90.00000	0.00000	0.00000	0 0 0 0 0 0
3	180.00000	0.00000	0.00000	0 0 1 0 0 0
4	300.00000	0.00000	0.00000	0 0 1 0 0 0
5	396.00000	0.00000	0.00000	1 1 1 1 1 1

*SAP2000 considers beams as three dimensional elements with six nodal coordinates at each joint

FRAME ELEMENT DATA

FRAME	JNT-1	JNT-2	SECTION	ANGLE	RELEASES	SEGMENTS	LENGTH
1	1	2	W14X82	0.000	000000	4	90.000
2	2	3	W14X82	0.000	000000	4	90.000
3	3	4	W14X82	0.000	000000	4	120.000
4	4	5	W14X82	0.000	000000	4	96.000

JOINT FORCES Load Case LOAD1

JOINT	GLOBAL-X	GLOBAL-Y	GLOBAL-Z	GLOBAL-XX	GLOBAL-YY	GLOBAL-ZZ
2	0.000	0.000	-10.000	0.000	0.000	0.000
3	0.000	0.000	0.000	0.000	50.000	0.000

JOINT DISPLACEMENT LOADS Load Case LOAD1

JOINT	GLOBAL-X	GLOBAL-Y	GLOBAL-Z	GLOBAL-XX	GLOBAL-YY	GLOBAL-ZZ
3	0.00000	0.00000	-1.00000	0.00000	0.00000	0.00000
4	0.00000	0.00000	-2.00000	0.00000	0.00000	0.00000

FRAME SPAN DISTRIBUTED LOADS Load Case LOAD1

FRAME	TYPE	DIRECTION	DISTANCE-A *	VALUE-A	DISTANCE-B *	VALUE-B
2	FORCE	LOCAL-2	0.0000	-0.1000	1.0000	-0.1000
3	FORCE	LOCAL-2	0.1667	-0.1000	0.7917	-0.2000
4	FORCE	LOCAL-2	0.0000	-0.0500	1.0000	-0.0500

*Distance A and distance B are the relative distances measured from the left node of the beam element respectively, to the beginning and to the end of the distributed load in the span.

FRAME SPAN POINT LOADS Load Case LOAD1

FRAME	TYPE	DIRECTION	DISTANCE*	VALUE
1	FORCE	LOCAL-2**	0.1111	30.0000
1	FORCE	LOCAL-2	0.2222	10.0000
4	MOMENT	LOCAL-3	0.5000	100.0000

*Relative distance measured from the left node of the beam element.

**Local-1, Local-2 and Local-3 refer to the element local axes *x*, *y* and *z*, respectively.

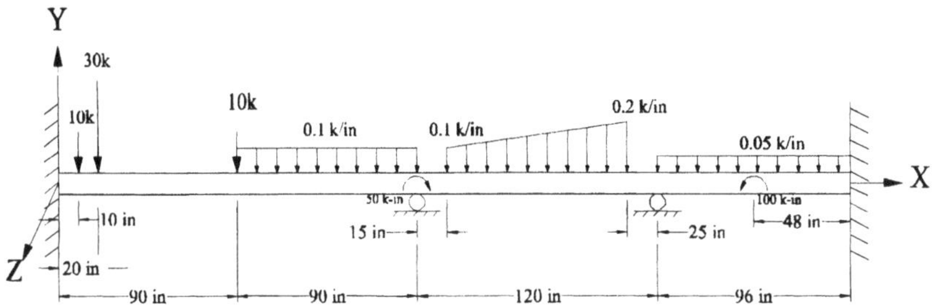

Fig. 2.1 Beam of Illustrative Example 2.1

Solution:

Begin: Load SAP2000 from the CD enclosed with this volume. Follow the installation instructions and the prompts in the software. After successful installation of the program, begin this example by opening the program.

Enter: "OK" to disable the "Tip of the Day" screen.

Hint: Maximize both screens for full views of all windows.

Select: In the lower right-hand corner of the screen use the drop-down menu to select "kip-in".

Select: From the Main Menu bar: FILE > NEW MODEL FROM TEMPLATE

Select: From the Model Template Screen click on the button displaying the drawing of a beam.

Edit: On the screen named "Beam", change the number of spans to 4 and change the span length to 90. Then OK.

Edit: Minimize the 3-D screen; it appears on the left of the split screen. Then maximize the 2-D screen. To have a full view of the beam drawing, click on the PAN icon on the toolbar and drag the figure to the center of the screen.

Select: Translate the origin of the Global Coordinate System to the left end of the beam using the following commands from the Main Menu:

SELECT>SELECT>ALL. This will mark all members of the beam. Then select EDIT>MOVE. In the screen labeled "Move Selected Points" change the Delta X to equal 180. Then OK.

Edit: Add grid lines to the last two supports of the beam with the following commands from the Main menu: DRAW>EDIT GRID.

When the pop-up menu appears, highlight the value -180 and change it to 270. Then click the button labeled "Move Grid Line".
Next, highlight the value -90 and change it to 360.

Click "Move Grid Line". The screen should now show values of 0, 90, 180, 270 and 360. Next click on the check box labeled "glue joints to grid lines". This check will allow to move the grid lines together with the joints. Then OK.

Select: To change the length of the last two beam elements edit the grid lines once again. Using the Main Menu: DRAW>EDIT GRID.

Highlight the value 270 and change it to 300, then click "Move grid lines". Highlight the value 360 and change it to 396. Click "Move Grid Lines". Then OK.

Define: Beam material: Using the Main Menu select DEFINE>MATERIALS. From the list on the left side of the pop-up screen select STEEL.

Click on SHOW>MODIFY PROPERTIES and change the Modulus of Elasticity to $E = 29000$. Then OK.

Define: Cross sections: From the Main Menu select the following: DEFINE>FRAME SECTIONS[1].

From the window on the right side of the pop-up screen select Import/Wide Flange; then OK. Another pop-up window will appear with the label C:\sap2000e. Select "sections.prop" then click OPEN.

Using the scroll bar locate the section "W14X82" and then OK. A pop-up window labeled "I/Wide Flange Section" will appear. Click on the button labeled "Modification Factors". Another menu will pop-up labeled "Analysis Property Modification Factors". In order not to include the shear deformation in the calculation of the stiffness coefficients, change the Shear Area in the 2 direction to 0 (zero). Then click OK on all three screens.

Assign: Assign frame sections: From the Main Menu bar SELECT>SELECT>ALL. Then from the Main Menu bar ASSIGN>FRAME>SECTIONS.
In the Define Frame Sections window select W14X82. Then OK.

---

[1]SAP2000 refers to the beam elements as frames

Label: For viewing convenience label the joints and elements of the beam: From the Main Menu VIEW>SET ELEMENTS. From the pop-up window check the boxes under joint labels and frame labels. Then OK.

Modify: Restraint settings: Click on the center of the "dot" that locate joints ① and ⑤. This command will mark the selected joints with an "X". From the Main Menu select: ASSIGN>JOINT>RESTRAINTS. In the window labeled "Joint Restraints" select restraints in all directions. Then OK.

Mark joint ②. Then from the Main Menu enter
ASSIGN>JOINT>RESTRAINTS.
In the pop-up window select no restraints with the symbol of a dot. Then OK.

Mark joints ③ and ④. Then from the Main Menu enter:
ASSIGN>JOINT>RESTRAINTS. Restrain only Translation-3. Then OK.

Assign: Loads: Using the Main menu enter:
DEFINE>STATIC LOAD CASES.

In the window labeled "Define Static Load Case Names" change the label DEAD to LIVE using the drop-down menu. Also change the self-weight multiplier to 0 (zero). Then click the button labeled "Change Load". Then OK.

Assign: Mark joint ②. Then from the Main Menu enter:
ASSIGN>JOINT STATIC LOAD>FORCES. In the Joint Forces menu change the force in the Global Z to -10.0. Then OK.

Warning: make certain to zero out any previous entries made on this screen!

Mark joint ③. Then from the Main Menu enter:
ASSIGN>JOINT STATIC LOAD>FORCES. Change the Moment Global YY to 50. Then OK.

Assign: Frame loads: Select Frame 1 by clicking on this frame element of the beam shown on the screen. Frame 1 will change from a continuous line to dashes. On the Main Menu select:

ASSIGN>FRAME STATIC LOADS>POINT AND UNIFORM LOADS. Select the button "forces" and direction Local 2 using the drop-down menu. Click on the button for "absolute distance".

Warning: make certain to zero out any previous entries made on this screen!
In the first two columns on the left enter the following:

Distance = 10.0 Distance = 20.0
Load = -30.0 Load = -10.0
Then OK.

Assign: Select Frame 4. On the Main Menu enter:
ASSIGN>FRAME STATIC LOADS>POINT AND UNIFORM LOADS.

Select the button "MOMENTS". In the Drop Down menu select "Local 3". Then click on the button for "absolute distance".

Warning: make certain to zero out any previous entries made on this screen!
In the left-hand column enter the following:

Distance = 48.0 Load[2] = 100.0 Then OK.

Assign: Select Frame 4 . On the Main Menu enter:
ASSIGN > FRAME STATIC LOADS > POINT AND UNIFORM LOADS
Select the button "Forces" and in the drop-down menu select Local 2.

Warning: make certain to zero out any previous entries made on this screen!
In the box that is located in the left-hand lower corner of the this window enter:

Uniform Load = - 0.05 Then OK.

Assign: Select Frame 2. From the Main Menu enter:
ASSIGN>FRAME STATIC LOAD>POINT AND UNIFORM LOAD.

Warning: make certain to zero out any previous entries made on this screen!
Select "Forces" and change the direction to "Local 2". Enter a uniform load of:

Uniform Load = - 0.10

Then OK.

Assign: Select Frame 3. From the Main Menu enter:
ASSIGN>FRAME STATIC LOADS>TRAPEZOIDAL.

Warning: make certain to zero out any previous entries made on this screen!
Then select "Forces", and change the direction to "Local 2". Click on the button "Absolute Distance". Enter the following in the two left-hand columns: Distance =20.0 Distance = 95.0
Load = - 0.10 Load = - 0.20
Then OK.

[2] Positive Moment in reference to the local coordinate axes for frame element 1 of the beam.

Set Analysis: From the Main Menu enter:
ANALYZE>SET OPTIONS.
In the box marked "Available DOF" make certain that the only check boxes marked are UZ and RY.

Final Analysis: On the Main Menu enter:
ANALYZE>RUN.
On the window requesting to name the model enter its name as "Example 2.1". Click on "SAVE". The calculations will appear on the screen as they take place. When the analysis is completed, a message will appear on the bottom of the pop-up window "The analysis is complete". Click OK.

Observe: The deformed shape of the beam will be plotted on the screen. On the Main Menu select DISPLAY>SHOW DEFORMED SHAPE.

Click on the selection for "Wire Shadow". Also click on the "XZ" icon to see the plot on that plane. Then OK. To see the displacement values at any node, right-click on the node and a pop-up window will show the nodal displacements.

Print Input Tables: From the Main Menu enter:
FILE>PRINT INPUT TABLES. The previous Table 2.1 contains the edited input tables for Illustrative Example 2.1.

Print Out put Tables: From the Main Menu enter:
FILE>PRINT OUTPUT TABLES The following Table 2.2 contains the edited output tables for Illustrative Example 2.1.:

Table 2.2 Edited Output tables for Illustrative Example 2.1 (Units: Kips, Inches)

JOINT DISPLACEMENTS

JOINT	LOAD	UX	UY	UZ	RX	RY	RZ
1	LOAD2	0.0000	0.0000	0.0000	0.0000	0.0000	0.0000
2	LOAD2	0.0000	0.0000	-0.0239	0.0000	4.548E-05	0.0000
3	LOAD2	0.0000	0.0000	0.0000	0.0000	-1.761E-04	0.0000
4	LOAD2	0.0000	0.0000	0.0000	0.0000	-7.099E-06	0.0000
5	LOAD2	0.0000	0.0000	0.0000	0.0000	0.0000	0.0000

JOINT REACTIONS

JNT	LOAD	F1	F2	F3	M1	M2	M3
1	LOAD2	0.0000	0.0000	46.9113	0.0000	-785.0463	0.0000
3	LOAD2	0.0000	0.0000	19.4328	0.0000	0.0000	0.0000
4	LOAD2	0.0000	0.0000	7.9866	0.0000	0.0000	0.0000
5	LOAD2	0.0000	0.0000	0.7193	0.0000	9.6171	0.0000

Table 2.2 Continued

FRAME ELEMENT FORCES

FRM	LOAD	LOC	P	V2	V3	T	M2	M3
1	LOAD2							
		0.00	0.00	-46.91	0.00	0.00	0.00	-785.05
		22.50	0.00	-6.91	0.00	0.00	0.00	-129.54
		45.00	0.00	-6.91	0.00	0.00	0.00	25.96
		67.50	0.00	-6.91	0.00	0.00	0.00	181.47
		90.00	0.00	-6.91	0.00	0.00	0.00	336.97
2	LOAD2							
		0.00	0.00	3.09	0.00	0.00	0.00	336.97
		22.50	0.00	5.34	0.00	0.00	0.00	242.16
		45.00	0.00	7.59	0.00	0.00	0.00	96.73
		67.50	0.00	9.84	0.00	0.00	0.00	-99.33
		90.00	0.00	12.09	0.00	0.00	0.00	-346.01
3	LOAD2							
		0.00	0.00	-7.34	0.00	0.00	0.00	-296.01
		30.00	0.00	-6.28	0.00	0.00	0.00	-98.24
		60.00	0.00	-2.28	0.00	0.00	0.00	33.08
		90.00	0.00	2.92	0.00	0.00	0.00	26.40
		120.00	0.00	3.91	0.00	0.00	0.00	-63.72
4	LOAD2							
		0.00	0.00	-4.08	0.00	0.00	0.00	-70.97
		24.00	0.00	-2.88	0.00	0.00	0.00	12.57
		48.00	0.00	-1.68	0.00	0.00	0.00	-32.69
		72.00	0.00	-4.807E-01	0.00	0.00	0.00	-6.75
		96.00	0.00	7.193E-01	0.00	0.00	0.00	-9.62

Plot Displacements: From the Main Menu enter:

DISPLAY>SHOW DEFORMED SHAPE, then enter
FILE>PRINT GRAPHICS
(The deformed shape shown on the screen is reproduced in Fig. 2.2)

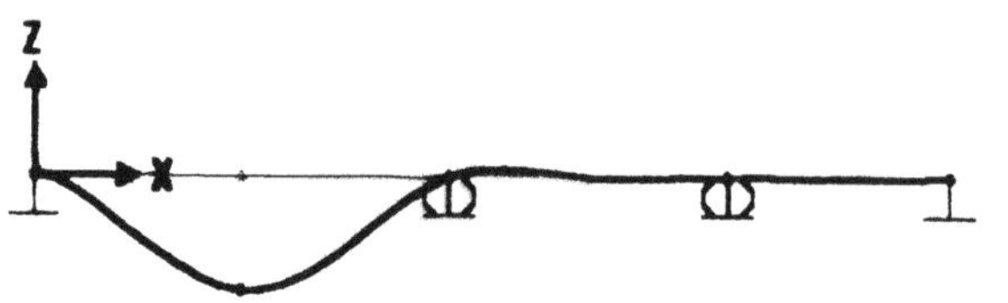

Fig. 2.2 Deformed shape for the beam of Illustrative Example 2.1

Plot Shear Forces: From the Main Menu enter:

DISPLAY>SHOW ELEMENT FORCES / STRESSES>FRAMES
Select Shear 2-2. Then
FILE>PRINT GRAPHICS
(The shear force diagram shown on the screen is depicted in Fig. 2.3)

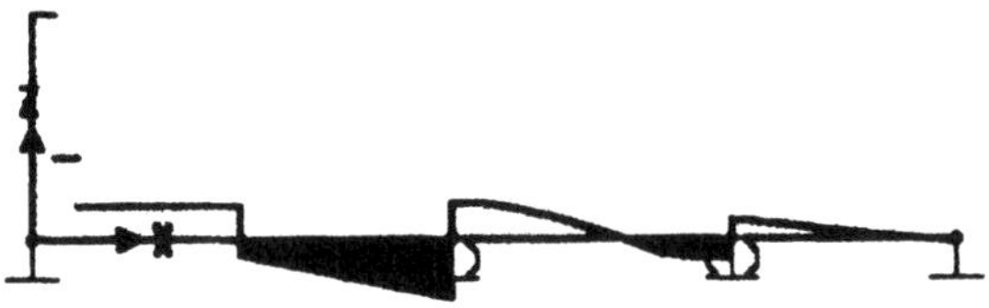

Fig. 2.3 Shear Force diagram for the beam of Illustrative Example 2.1

Plot the Bending Moment: From the Main Menu enter:
DISPLAY>SHOW FORCES / STRESSES>FRAME
Select Moment 3-3 and OK. Then
FILE>PRINT GRAPHICS
(The bending moment diagram shown on the screen is depicted in Fig. 2.4)

Note: To plot the Moment Diagram on the compression side, from the Main Menu enter: OPTIONS then remove check on "Moment Diagram on the Tension Side"

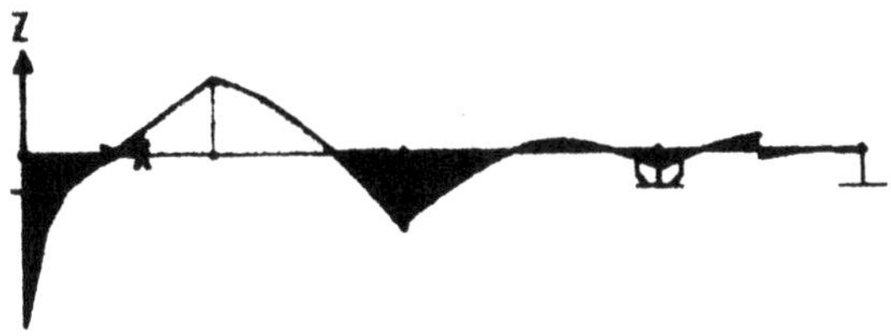

Fig 2.4 Bending Moment Diagram for the beam of Illustrative Example 2.1

Note: To view plots of either the shear force or bending moment diagrams for any element of the beam, right-click on the element and a pop-up window will depict the diagram for the selected element. It should be observed that as the cursor is moved across the plot in this window, values of shear force or of the bending moment will be displayed.

Illustrative Example 2.2

Use SAP2000 to solve the beam of Illustrative Example 2.2, which is shown in Fig.2.1. Assume that the supports ③ and ④ have undergone downward vertical displacements of 1.0 in and 2.0 in, respectively.

Solution:

The input data for this problem is the same as the input data for Illustrative Example 2.1, except for the additional data on the imposed displacements at nodes ③ and ④. The following data should be added to the file containing data for Illustrative Example 2.1.

Note: To use the same data file of Illustrative Example 2.1, it is necessary to unlock this file by clicking on the icon showing a padlock (Main Menu Bar). Then OK.

Select: Click on node ③. Then enter
ASSIGN>JOINT STATIC LOADS>DISPLACEMENTS and enter translation Z=-1.0, then OK.

Analogously, select node ④ and enter
ASSIGN>JOINT STATIC LOADS>DISPLACEMENTS and enter translation Z=-2.0 then OK.

Set Analysis: Select from the Main Menu:
ANALYZE>RUN> on the window requesting to name the model enter its new name "Example 2.2".
Click on "SAVE", then OK.

Observe: The deformed shape of the beam will be plotted on the screen. On the Main Menu select DISPLAY>SHOW DEFORMED SHAPE. Click on the selection for "Wire Shadow". Also click on the "XZ" icon to see the plot on that plane. Then OK.

Results: To see the displacement values at any node, right-click on the node and a pop-up window will show the nodal displacements.

Print Input Tables: From the Main Menu enter:
FILE>PRINT INPUT TABLES. Table 2.3 contains the edited input tables for Illustrative Example 2.2.

Table 2.3 Edited Input Tables for Illustrative Example 2.2 (Units: kips, inches)

JOINT DATA

JNT	GLBL-X	GLBL-Y	GLBL-Z	RESTRAINTS
1	0.00000	0.00000	0.00000	1 1 1 1 1 1
2	90.00000	0.00000	0.00000	0 0 0 0 0 0
3	180.00000	0.00000	0.00000	0 0 1 0 0 0
4	300.00000	0.00000	0.00000	0 0 1 0 0 0
5	396.00000	0.00000	0.00000	1 1 1 1 1 1

Table 2.3 Continued

FRAME ELEMENT DATA

FRM	JNT-1	JNT-2	SEC	ANGL	RELEASES	SEGTS	R1	R2	FACTR	LENGTH
1	1	2	W14X82	0.000	000000	4	0.000	0.000	1.000	90.000
2	2	3	W14X82	0.000	000000	4	0.000	0.000	1.000	90.000
3	3	4	W14X82	0.000	000000	4	0.000	0.000	1.000	120.000
4	4	5	W14X82	0.000	000000	4	0.000	0.000	1.000	96.000

JOINT FORCES Load Case LOAD2

JNT	GLBL-X	GLBL-Y	GLBL-Z	GLBL-XX	GLBL-YY	GLBL-ZZ
3	0.000	0.000	0.000	0.000	50.000	0.000
2	0.000	0.000	-10.000	0.000	0.000	0.000

JOINT DISPLACEMENT LOADS Load Case LOAD2

JNT	GLBL-X	GLBL-Y	GLBL-Z	GLBL-XX	GLBL-YY	GLBL-ZZ
3	0.00000	0.00000	-1.00000	0.00000	0.00000	0.00000
4	0.00000	0.00000	-2.00000	0.00000	0.00000	0.00000

FRAME SPAN DISTRIBUTED LOADS Load Case LOAD2

FRM	TYPE	DIRECTION	DIST-A	VALUE-A	DIST-B	VALUE-B
4	FORCE	LOCAL-2	0.0000	-0.0500	1.0000	-0.0500
2	FORCE	LOCAL-2	0.0000	-0.1000	1.0000	-0.1000
3	FORCE	LOCAL-2	0.1667	-0.1000	0.7917	-0.2000

FRAME SPAN POINT LOADS Load Case LOAD2

FRM	TYPE	DIRECTION	DIST	VALUE
4	MOMENT	LOCAL-3	0.5000	100.0000
1	FORCE	LOCAL-2	0.1111	-30.0000
1	FORCE	LOCAL-2	0.2222	-10.0000

Print Out put Tables: From the Main Menu enter:

FILE>PRINT OUTPUT TABLES

(Table 2.4 contains the output tables for Illustrative Example 2.2)

Table 2.4 Edited Output tables for Illustrative Example 2.2 (Units: Kips, inches)

JOINT DISPLACEMENTS

JOINT	LOAD	UX	UY	UZ	RX	RY	RZ
1	LOAD2	0.0000	0.0000	0.0000	0.0000	0.0000	0.0000
2	LOAD2	0.0000	0.0000	-0.1774	0.0000	4.528E-03	0.0000
3	LOAD2	0.0000	0.0000	-1.0000	0.0000	0.0152	0.0000
4	LOAD2	0.0000	0.0000	-2.0000	0.0000	-0.0152	0.0000
5	LOAD2	0.0000	0.0000	0.0000	0.0000	0.0000	0.0000

Table 2.4 Continued

JOINT REACTIONS

JOINT	LOAD	F1	F2	F3	M1	M2	M3
1	LOAD2	0.0000	0.0000	26.5878	0.0000	-1144.5255	0.0000
3	LOAD2	0.0000	0.0000	215.5311	0.0000	0.0000	0.0000
4	LOAD2	0.0000	0.0000	-608.0510	0.0000	0.0000	0.0000
5	LOAD2	0.0000	0.0000	440.9821	0.0000	25199.5918	0.0000

FRAME ELEMENT FORCES

FRAME	LOAD	LOC	P	V2	V3	T	M2	M3
1	LOAD2							
		0.00	0.00	-26.59	0.00	0.00	0.00	-1144.53
		22.50	0.00	13.41	0.00	0.00	0.00	-946.30
		45.00	0.00	13.41	0.00	0.00	0.00	-1248.07
		67.50	0.00	13.41	0.00	0.00	0.00	-1549.85
		90.00	0.00	13.41	0.00	0.00	0.00	-1851.62
2	LOAD2							
		0.00	0.00	23.41	0.00	0.00	0.00	-1851.62
		22.50	0.00	25.66	0.00	0.00	0.00	-2403.71
		45.00	0.00	27.91	0.00	0.00	0.00	-3006.42
		67.50	0.00	30.16	0.00	0.00	0.00	-3659.76
		90.00	0.00	32.41	0.00	0.00	0.00	-4363.72
3	LOAD2							
		0.00	0.00	-183.12	0.00	0.00	0.00	-4313.72
		30.00	0.00	-182.05	0.00	0.00	0.00	1157.29
		60.00	0.00	-178.05	0.00	0.00	0.00	6561.86
		90.00	0.00	-172.85	0.00	0.00	0.00	11828.42
		120.00	0.00	-171.87	0.00	0.00	0.00	17011.54
4	LOAD2							
		0.00	0.00	436.18	0.00	0.00	0.00	17004.29
		24.00	0.00	437.38	0.00	0.00	0.00	6521.52
		48.00	0.00	438.58	0.00	0.00	0.00	-4090.05
		72.00	0.00	439.78	0.00	0.00	0.00	-14630.42
		96.00	0.00	440.98	0.00	0.00	0.00	-25199.59

Plot Displacements: From the Main Menu enter:

DISPLAY>SHOW DEFORMED SHAPE, then enter
FILE>PRINT GRAPHICS
(The deformed shape shown on the screen is reproduced in Fig. 2.5)

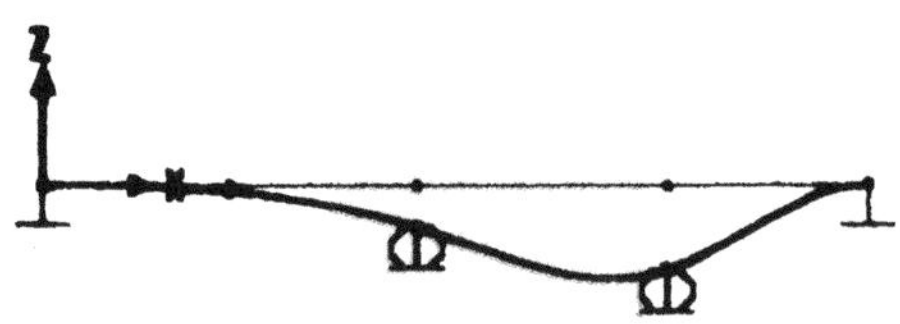

Fig. 2.5 Deformed shape for the beam in Illustrative Example 2.2

Plot Shear Forces: From the Main Menu enter:

DISPLAY>SHOW ELEMENT FORCES / STRESSES>FRAMES
Select Shear 2-2. Then
FILE>PRINT GRAPHICS
(The shear force diagram shown on the screen is depicted in Fig. 2.6)

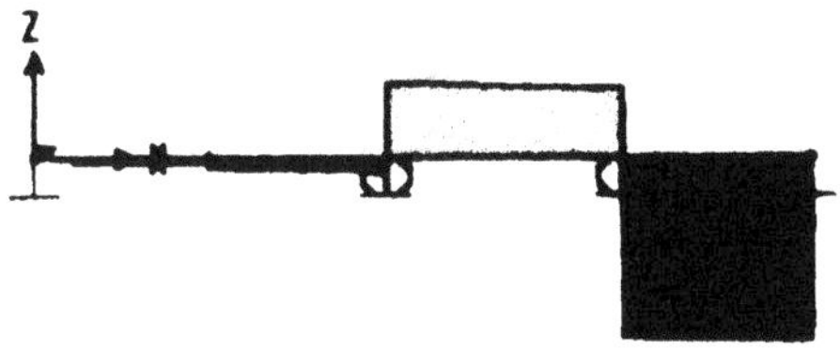

Fig. 2.6 Shear Force diagram for the beam of Illustrative Example 2.2

Plot the Bending Moment: From the Main Menu enter:
DISPLAY>SHOW FORCES / STRESSES>FRAME
Select Moment 3-3 and OK. Then
FILE>PRINT GRAPHICS
(The bending moment diagram shown on the screen is depicted in Fig. 2.7)

Note: To plot the Moment Diagram on the compression side enter from the Main Menu:
OPTIONS then remove check on "Moment Diagram on the Tension Side"

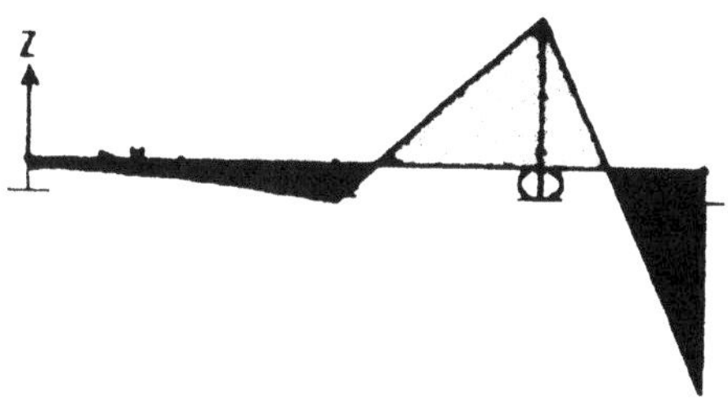

Fig 2.7 Bending Moment Diagram for the beam of Illustrative Example 2.2

Note: To view plots of either the shear force or bending moment diagrams for any element of the beam, right-click on the element and a pop-up window will depict the diagram for the selected element. It should be observed that as the cursor is moved across the plot in this window, values of shear force or of the bending moment will be displayed.

Illustrative Example 2.3

Use SAP2000 to solve the beam of Illustrative Example 2.1, which is shown in Fig.2.1. Assume that beam element 3 has a hinge (released for bending) at its left end.

Solution:

The input data for this example is the same as the input data for Illustrative Example 2.1, except for additional data to implement the release at the left end of element 3.

The following data should be added to the file containing data for Illustrative Example 2.1:

Select: Click on element 3. Then enter
ASSIGN>FRAME>RELEASES and check Start Moment 3-3 (major axis of the cross-sectional area), then OK.

Set Analysis: Select from the Main Menu:
ANALYZE>RUN> on the window requesting to name the model enter its new name: "Example 2.3". Click on "SAVE".

Observe: The deformed shape of the beam will be shown on the screen.
From the Main Menu select:
DISPLAY>SHOW DEFORMED SHAPE. Click on the selection for "Wire Shadow". Also click on the "XZ" icon to see the plot on that plane. Then OK.

Results: To see the displacement values at any node, right-click on the node and a pop-up window will show the nodal displacements.

Print Out put Tables: From the Main Menu enter:
FILE>PRINT OUTPUT TABLES
(Table 2.5 contains the edited output tables for Illustrative Example 2.3)

Table 2.5 Edited Output tables for Illustrative Example 2.3 (Units: Kips, inches)

JOINT DISPLACEMENTS

JOINT	LOAD	UX	UY	UZ	RX	RY	RZ
1	LOAD2	0.0000	0.0000	0.0000	0.0000	0.0000	0.0000
2	LOAD2	0.0000	0.0000	-0.0356	0.0000	1.757E-04	0.0000
3	LOAD2	0.0000	0.0000	0.0000	0.0000	-6.969E-04	0.0000
4	LOAD2	0.0000	0.0000	0.0000	0.0000	-9.390E-05	0.0000
5	LOAD2	0.0000	0.0000	0.0000	0.0000	0.0000	0.0000

Table 2.5 Continued

JOINT REACTIONS

JOINT	LOAD	F1	F2	F3	M1	M2	M3
1	LOAD2	0.0000	0.0000	49.3781	0.0000	-933.0516	0.0000
3	LOAD2	0.0000	0.0000	13.7284	0.0000	0.0000	0.0000
4	LOAD2	0.0000	0.0000	12.6696	0.0000	0.0000	0.0000
5	LOAD2	0.0000	0.0000	-0.7261	0.0000	-36.6346	0.0000

FRAME ELEMENT FORCES

FRAME	LOAD	LOC	P	V2	V3	T	M2	M3
1	LOAD2							
		0.00	0.00	-49.38	0.00	0.00	0.00	-933.05
		22.50	0.00	-9.38	0.00	0.00	0.00	-222.05
		45.00	0.00	-9.38	0.00	0.00	0.00	-11.04
		67.50	0.00	-9.38	0.00	0.00	0.00	199.97
		90.00	0.00	-9.38	0.00	0.00	0.00	410.97
2	LOAD2							
		0.00	0.00	6.219E-01	0.00	0.00	0.00	410.97
		22.50	0.00	2.87	0.00	0.00	0.00	371.67
		45.00	0.00	5.12	0.00	0.00	0.00	281.74
		67.50	0.00	7.37	0.00	0.00	0.00	141.18
		90.00	0.00	9.62	0.00	0.00	0.00	-50.00
3	LOAD2							
		0.00	0.00	-4.11	0.00	0.00	0.00	0.00
		30.00	0.00	-3.04	0.00	0.00	0.00	100.64
		60.00	0.00	9.602E-01	0.00	0.00	0.00	134.83
		90.00	0.00	6.16	0.00	0.00	0.00	31.03
		120.00	0.00	7.14	0.00	0.00	0.00	-156.22
4	LOAD2							
		0.00	0.00	-5.53	0.00	0.00	0.00	-163.47
		24.00	0.00	-4.33	0.00	0.00	0.00	-45.24
		48.00	0.00	-3.13	0.00	0.00	0.00	-55.82
		72.00	0.00	-1.93	0.00	0.00	0.00	4.81
		96.00	0.00	-7.261E-01	0.00	0.00	0.00	36.63

Plot Displacements: From the Main Menu enter:

DISPLAY>SHOW DEFORMED SHAPE, then enter
FILE>PRINT GRAPHICS
(The deformed shape shown on the screen is reproduced in Fig.2.8)

Fig. 2.8 Deformed shape for the beam of Illustrative Example 2.3

Plot Shear Forces: From the Main Menu enter:

DISPLAY>SHOW ELEMENT FORCES / STRESSES>FRAMES
Select Shear 2-2. Then to obtain a hard copy enter:
FILE>PRINT GRAPHICS

(The shear force diagram shown on the screen is depicted in Fig. 2.9)

Fig. 2.9 Shear Force diagram for the beam of Illustrative Example 2.3

Plot the Bending Moment: From the Main Menu enter:
DISPLAY>SHOW FORCES / STRESSES>FRAME
Select Moment 3-3 and OK. Then to obtain a hard copy enter:
FILE>PRINT GRAPHICS
(The bending moment diagram shown on the screen is depicted in Fig. 2.10)

Note: To plot the Moment Diagram on the compression side enter from the Main Menu:
OPTIONS then remove check on "Moment Diagram on the Tension Side"

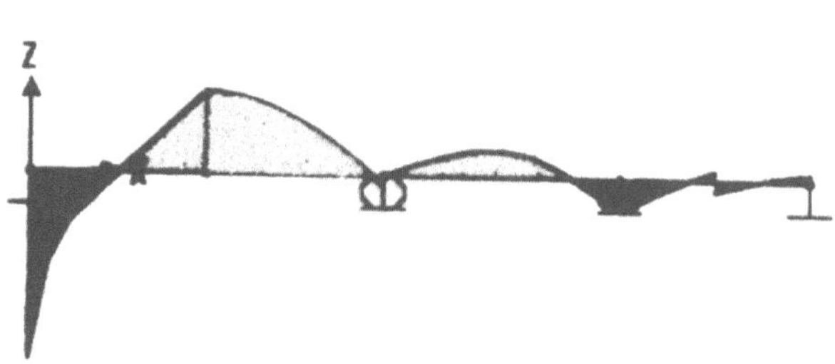

Fig 2.10 Bending Moment Diagram for the beam of Illustrative Example 2.3

Note: To view plots of either the shear force or bending moment diagrams for any element of the beam, right-click on the element and a pop-up window will depict the diagram for the selected element. It should be observed that as the cursor is moved across the plot in this window, values of shear force or of the bending moment will be displayed.

Illustrative Example 2.4

Use SAP2000 to analyze the loaded beam shown in Fig.2.1, assuming that beam element 3 is subjected to a linear change in temperature with $T_2 = 60\ \mathrm{F}^0$ at the top and $T_1 = 120\ \mathrm{F}^0$ at the bottom. The coefficient of thermal expansion is $\alpha = 6.5\mathrm{xE}\text{-}6\ /\mathrm{F}^o$ and the height of the cross section area is $h = 14.31$ in.

Solution:

The temperature gradient across the section of the beam is

$$\text{Temp. Gradiant } = \frac{T_2 - T_1}{h} = \frac{60 - 120}{14.31} = -4.19\ \mathrm{F}^0/\mathrm{in}$$

The beam and the load for this example are the same as in Illustrative Example 2.1, with the exception of the thermal load applied at element 3. Therefore, the stored data file for Illustrative Example 2.1 may be used by simply adding the additional equivalent load due to change in temperature in element 3.

The following commands are then implemented in SAP2000:

FILE>OPEN then select the data file for Illustrative Example 2.1.

Unlock: Since the program has locked the data file used in the solution of Illustrative Example 2.1, it is necessary to unlock this file. To unlock click the icon showing the symbol of a padlock and responding YES to the question that follows.

Thermal Load: Click on element 3 of the beam to select.
Then from the Main Menu enter:

ASSIGN>FRAME STATIC LOADS>TEMPERATURE
Click on Temperature Gradient 2-2
Enter Temperature Gradient = - 4.19.
Then OK.

Analyze: From the Main Menu enter:
RUN then OK.
The screen will show the deformed shape of the beam.

Print Input Tables: From the Main Menu enter:
FILE>PRINT>INPUT TABLE

(Table 2.6 contains the edited input tables for Illustrative Example 2.4)

Table 2.6 Edited Input Table for Illustrative Example 2.4 (Units: Kips, inches)

JOINT DATA

JOINT	GLBL-X	GLBL-Y	GLBL-Z	RESTRAINTS
1	0.00000	0.00000	0.00000	1 1 1 1 1 1
2	90.00000	0.00000	0.00000	0 0 0 0 0 0
3	180.00000	0.00000	0.00000	0 0 1 0 0 0
4	300.00000	0.00000	0.00000	0 0 1 0 0 0
5	396.00000	0.00000	0.00000	1 1 1 1 1 1

FRAME ELEMENT DATA

FRM	JNT-1	JNT-2	SEC	ANGL	RELEASES	SEG	R1	R2	FACTOR	LTH
1	1	2	W14X82	0.000	000000	4	0.000	0.000	1.000	90.000
2	2	3	W14X82	0.000	000000	4	0.000	0.000	1.000	90.000
3	3	4	W14X82	0.000	000000	4	0.000	0.000	1.000	120.000
4	4	5	W14X82	0.000	000000	4	0.000	0.000	1.000	96.000

JOINT FORCES Load Case LOAD2

JOINT	GLBL-X	GLBL-Y	GLBL-Z	GLBL-XX	GLBL-YY	GLBL-ZZ
3	0.000	0.000	0.000	0.000	50.000	0.000
2	0.000	0.000	-10.000	0.000	0.000	0.000

FRAME SPAN DISTRIBUTED LOADS Load Case LOAD2

FRAME	TYPE	DIRECTION	DIST-A	VALUE-A	DIST-B	VALUE-B
4	FORCE	LOCAL-2	0.0000	-0.0500	1.0000	-0.0500
2	FORCE	LOCAL-2	0.0000	-0.1000	1.0000	-0.1000
3	FORCE	LOCAL-2	0.1667	-0.1000	0.7917	-0.2000

FRAME SPAN POINT LOADS Load Case LOAD2

FRAME	TYPE	DIRECTION	DISTANCE	VALUE
4	MOMENT	LOCAL-3	0.5000	100.0000
1	FORCE	LOCAL-2	0.1111	-30.0000
1	FORCE	LOCAL-2	0.2222	-10.0000

FRAME THERMAL LOADS Load Case LOAD2

FRAME	TYPE	VALUE
3	GRAD 2-2	-4.1900

Print Output Tables: From the Main Menu enter:
FILE > PRINT > OUTPUT TABLE

(Table 2.7 contains the edited output tables for Illustrative Example 2.4)

Table 2.7 Output tables for Illustrative Example 2.4 (Units: kips,inches)

JOINT DISPLACEMENTS

JOINT	LOAD	UX	UY	UZ	RX	RY	RZ
1	LOAD2	0.0000	0.0000	0.0000	0.0000	0.0000	0.0000
2	LOAD2	0.0000	0.0000	-9.458E-03	0.0000	-1.150E-04	0.0000
3	LOAD2	0.0000	0.0000	0.0000	0.0000	4.658E-04	0.0000
4	LOAD2	0.0000	0.0000	0.0000	0.0000	-5.129E-04	0.0000
5	LOAD2	0.0000	0.0000	0.0000	0.0000	0.0000	0.0000

JOINT REACTIONS

JOINT	LOAD	F1	F2	F3	M1	M2	M3
1	LOAD2	0.0000	0.0000	43.8705	0.0000	- 602.5990	0.0000
3	LOAD2	0.0000	0.0000	21.0223	0.0000	0.0000	0.0000
4	LOAD2	0.0000	0.0000	17.8605	0.0000	0.0000	0.0000
5	LOAD2	0.0000	0.0000	-7.7034	0.0000	-259.9073	0.0000

FRAME ELEMENT FORCES

FRAME	LOAD	LOC	P	V2	V3	T	M2	M3
1	LOAD2							
		0.00	0.00	-43.87	0.00	0.00	0.00	-602.60
		22.50	0.00	-3.87	0.00	0.00	0.00	-15.51
		45.00	0.00	-3.87	0.00	0.00	0.00	71.57
		67.50	0.00	-3.87	0.00	0.00	0.00	158.66
		90.00	0.00	-3.87	0.00	0.00	0.00	245.75
2	LOAD2							
		0.00	0.00	6.13	0.00	0.00	0.00	245.75
		22.50	0.00	8.38	0.00	0.00	0.00	82.52
		45.00	0.00	10.63	0.00	0.00	0.00	-131.33
		67.50	0.00	12.88	0.00	0.00	0.00	-395.80
		90.00	0.00	15.13	0.00	0.00	0.00	-710.91
3	LOAD2							
		0.00	0.00	-5.89	0.00	0.00	0.00	-660.91
		30.00	0.00	-4.83	0.00	0.00	0.00	-506.68
		60.00	0.00	-8.262E-01	0.00	0.00	0.00	-418.89
		90.00	0.00	4.37	0.00	0.00	0.00	-469.11
		120.00	0.00	5.36	0.00	0.00	0.00	-602.76
4	LOAD2							
		0.00	0.00	-12.50	0.00	0.00	0.00	-610.01
		24.00	0.00	-11.30	0.00	0.00	0.00	-324.33
		48.00	0.00	-10.10	0.00	0.00	0.00	-167.45
		72.00	0.00	-8.90	0.00	0.00	0.00	60.63
		96.00	0.00	-7.70	0.00	0.00	0.00	259.91

Plot Displacements: From the Main Menu enter:

DISPLAY>SHOW DEFORMED SHAPE, then enter
FILE>PRINT GRAPHICS

(The deformed shape shown on the screen is reproduced in Fig.2.11)

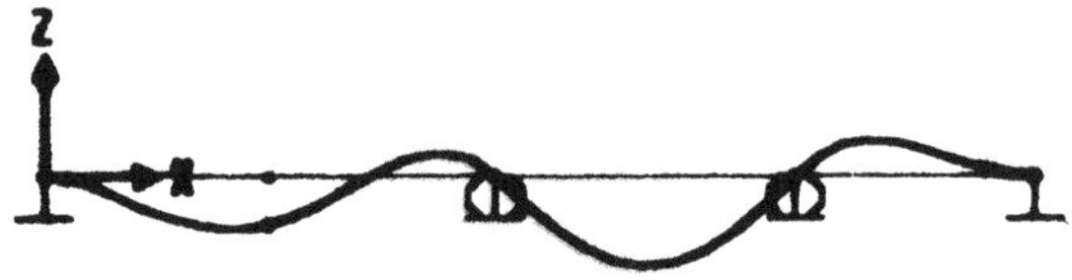

Fig. 2.11 Deformed shape for the beam of Illustrative Example 2.4

Plot Shear Forces: From the Main Menu enter:

DISPLAY>SHOW ELEMENT FORCES / STRESSES>FRAMES
Select Shear 2-2. Then enter:
FILE>PRINT GRAPHICS

(The shear force diagram shown on the screen is depicted in Fig.2.12)

Fig. 2.12 Shear Force diagram for the beam of Illustrative Example 2.4

Plot the Bending Moment: From the Main Menu enter:
DISPLAY>SHOW FORCES / STRESSES>FRAME
Select Moment 3-3 and OK. Then enter:
FILE>PRINT GRAPHICS
(The bending moment diagram on the screen is depicted in Fig. 2.13)

Fig 2.13 Bending Moment Diagram for the beam of Illustrative Example 2.4

Note: To plot the Moment Diagram on the compression side from the Main Menu enter: OPTIONS then remove check on "Moment Diagram on the Tension Side"

Note: To view plots of either the shear force or bending moment diagrams for any element of the beam, right-click on the element and a pop-up window will depict the diagram for the selected element. It should be observed that as the cursor is moved across the plot in this window, values of shear force or of the bending moment will be displayed.

Illustrative Example 2.5

Consider in Fig.2.14(a)) a loaded concrete continuous beam of two spans and in Fig.2.14(b) the selected analytical model with 5 elements and 6 nodes. Use the program SAP2000 to perform the structural analysis to determine displacements, reactions and the shear force and bending moment diagrams.

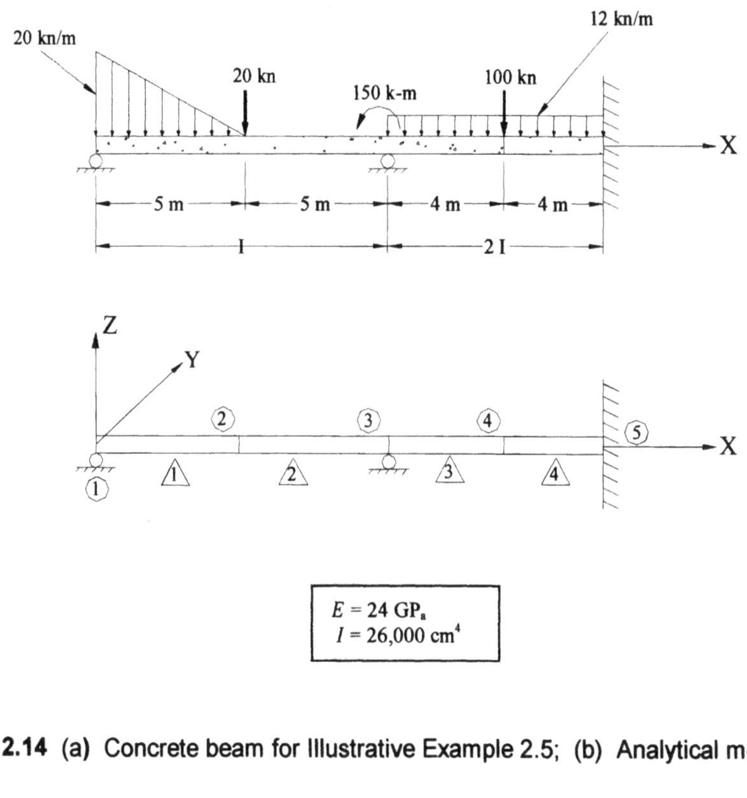

Fig. 2.14 (a) Concrete beam for Illustrative Example 2.5; (b) Analytical model

Table 2.8 Edited Input Data for Illustrative Example 2.5 (Units: KN, m)

JOINT DATA

JNT	GLBL-X	GLBL-Y	GLBL-Z	RESTRAINTS	AN-A	AN-B	AN-C
1	0.00000	0.00000	0.00000	0 0 1 0 0 0	0.000	0.000	0.000
2	5.00000	0.00000	0.00000	0 0 0 0 0 0	0.000	0.000	0.000
3	10.00000	0.00000	0.00000	0 0 1 0 0 0	0.000	0.000	0.000
4	14.00000	0.00000	0.00000	0 0 0 0 0 0	0.000	0.000	0.000
5	18.00000	0.00000	0.00000	1 1 1 1 1 1	0.000	0.000	0.000

FRAME ELEMENT DATA

FRM	JNT-1	JNT-2	SEC	ANGL	RELEASES	SGMNTS	R1	R2	FACTOR	LENGTH
1	1	2	FSEC3	0.000	000000	4	0.000	0.000	1.000	5.000
2	2	3	FSEC3	0.000	000000	4	0.000	0.000	1.000	5.000
3	3	4	FSEC2	0.000	000000	4	0.000	0.000	1.000	4.000
4	4	5	FSEC2	0.000	000000	4	0.000	0.000	1.000	4.000

JOINT FORCES Load Case LOAD1

JOINT	GLBL-X	GLBL-Y	GLBL-Z	GLBL-XX	GLBL-YY	GLBL-ZZ
3	0.000	0.000	0.000	0.000	-150.000	0.000
4	0.000	0.000	-100.000	0.000	0.000	0.000
2	0.000	0.000	-20.000	0.000	0.000	0.000

FRAME SPAN DISTRIBUTED LOADS Load Case LOAD1

FRM	TYPE	DIRECTION	DIST-A	VALUE-A	DIST-B	VALUE-B
3	FORCE	GLOBAL-Z	0.0000	-12.0000	1.0000	-12.0000
4	FORCE	GLOBAL-Z	0.0000	-12.0000	1.0000	-12.0000
1	FORCE	GLOBAL-Z	0.0000	-20.0000	1.0000	0.0000

Solution:

Begin: Open SAP2000.

Hint: Maximize both screens for a full views of all windows.

Units: Select kn-m (Kilo Newtons-Meters) in the drop-down menu located on the lower right hand corner of the screen.

Model: FILE>NEW MODEL FROM TEMPLATE
Select "Beam"
Change the number of spans to 4 and span length to 5, then OK

Edit: Minimize the 3-D screen.
Maximize the 2-D screen (X-Z screen). Then drag the figure to the center of the screen using the PAN icon on the toolbar.

Translate: Enter: SELECT>SELECT ALL
EDIT>MOVE
Delta X = 10. Then OK. Use icon PAN to center figure.

Grid: Enter: DRAW>EDIT GRID
Highlight –5 value and change to 15.
Click "Move Grid Line", then OK
Highlight –10 value and change to 20. Then OK

Labels: Enter: VIEW>SET ELEMENTS
Check: Joint Labels
Check: Frame Labels, then OK.

Restraints: Click (mark): Joints ① and ③ and enter:
ASSIGN>JOINT>RESTRANTS
Check: Restraint translation 3. Then OK.
Select: joint ⑤
Enter: ASSIGN>JOINTS>RESTRAINTS
Check: Restrain all directions; then OK.

Material: Enter:DEFINE MATERIAL>CONC
Click: Modify>Show Materials
Set: Modulus of Elasticity = 2.4E6; then OK

Sections: Enter: DEFINE>FRAME SECTIONS
Click: Add/Wide Flange
Select: Add General (Section)
Set: Moment of Inertia about axis 3 = 5.2E-3
Shear area in 2 direction = 0
Scroll: Material "CONC" Then OK, OK.

Enter: DEFINE >FRAME SECTIONS
Click: Add/Wide Flange
Select: Add General (Section)
Set: Moment of Inertia about axis 3 = 2.6E-3
Shears are in 2 direction = 0; then OK
Material: Scroll to CONC. Then OK, OK.

Loads: Enter: DEFINE>STATIC LOAD CASES
Change: DEAD to LIVE
Self weight multiplier = 0
Click: Change Load; then OK
Mark: Joint 2
Enter: ASSIGN>JOINT STATIC LOAD>FORCES
Check: Forces Global Z = -20; then OK

Warning: Be certain to zero out the values of previous entries.

Mark: Joint ④

Enter: ASSIGN>JOINT STATIC LOADS>FORCES
Check: Forces Global Z = -100

Assign Frame Loads: Mark: Frame 1

Enter: ASSIGN>FRAME STATIC LOADS>TRAPEZOIDAL
Verify: Checks on Forces and on Global Z
Enter: Distance 0.0 1.0
Load -20 0 Then OK

Mark: Frames 3 and 4

Enter: ASSIGN>FRAME STATIC LOADS>POINT AND UNIFORM
Uniform Load = -12
Warning: Be certain to zero out the values of previous entries. Then OK

Options: Enter: ANALYZE>SET OPTIONS

Mark: Available Degrees of Freedom UX, UZ, RY. Then OK.

Analyze: Enter: ANALYZE>RUN

Filename "Example 2.5" , SAVE
At the conclusion of the solution process enter OK.

Print Tables: Enter: FILE>PRINT INPUT TABLES

Check: Joint Data: Coordinates
Element data: Frames
Static Loads: Joints, Frames; then OK

(Table 2.8 reproduces the edited Input Data for Illustrative Example 2.5)

Enter: FILE>PRINT OUTPUT TABLE
Check: Displacements
Reactions
Frame Forces, then OK

Table 2.9 reproduces the edited output table for Illustrative Example 2.5

Table 2.9 Edited Output tables for Illustrative Example 2.5 (Units: kN, m)

JOINT DISPLACEMENTS

JNT	LOAD	UX	UY	UZ	RX	RY	RZ
1	LOAD1	0.0000	0.0000	0.0000	0.0000	3.304E-03	0.0000
2	LOAD1	0.0000	0.0000	-8.007E-03	0.0000	-8.017E-04	0.0000
3	LOAD1	0.0000	0.0000	0.0000	0.0000	-7.648E-04	0.0000
4	LOAD1	0.0000	0.0000	-2.398E-03	0.0000	1.912E-04	0.0000
5	LOAD1	0.0000	0.0000	0.0000	0.0000	0.0000	0.0000

Table 2.9 Continued

JOINT REACTIONS

JNT	LOAD	F1	F2	F3	M1	M2	M3
1	LOAD1	0.0000	0.0000	45.4942	0.0000	0.0000	0.0000
3	LOAD1	0.0000	0.0000	131.4541	0.0000	0.0000	0.0000
5	LOAD1	0.0000	0.0000	89.0517	0.0000	140.1378	0.0000

FRAME ELEMENT FORCES

FRAME	LOAD	LOC	P	V2	V3	T	M2	M3
1	LOAD1	0.00	0.00	-45.49	0.00	0.00	0.00	0.00
		1.25	0.00	-23.62	0.00	0.00	0.00	42.54
		2.50	0.00	-7.99	0.00	0.00	0.00	61.65
		3.75	0.00	1.38	0.00	0.00	0.00	65.13
		5.00	0.00	4.51	0.00	0.00	0.00	60.80
2	LOAD1	0.00	0.00	24.51	0.00	0.00	0.00	60.80
		1.25	0.00	24.51	0.00	0.00	0.00	30.17
		2.50	0.00	24.51	0.00	0.00	0.00	-0.46
		3.75	0.00	24.51	0.00	0.00	0.00	-31.09
		5.00	0.00	24.51	0.00	0.00	0.00	-61.72
3	LOAD1	0.00	0.00	-106.95	0.00	0.00	0.00	-211.72
		1.00	0.00	-94.95	0.00	0.00	0.00	-110.78
		2.00	0.00	-82.95	0.00	0.00	0.00	-21.83
		3.00	0.00	-70.95	0.00	0.00	0.00	55.12
		4.00	0.00	-58.95	0.00	0.00	0.00	120.07
4	LOAD1	0.00	0.00	41.05	0.00	0.00	0.00	120.07
		1.00	0.00	53.05	0.00	0.00	0.00	73.02
		2.00	0.00	65.05	0.00	0.00	0.00	13.97
		3.00	0.00	77.05	0.00	0.00	0.00	-57.09
		4.00	0.00	89.05	0.00	0.00	0.00	-140.14

Plot Output:

Enter: DISPLAY > SHOW DEFORMED SHAPE
Check: Wire shadow. Then O.K.

Enter : FILE > PRINT GRAPHICS
[The deformed shape is reproduced in Fig. 2.15(a)]

Enter: DISPLAY > SHOW ELEMENT FORCES/STRESSES > FRAMES
Click: Shear 2-2. Then O.K.

Enter: FILE > PRINT GRAPHICS
[The shear force diagram is reproduced in Fig, 2.15(b)]

Enter: DISPLAY > SHOW ELEMENT FORCES/STRESSES > FRAMES
Click: Moment 3 - 3. Then O.K.

Enter: FILE > PRINT GRAPHICS
[The bending moment diagram is reproduced in Fig. 2.15(c)]

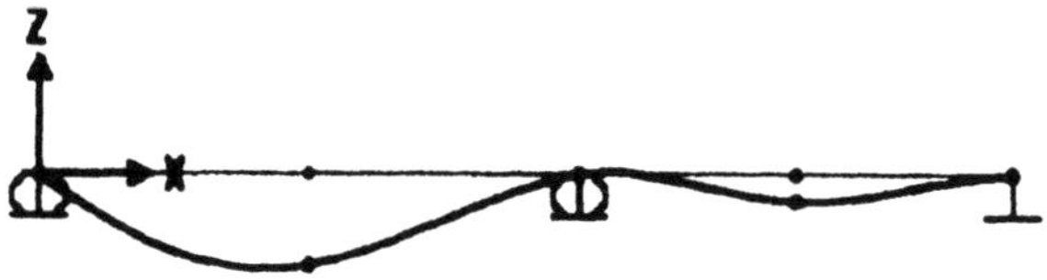

(a)

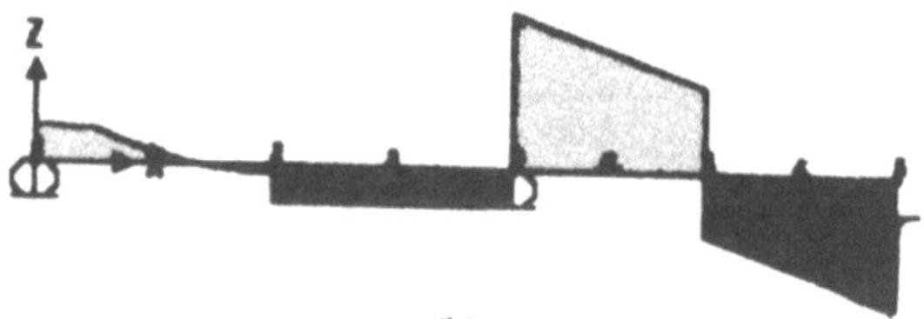

(b)

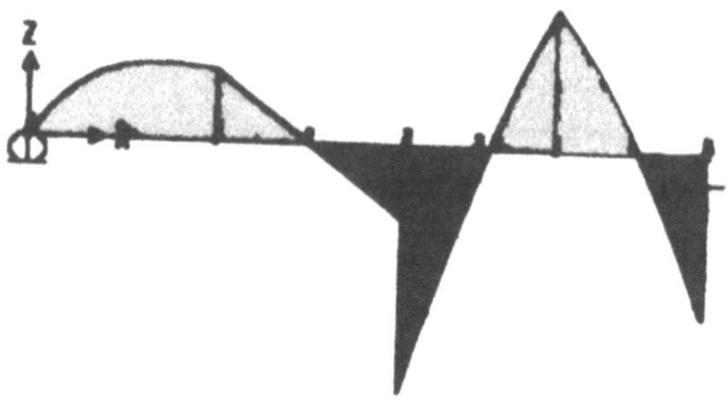

(c)

Fig. 2.15 (a) Deformed shape, (b) Shear Force diagram, and (c) Bending Moment Diagram for the beam of Illustrative Example 2.5.

Illustrative Example 2.6

Use SAP2000 to analyze the beam shown in Fig. 1.16 reproduced here for convenience as Fig. 2.16.

$E = 10{,}000$ kip
$I = 100$ in^4
$L = 100$ in
$k_a = 100$ kip/in
$k_t = 1000$ kip-in/rad

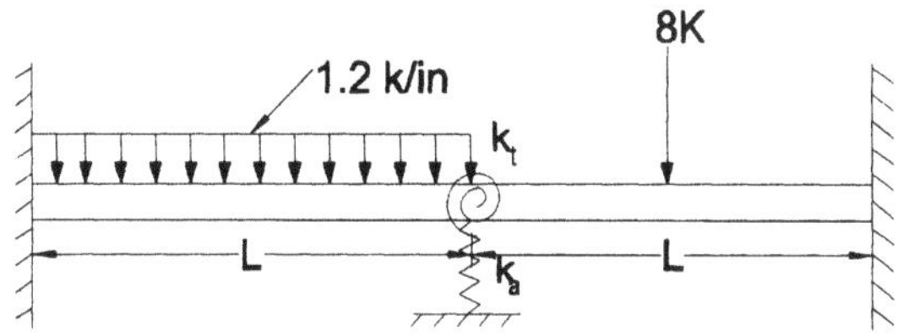

Fig. 2.16 Beam for Illustrative Example 2.6

Edited Input data for Illustrative Example 2.6 is given in Table 2.10.

Table 2.10 Edited Input Data for Illustrative Example 2.6 (Units: kips, inches)

JOINT DATA

JOINT	GLOBAL-X	GLOBAL-Y	GLOBAL-Z	RESTRAINTS
1	0.00000	0.00000	0.00000	1 1 1 1 1 1
2	100.00000	0.00000	0.00000	0 0 0 0 0 0
3	200.00000	0.00000	0.00000	1 1 1 1 1 1

JOINT SPRING DATA

JOINT	K-U1	K-U2	K-U3	K-R1	K-R2	K-R3
2	0.000	0.000	100.000	0.000	1000.000	0.000

FRAME ELEMENT DATA

FRAME	JNT-1	JNT-2	SECTION	RELEASES	SEGMENTS	LENGTH
1	1	2	FSEC2	0.000	4	100.000
2	2	3	FSEC2	0.000	4	100.000

FRAME SPAN DISTRIBUTED LOADS Load Case LOAD1

FRAME	TYPE	DIRECTION	DISTANCE-A	VALUE-A	DISTANCE-B	VALUE-B
1	FORCE	GLOBAL-Z	0.0000	-1.2000	1.0000	-1.2000

FRAME SPAN POINT LOADS

Load Case LOAD1

FRAME	TYPE	DIRECTION	DISTANCE	VALUE
2	FORCE	GLOBAL-Z	0.5000	-8.0000

Solution:

Begin: Open SAP2000.

Hint: Maximize both screens for a full views of all windows.

Units: Select kn-m in the drop-down menu located on the lower right hand corner of the screen.

Model: FILE>NEW MODEL FROM TEMPLATE
Select "Beam"
Change the number of spans to 2 and span length to 100, then OK

Edit: Minimize the 3-D screen.
Maximize the 2-D screen (X-Z screen). Then drag the figure to the center of the screen using the PAN icon on the toolbar.

Translate: Enter: SELECT>SELECT ALL
EDIT>MOVE
Delta X = 150. Then OK. Use icon PAN to center figure.

Grid: Enter: DRAW>EDIT GRID
Highlight –100 value and change to 200.
Click "Move Grid Line", then OK
VIEW
Remove check on "Show Grid"

Labels: Enter: VIEW>SET ELEMENTS
Check: Joint Labels
Check: Frame Labels, then OK.

Restraints:

Click (mark): Joints ① and ③ and enter:
ASSIGN>JOINT>RESTRANTS
Check: Restrain all directions. Then OK.

Click: Joint ②
Enter: ASSIGN>JOINT> RESTRAINTS
Check restraints in the Y direction only; then OK

Click: Joint ②
ASSIGN>JOINT>RESTRAINTS
No restraints in all directions; then OK

Material: Enter:DEFINE MATERIAL>OTHER
Click: Modify>Show Materials
Set: Modulus of Elasticity = 10E3; then OK

Sections: Enter: DEFINE>FRAME SECTIONS
Click: Add/Wide Flange
Select: Add General (Section)
Set: Moment of Inertia about axis 3 = 100
Shear area in 2 direction = 0
Scroll: Material "Other. Then OK, OK.

Enter: DEFINE >FRAME SECTIONS
Click: Add/Wide Flange
Select: Add General (Section)
Set: Moment of Inertia about axis 3 = 200
Shears are in 2 direction = 0
Scroll: Material "Other". Then OK, OK

Click (mark) Frames 1 and 2
Enter: ASSIGN>FRAME>SECTIONS
Select: FSFC2; then OK

Loads: Enter: DEFINE>STATIC LOAD CASES
Change: DEAD to LIVE
Self weight multiplier = 0
Click: Change Load; then OK

Click on Frame 1
Enter: ASSIGN>FRAME STATIC LOAD>POINT LOAD AND UNIFORM
Warning: Be certain to zero out the values of previous entries.
Check: Forces Z direction
Uniform Load = -1.2; then OK.

Click on Frame 2
Enter: ASSIGN>FRAME STATIC LOADS>POINT AND UNIFORM LOADS
Warning: Be certain to zero out the values of previous entries.
Check: Forces in Z Direction.
Enter: Distance 0.5
Load –8.0

Options: Enter: ANALYZE>SET OPTIONS
Mark: Available Degrees of Freedom UX, UZ, RY. Then OK.

Analyze: Enter: ANALYZE>RUN
Filename "Example 2.6", SAVE
At the conclusion of the solution process enter OK.

Print Tables: Enter: FILE>PRINT INPUT TABLES
Check: Joint Data: Coordinates
Element data: Frames
Static Loads: Joints, Frames; then OK

(Table 2.10 reproduces the edited Input Data for Illustrative Example 2.6)*

Enter: FILE>PRINT OUTPUT TABLE
Check: Displacements
Reactions
Frame Forces
Spring Forces
Then OK

(Table 2.11 reproduces the edited Output Table for Illustrative Example 2.6)

Table 2.11 Edited Output for Illustrative Example 2.6 (Units: kips, inches)*

JOINT DISPLACEMENTS

JOINT	LOAD	UX	UY	UZ	RX	RY	RZ
1	LOAD1	0.0000	0.0000	0.0000	0.0000	0.0000	0.0000
2	LOAD1	0.0000	0.0000	-0.5161	0.0000	-0.0111	0.0000

JOINT REACTIONS

JOINT	LOAD	F1	F2	F3	M1	M2	M3
1	LOAD1	0.0000	0.0000	72.8602	0.0000	-1531.8997	0.0000
3	LOAD1	0.0000	0.0000	3.5269	0.0000	187.4552	0.0000

FRAME ELEMENT FORCES

FRAME	LOAD	LOC	P	V2	V3	T	M2	M3
1	LOAD1							
		0.00	0.00	-72.86	0.00	0.00	0.00	-1531.90
		25.00	0.00	-42.86	0.00	0.00	0.00	-85.39
		50.00	0.00	-12.86	0.00	0.00	0.00	611.11
		75.00	0.00	17.14	0.00	0.00	0.00	557.62
		100.00	0.00	47.14	0.00	0.00	0.00	-245.88

* This table has been edited to conform to the format requirements of this text.

Table 2.11 Continued

FRAME	LOAD	LOC	P	V2	V3	T	M2	M3
2	LOAD1							
		0.00	0.00	-4.47	0.00	0.00	0.00	-234.77
		25.00	0.00	-4.47	0.00	0.00	0.00	122.94
		50.00	0.00	3.53	0.00	0.00	0.00	-11.11
		75.00	0.00	3.53	0.00	0.00	0.00	-99.28
		100.00	0.00	3.53	0.00	0.00	0.00	-187.46

JOINT SPRING FORCES

JOINT	LOAD	F1	F2	F3	M1	M2	M3
2	LOAD1	0.0000	0.0000	51.6129	0.0000	11.1111	0.0000

Deformed Shape: Enter: DISPLAY>SHOW DEFORMED SHAPE
Check: Wire Shadow; then OK
Enter: FILE>PRINT GRAPHICS

(The deformed shape is reproduced in Fig. 2.17)

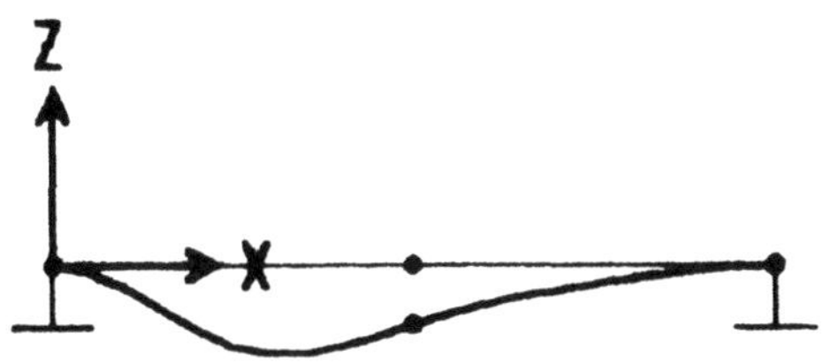

Fig. 2.17 Deformed shape of the beam of Illustrative Example 2.6

Shear Force Diagram: Enter: DISPLAY>SHOW ELEMENT FORCES/STRESSES>FRAMES
Check: Shear 2-2; then OK.
(The Shear Force Diagram is reproduced in Fig. 2.18.)

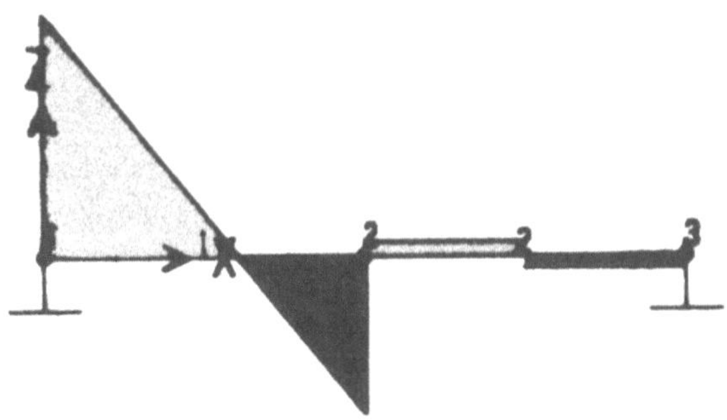

Fig. 2.18 Shear Force Diagram of the beam in Illustrative Example 2.6.

Bending Moment Diagram: Enter:

DISPLAY>SHOW ELEMENT FORCES/STRESSES>FRAMES

Check: Moment 3-3, then OK

(The Bending Moment diagram is reproduced in Fig. 2.19.)

Fig. 2.19 Bending Moment diagram of the beam in Illustrative Example 2.6.

2.2 Problems

Use SAP2000 to solve the problems in Chapter 1, repeated here for convenience.

Problem 2.1

For the beam shown in Fig. P2.1 determine:

(a) Displacements at nodes ① and ②

(b) Reactions at the support

E = 30,000 ksi I = 100 in^4

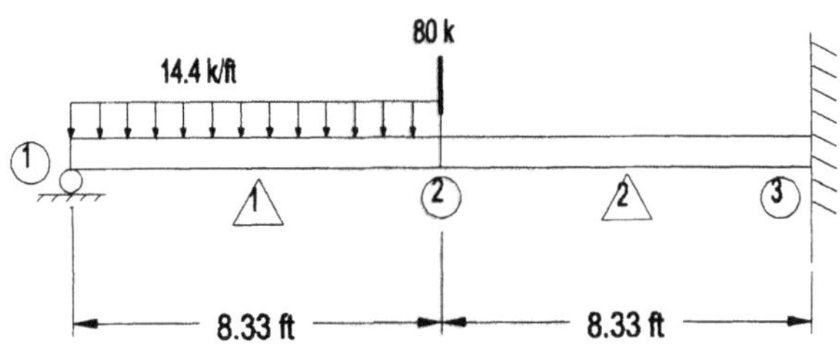

Fig. P2.1

Problem 2.2

For the beam shown in Fig. P2.2 determine:

(a) Displacements at node ②
(b) End forces on the beam elements
(c) Reactions at the supports

$E = 30{,}000$ ksi
$I = 200$ in^4

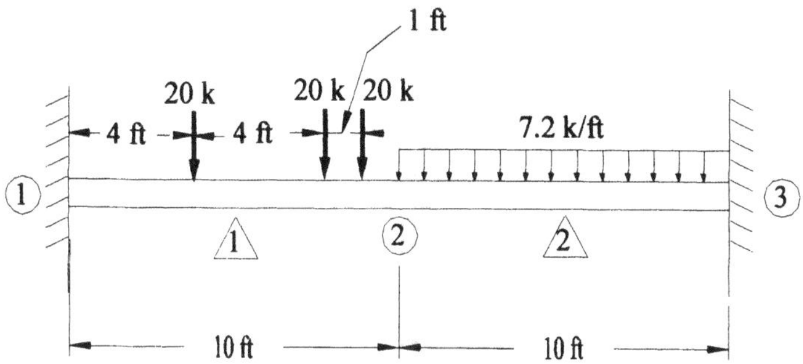

Fig P2.2

Problem 2.3

For the beam shown in Fig. P2.3 determine:

(a) Displacement at the nodes ② and ③,
(b) End-forces on the beam elements
(c) Reactions at the supports

$E = 200$ GP$_a$
$I = 40{,}000$ cm^4

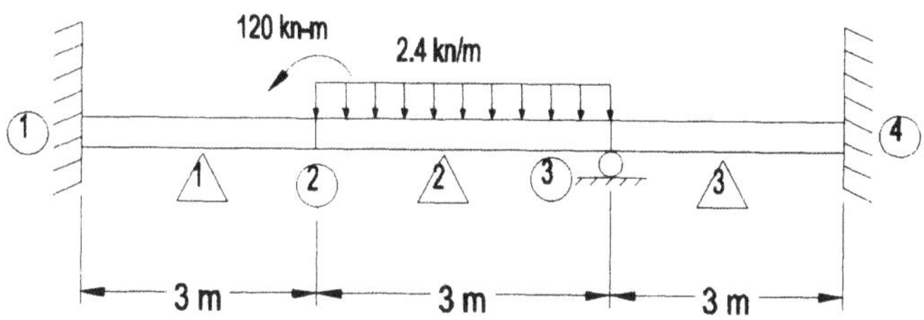

Fig. P2.3

Problem 2.4

For the beam shown in Fig. P2.4 determine:

(a) Displacement at node ②
(b) End-forces on the beam elements
(c) Reactions at the supports

$E = 4500$ ksi
$I = 1200$ in^4

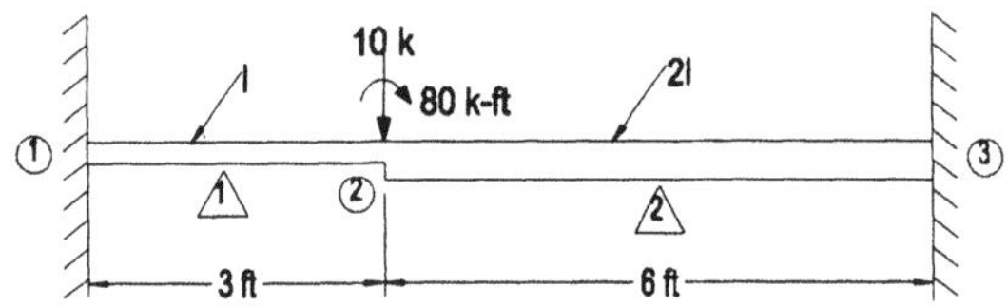

Fig. P2.4

Problem 2.5

For the beam shown in Fig. P2.5 determine :

(a) displacements at nodes ② and ③,
(b) End-forces on the beam elements, and
(c) Reactions at the supports.

$E = 200$ GP$_a$
$I = 40{,}000$ cm^4

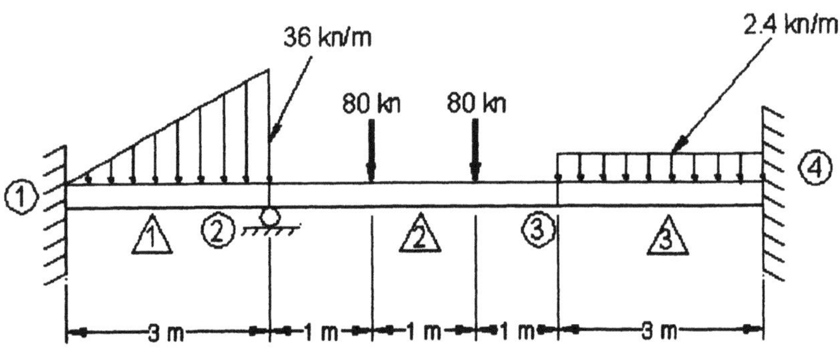

Fig. P2.5

Problem 2.6

For the beam shown in Fig. P2.6 determine:

(a) Displacements at nodes ① ② and ③,
(b) End-forces on the beam elements, and
(c) Reactions at the supports.

$E = 30{,}000$ ksi $I = 120$ in^4

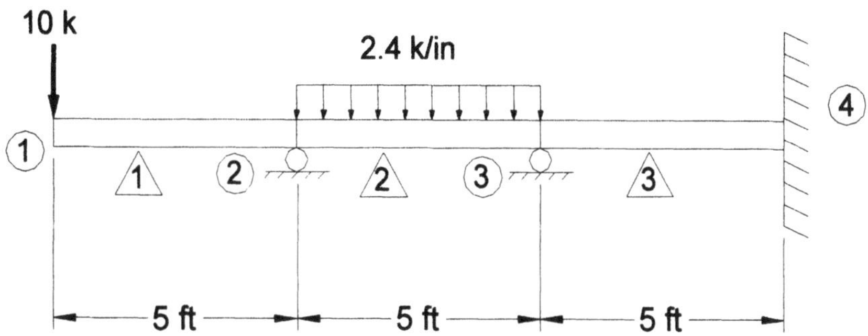

Fig. P2.6

Problem 2.7
Solve Problem 2.1 assuming that a hinge exists at the right end of beam element 1.

Problem 2.8
Solve Problem 2.2 assuming a hinge at the left end of beam element 2.

Problem 2.9
Solve Problem 2.3 assuming a hinge at the right end of beam element 2.

Problem 2.10

Solve Problem 2.4 assuming a hinge at node ② (right end of beam element 1 and left end of element 2).

Problem 2.11
Solve Problem 2.5 assuming a hinge at the left end of beam element 2.

Problem 2.12
Solve Problem 2.6 assuming hinges at the two ends of beam element 2.

Problem 2.13

Solve Problem 2.7 assuming that the support at node ① underwent a downward displacement of 1.0 in and rotational displacement of 5 degrees in the clockwise direction.

Problem 2.14

Solve Problem 2.8 assuming that the support at node ③ underwent a downward displacement of 2.0 in.

Problem 2.15

Solve Problem 2.9 assuming that the support at node ① underwent a downward displacement of 0.5 in and clockwise rotation of 10 degrees.

Problem 2.16

Solve Problem 2.10 assuming that the temperature at the top side of beam element 2 is 30°F and at the bottom side is 80°F.

Problem 2.17

Solve Problem 2.11 knowing that the temperature along the top of the beam is 40°F and at its bottom side is 100°F.

Problem 2.18

For the beam shown in Fig. P2.18 determine:

(a) Displacements at nodes ① and ②.

(b) The force and the moment of the spring at node ②.

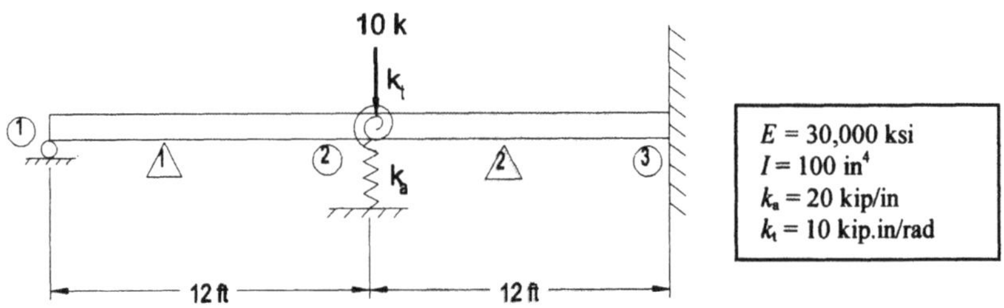

Fig. P2.18

Problem 2.19

For the beam shown in Fig. P2.19 determine:

(a) Displacements at nodes ② and ③.

(b) Force in the spring at node ②.

E = 30,000 ksi
I = 400 in^4
k_a = 10 kip/in
L = 100 in

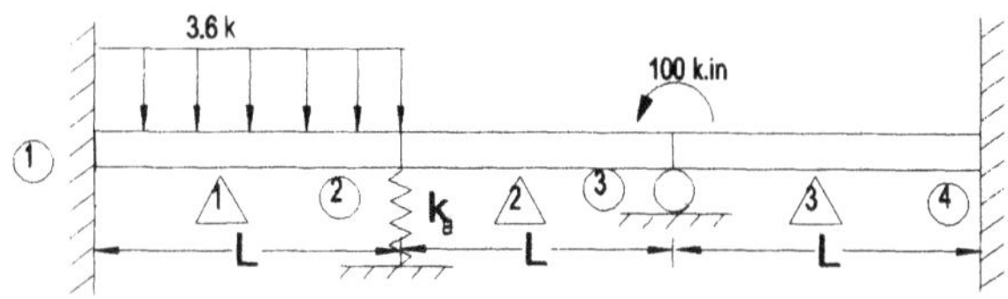

Fig. P2.19

3 Plane Frames

3.1 Introduction

Structural analysis using the matrix stiffness method for structures modeled as beams has been presented in the preceding two chapters. This method of analysis when applied to structures modeled as plane frames with loads applied in the plane of the frame, requires the inclusion of the axial effect in the element stiffness matrices; hence, the inclusion of axial effect in the system stiffness matrix. The analysis also requires a coordinate transformation of element end forces and displacements from element or local coordinate axes to the system or global coordinate axes. Except for the consideration of the axial effect and the need to transform the element end forces and displacements, the matrix stiffness method when applied to plane frames is identical to the analysis of beams presented in Chapters 1 and 2.

3.2 Stiffness Coefficients for Axial Forces.

The inclusion of axial forces and axial deformations in the stiffness matrix of a flexural beam requires the determination of the stiffness coefficients for axial loads. To derive the stiffness matrix for an axially loaded member, consider in Fig.3.1 a beam element acted on by the axial forces P_1 and P_2 producing axial displacements δ_1 and δ_2 at the nodes of the element. For a prismatic and uniform beam segment of length L and cross-sectional A, it is relatively simple to obtain the stiffness relation for axial effects by the application of Hooke's law. In relation to the beam shown in Fig.3.1 the displacements δ_1 produced by the force P_1 acting at node ① while node ② is maintained fixed (δ_2 =0) is given by

$$\delta_1 = \frac{P_1 L}{AE} \tag{3.1}$$

From eq.(3.1) and the definition of the stiffness coefficient k_{11} (force at node ① to produce a unit displacement $\delta_1 = 1.0$), we obtain

$$k_{11} = \frac{P_1}{\delta_1} = \frac{AE}{L} \tag{3.2a}$$

The equilibrium of the beam element acted upon by the force k_{11} requires an opposite force k_{21} at the other end, namely

$$k_{21} = -k_{11} = -\frac{AE}{L} \tag{3.2b}$$

Analogously, the other stiffness coefficients are

$$k_{22} = \frac{AE}{L} \tag{3.2c}$$

and

$$k_{12} = -\frac{AE}{L} \tag{3.2d}$$

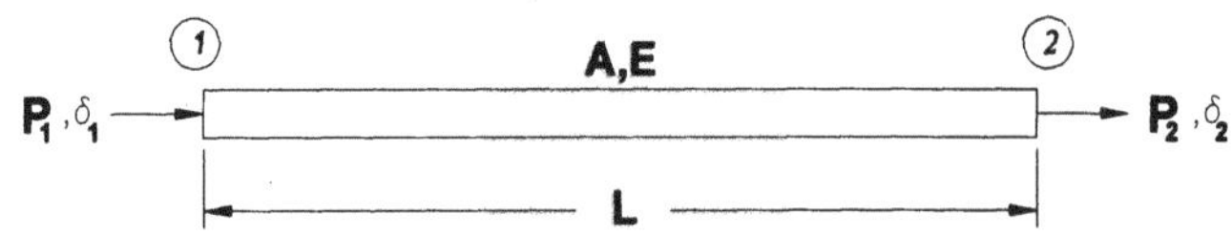

Fig. 3.1 Beam element with nodal axial loads P_1, P_2, and corresponding nodal displacements δ_1, δ_2

The stiffness coefficients as given by eqs.(3.2) are conveniently arranged in the stiffness matrix relating axial forces $\{P\}$ and displacements $\{\delta\}$ for a prismatic beam element, that is

$$\begin{Bmatrix} P_1 \\ P_2 \end{Bmatrix} = \frac{AE}{L} \begin{bmatrix} 1 & -1 \\ -1 & 1 \end{bmatrix} \begin{Bmatrix} \delta_1 \\ \delta_2 \end{Bmatrix} \tag{3.3}$$

3.3 Displacement Functions for Axially Loaded Beams.

Consider in Fig. 3.2 a beam element undergoing a unit displacement $\delta_1 = 1$ at its left end.

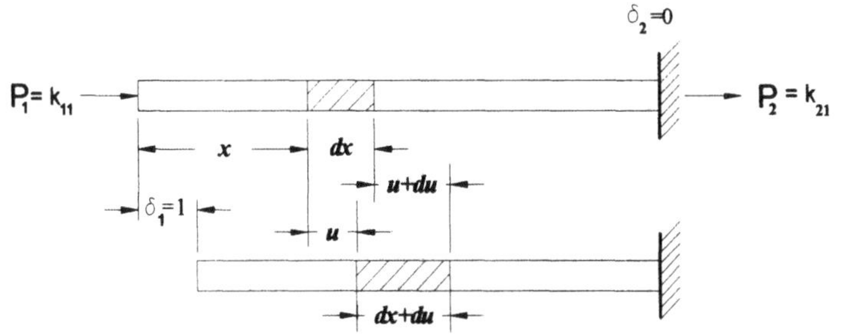

Fig. 3.2 Beam element showing deformation due to a unit displacement of $\delta_1 = 1$.

As shown in Fig. 3.2, if $u = u(x)$ is the displacement at section x , the displacement at section $x + dx$ will be $u + du$. The element dx in this new position has changed in length by the amount du, and thus, the strain is du/dx. Since by Hooke's law, the ratio of stress to strain is equal to the modulus of elasticity E, we can write

$$\frac{du}{dx} = \frac{P}{AE} \tag{3.4}$$

Integration of eq.(3.4) with respect to x yields

$$u\,(x) = \frac{P}{AE}x + C$$

in which C is a constant of integration. Introducing the boundary conditions $u = 1$ at $x = 0$ and $u = 0$ at $x = L$, we obtain the displacement function $u_1 = u_1(x)$ corresponding to $\delta_1 = 1$ as

$$u_1(x) = 1 - \frac{x}{L} \tag{3.5}$$

Analogously, the displacement function $u_2(x)$ corresponding to a unit displacement $\delta_2 = 1$ is

$$u_2(x) = \frac{x}{L} \tag{3.6}$$

The general expression for the stiffness coefficient k_{ij} for axial effect may be obtained by application of the Principle for Virtual Work (see Problem 3.1) as

$$k_{ij} = \int_0^L AE\, u_i'(x)\;\; u_j'(x)\, dx \tag{3.7}$$

where u_i' or u_j' is the derivative with respect to x of the displacement function $u_1(x)$ or $u_2(x)$ given by eq.(3.5) or (3.6).

Using eq.(3.7) the reader may check the results obtained in eqs.(3.2) for a uniform beam element. However, eq.(3.7) could as well be used for non-uniform elements in which, in general, AE would be a function of x. In practice, the same displacement functions $u_1(x)$ and $u_2(x)$ obtained for a uniform beam element are also used in eq.(3.7) for a non-uniform element.

3.4 Element Stiffness Matrix for Plane Frame Elements

The stiffness matrix corresponding to the nodal coordinates for an element of a plane frame (Fig. 3.3) is obtained by combining in a single matrix the stiffness matrix for axial effects [eq.(3.3)] and the stiffness matrix for flexural effects [eq.(1.11)]. The matrix resulting from this combination relates the forces P_i and the displacements δ_i at the nodal coordinates indicated in Fig 3.3 as

$$\begin{Bmatrix} P_1 \\ P_2 \\ P_3 \\ P_4 \\ P_5 \\ P_6 \end{Bmatrix} = \frac{EI}{L^3} \begin{bmatrix} AL^2/I & 0 & 0 & -AL^2/I & 0 & 0 \\ 0 & 12 & 6L & 0 & -12 & 6L \\ 0 & 6L & 4L^2 0 & 0 & -6L & 2L^2 \\ -AL^2/I & 0 & 0 & AL^2/I & 0 & 0 \\ 0 & -12 & -6L & 0 & 12 & -6L \\ 0 & 6L & 2L^2 & 0 & -6L & 4L^2 \end{bmatrix} \begin{Bmatrix} \delta_1 \\ \delta_2 \\ \delta_3 \\ \delta_4 \\ \delta_5 \\ \delta_6 \end{Bmatrix} \tag{3.8}$$

or, in concise notation,

$$\{P\} = [k]\{\delta\} \tag{3.9}$$

in which $[k]$ is the stiffness matrix in reference to local coordinate axes of an element of a plane frame and $\{P\}$ and $\{\delta\}$ are, respectively, the nodal force and the nodal displacement vectors of the element.

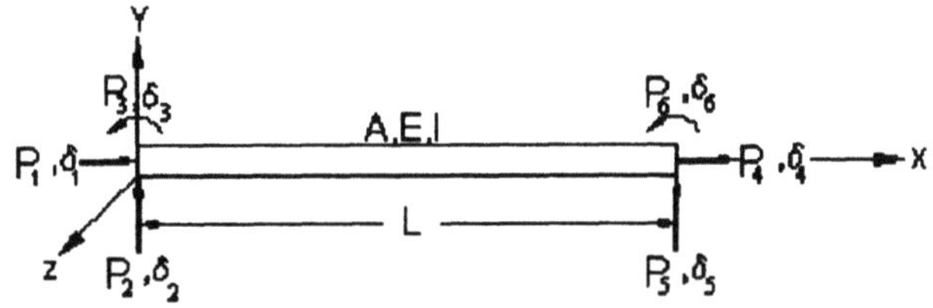

Fig. 3.3 Beam element showing flexural and axial nodal forces and displacements.

3.5 Coordinate Transformation

The stiffness matrix for the beam element of a plane frame in eq.(3.8) refers to nodal coordinates defined by coordinate axes fixed on the beam element. These axes (x, y, z) are called local or element axes while the coordinate axes (X, Y, Z) for the whole structure are known as global or system axes. Figure 3.4 shows a beam element of a plane frame with nodal forces (P_1, P_2,.....P_6) referring to the local coordinate axes x, y, z, and $(\overline{P}_1, \overline{P}_2, \ldots\ldots \overline{P}_6)$ referring to the global coordinate axes X, Y, Z. The objective is to transform the element stiffness matrix from the reference of local coordinate axes to the global coordinate axes. This transformation is required in order that the matrices for all the elements refer to the same set of coordinates; hence, the matrices become compatible for assemblage into the system matrix for the structure. We begin by expressing the forces (P_1, P_2, P_3) in terms of the forces $(\overline{P}_1, \overline{P}_2, \overline{P}_3)$. Since these two sets of forces are equivalent, we obtain from Fig. 3.4 the following relationships:

$$
\begin{aligned}
P_1 &= \overline{P}_1 \cos\theta + \overline{P}_2 \sin\theta \\
P_2 &= -\overline{P}_1 \sin\theta + \overline{P}_2 \cos\theta \\
P_3 &= \overline{P}_3
\end{aligned}
\qquad (3.10)
$$

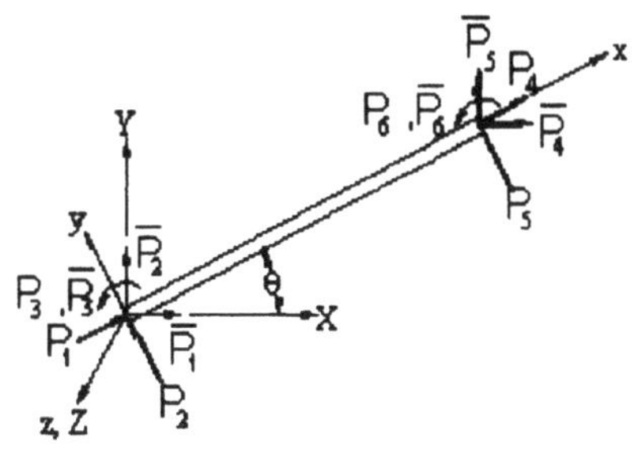

Fig. 3.4 Beam element showing nodal forces P_i in local (x, y, z) and nodal forces $\overline{P}_i$ in global coordinate axes (X, Y, Z)

Analogously, we obtain for the forces P_4, P_5 and P_6 on the other node, the relationships:

$$
\begin{aligned}
P_4 &= \overline{P}_4 \cos\theta + \overline{P}_5 \sin\theta \\
P_5 &= -\overline{P}_4 \sin\theta + \overline{P}_5 \cos\theta \\
P_6 &= \overline{P}_6
\end{aligned}
\qquad (3.11)
$$

Equations (3.10) and (3.11) may conveniently be arranged in matrix form as

$$\begin{Bmatrix} P_1 \\ P_2 \\ P_3 \\ P_4 \\ P_5 \\ P_6 \end{Bmatrix} = \begin{bmatrix} \cos\theta & \sin\theta & 0 & 0 & 0 & 0 \\ -\sin\theta & \cos\theta & 0 & 0 & 0 & 0 \\ 0 & 0 & 1 & 0 & 0 & 0 \\ 0 & 0 & 0 & \cos\theta & \sin\theta & 0 \\ 0 & 0 & 0 & -\sin\theta & \cos\theta & 0 \\ 0 & 0 & 0 & 0 & 0 & 1 \end{bmatrix} \begin{Bmatrix} \bar{P}_1 \\ \bar{P}_2 \\ \bar{P}_3 \\ \bar{P}_4 \\ \bar{P}_5 \\ \bar{P}_6 \end{Bmatrix} \tag{3.12}$$

or in condensed notation

$$\{P\} = [T]\{\bar{P}\} \tag{3.13}$$

in which $\{P\}$ and $\{\bar{P}\}$ are, respectively, the vectors of the element nodal forces in local and global coordinates and $[T]$ is the transformation matrix given by the square matrix in eq.(3.12). Repeating the same procedure, we obtain the relationship between the nodal displacements (δ_1, $\delta_{2,\ldots\ldots}$ δ_6) in the local coordinate system and the nodal displacements ($\bar{\delta}_1, \bar{\delta}_2, \ldots\ldots \bar{\delta}_6$) in the global coordinate system, namely

$$\begin{Bmatrix} \delta_1 \\ \delta_2 \\ \delta_3 \\ \delta_4 \\ \delta_5 \\ \delta_6 \end{Bmatrix} = \begin{bmatrix} \cos\theta & \sin\theta & 0 & 0 & 0 & 0 \\ -\sin\theta & \cos\theta & 0 & 0 & 0 & 0 \\ 0 & 0 & 1 & 0 & 0 & 0 \\ 0 & 0 & 0 & \cos\theta & \sin\theta & 0 \\ 0 & 0 & 0 & -\sin\theta & \cos\theta & 0 \\ 0 & 0 & 0 & 0 & 0 & 1 \end{bmatrix} \begin{Bmatrix} \bar{\delta}_1 \\ \bar{\delta}_2 \\ \bar{\delta}_3 \\ \bar{\delta}_4 \\ \bar{\delta}_5 \\ \bar{\delta}_6 \end{Bmatrix} \tag{3.14}$$

or

$$\{\delta\} = [T]\{\bar{\delta}\} \tag{3.15}$$

Now, the substitution of $\{P\}$ from eq.(3.13) and of $\{\delta\}$ from eq.(3.15) into the stiffness equation referred to local axes $\{P\} = [k]\{\delta\}$ results in

$$[T]\{\bar{P}\} = [k][T]\{\bar{\delta}\}$$

or

$$\{\bar{P}\} = [T]^{-1}[k][T]\{\bar{\delta}\}$$

where $[T]^{-1}$ is the inverse of matrix $[T]$. The reader may verify that the transformation matrix $[T]$ in eq.(3.12)is an orthogonal matrix, that is, $[T]^{-1} = [T]^{T}$. Hence

$$\{\bar{P}\} = [T]^T [k][T]\{\bar{\delta}\} \tag{3.16}$$

or, in a more convenient notation,

$$\{\bar{P}\} = [\bar{k}]\{\bar{\delta}\} \tag{3.17}$$

in which

$$[\bar{k}] = [T]^T [k]\ [T] \tag{3.18}$$

is the stiffness matrix for a beam element in reference to the global system of coordinates.

3.6 Equivalent Nodal Force for an Axially Loaded Beam Element

The application of the Principle of Virtual Work (Problem 3.2) provides the general expression to calculate the equivalent nodal forces for an axially loaded beam as

$$Q_i = \int_0^L p(x)\, u_i(x)\, dx \tag{3.19}$$

in which $p(x)$ is the applied distributed axial force on the beam element and $u_i(x)$ (i = 1,2) is the displacement function given by eq.(3.5) or (3.6). Appendix I Case (e) and Case (g) provide, respectively, the expressions for the equivalent nodal forces for a beam element loaded axially with a concentrated force or with a uniform distributed force.

3.7 Element End Forces

The element end forces in reference to local coordinates are calculated as in the case of beams [eq.(1.20)] by

$$\{P\} = [k]\{\delta\} - \{Q\} \tag{3.20}$$

in which

$\{P\}$ = element end forces
$[k]$ = element stiffness matrix
$\{\delta\}$ = element nodal displacements
(Q) = equivalent forces at the nodal coordinates for the element applied loads

Illustrative Example 3.1

For the loaded plane frame shown in Fig. 3.5 determine:

(a) Displacements at node ②
(b) End-forces on the elements
(c) Reactions at the supports

All Members
$E = 30{,}000$ ksi
$I = 882$ in^4
$A = 24.1$ in^2

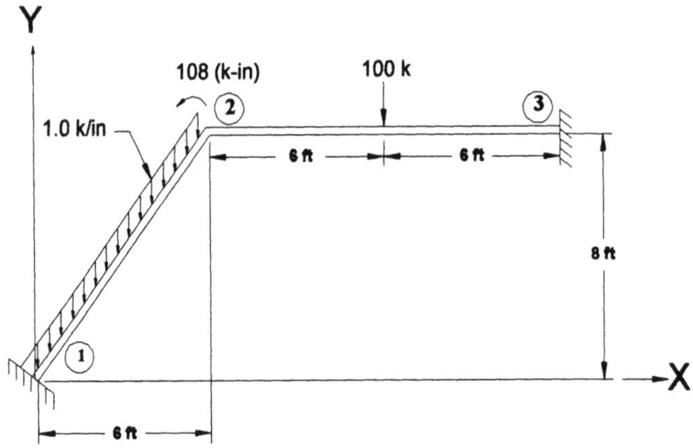

Fig. 3.5. Plane frame for Illustrative Example 3.1

Solution:

1. Modeling the structure.

The plane frame shown in Fig. 3.5 has been modeled into two beam elements, three nodes and nine system nodal coordinates. The first three system nodal coordinates correspond to the free nodal coordinates and the last six to the fixed nodal coordinates as shown in Fig. 3.6.

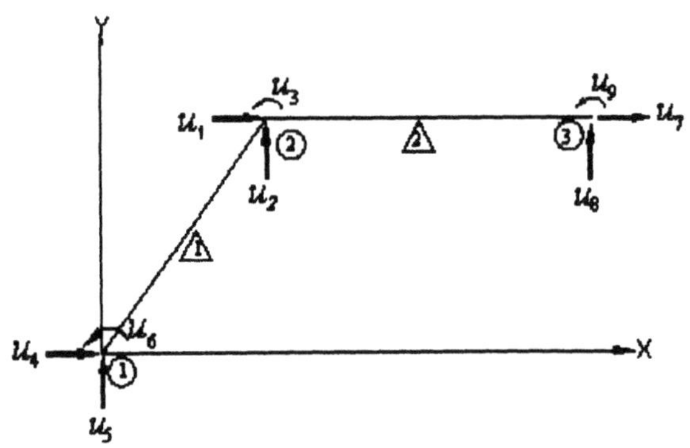

Fig. 3.6 Modeled plane frame of Illustrative Example 3.1 showing the system nodal coordinates u_1 through u_9.

2. Element stiffness matrices (local coordinates).

ELEMENT 1

Substituting numerical values into the stiffness matrix in eq.(3.8):

$$[k]_1 = \frac{30\text{x}10^3\text{x}\,882}{120^3}\begin{bmatrix} \frac{24.1\text{x}120^2}{882} & 0 & 0 & -\frac{24.1\text{x}120^2}{882} & 0 & 0 \\ 0 & 12 & 6\text{x}120 & 0 & -12 & 6\text{x}120 \\ 0 & 6\text{x}120 & 4\text{x}120^2 & 0 & -6\text{x}120 & 2\text{x}120^2 \\ -\frac{24.1\text{x}120^2}{882} & 0 & 0 & \frac{24.1\text{x}120^2}{882} & 0 & 0 \\ 0 & -12 & -6\text{x}120 & 0 & 12 & -6\text{x}120 \\ 0 & 6\text{x}120 & 2\text{x}120^2 & 0 & -6\text{x}120 & 4\text{x}120^2 \end{bmatrix}$$

or

$$[k]_1 = \begin{bmatrix} 6.025\text{E}3 & 0 & 0 & -6.025\text{E}6 & 0 & 0 \\ 0 & 1.838\text{E}2 & 1.103\,E4 & 0 & -1.838\,E2 & 1.103\text{E}4 \\ 0 & 1.103\text{E}4 & 8.820\text{E}5 & 0 & -1.103\text{E}4 & 4.410\text{E}5 \\ -6.025\text{E}6 & 0 & 0 & 6.025\text{E}3 & 0 & 0 \\ 0 & -1.838\text{E}2 & -1.103\text{E}4 & 0 & 1.838\text{E}2 & -1.103\text{E}4 \\ 0 & 1.103\text{E}4 & 4.410\text{E}5 & 0 & -1.103\text{E}4 & 8.820\text{E}5 \end{bmatrix} \quad \text{(a)}$$

ELEMENT 2

$$[k]_2 = \begin{array}{c} \begin{array}{cccccc} 1 & 2 & 3 & 7 & 8 & 9 \end{array} \\ \begin{bmatrix} 5.021\text{E}3 & 0 & 0 & -5.021\text{E}3 & 0 & 0 \\ 0 & 1.063\text{E}2 & 7.656\text{E}3 & 0 & -1.063\text{E}2 & 7.656\text{E}3 \\ 0 & 7.656\text{E}3 & 7.350\text{E}5 & 0 & -7.656\text{E}3 & 3.67\text{E}5 \\ -5.021\text{E}3 & 0 & 0 & 5.021\text{E}3 & 0 & 0 \\ 0 & -1.063\text{E}2 & -7.656\text{E}3 & 0 & 1.063\text{E}2 & -7.656\text{E}3 \\ 0 & 7.656\text{E}3 & 3.675\text{E}5 & 0 & -7.656\text{E}3 & 7.350\text{E}5 \end{bmatrix} \end{array} \begin{array}{c} 1 \\ 2 \\ 3 \\ 7 \\ 8 \\ 9 \end{array} \quad \text{(b)}$$

3. Element transformation matrices.

ELEMENT 1

From eq.(3.14) with $\cos\theta = 0.60$ and $\sin\theta = 0.80$

$$[T]_1 = \begin{bmatrix} 0.60 & 0.80 & 0 & 0 & 0 & 0 \\ -0.80 & 0.60 & 0 & 0 & 0 & 0 \\ 0 & 0 & 1 & 0 & 0 & 0 \\ 0 & 0 & 0 & 0.60 & 0.80 & 0 \\ 0 & 0 & 0 & -0.80 & 0.60 & 0 \\ 0 & 0 & 0 & 0 & 0 & 1 \end{bmatrix} \qquad \text{(c)}$$

ELEMENT 2

The transformation matrix for this element is the unitary matrix $[T]_2 = [I]$ since $\theta = 0$.

4. Element stiffness matrices (global coordinates).

ELEMENT 1

The element stiffness matrix in reference to the global system of coordinates is obtained by substituting into eq.(3.18) the element stiffness matrix and the element transformation matrix, respectively, from eqs.(a) and (c). After performing the multiplication of these matrices as indicated in eq.(3.18), we obtain

$$[\bar{k}]_1 = \begin{matrix} 4 & 5 & 6 & 1 & 2 & 3 & \\ \end{matrix}$$

$$[\bar{k}]_1 = \left[\begin{matrix} 2.287E3 & 2.804E3 & -8.820E3 & -2.287E3 & -2.804E3 & -8.820E3 \\ 2.804E3 & 3.922E3 & 6.615E3 & -2.804E3 & -3.922E3 & 6.665E3 \\ -8.820E3 & 6.615E3 & 8.820E3 & 8.820E3 & -6.615E3 & 4.410E3 \\ -2.287E3 & -2.804E3 & 8.820E3 & 2.287E3 & 2.804E3 & 8.820E5 \\ -2.804E3 & -3.922E3 & -6.615E3 & 2.804E3 & 3.922E3 & -6.615E3 \\ -8.820E3 & 6.615E3 & 4.410E3 & 8.820E5 & -6.615E3 & 8.820E5 \end{matrix}\right] \begin{matrix} 4 \\ 5 \\ 6 \\ 1 \\ 2 \\ 3 \end{matrix} \qquad \text{(d)}$$

ELEMENT 2

Since the transformation matrix for this element is the unitary matrix $[I]$, its stiffness matrix in global coordinates is identical to the stiffness matrix in the local coordinate as given by eq.(b).

5. Assemblage of the reduced system matrix.

To assemble the reduced system stiffness matrix for the structure, we proceed to transfer each coefficient from the two element matrices eqs.(d) and (b) into the system stiffness matrix. To this objective we write at the top and on the right side of these two matrices the coded nodal coordinates assigned for these elements in global nodal coordinates as labeled in Fig. 3.6.

Proceeding systematically to transfer the entries in the element stiffness matrices, eqs.(d) and (b), according to the rows and columns indicated on the right and at the top of these two matrices, we obtain the reduced system stiffness matrix by transferring only rows and columns 1, 2 and 3 which correspond to the free nodal coordinates.

$$[K]_R = \begin{bmatrix} 2.287\,E3 + 5.021\,E3 & 2.804\,E3 + 0 & 8.820\,E5 + 0 \\ 2.804\,E3 + 0 & 3.922\,E3 + 1.063\,E2 & -6.615\,E3 + 7.656\,E3 \\ 8.820\,E3 + 0 & -6.615\,E3 + 7.656\,E3 & 8.820\,E5 + 7.350\,E5 \end{bmatrix}$$

or

$$[K]_R = \begin{bmatrix} 0.7307E4 & 0.2804E4 & 0.8820E4 \\ 0.2804E4 & 0.4028E4 & 0.1041E4 \\ 0.8820E4 & 0.1041E4 & 0.1617E7 \end{bmatrix} \tag{e}$$

6. Equivalent element nodal forces (local coordinates).

ELEMENT 1

Figure 3.7 shows the vertically distributed load on element 1 equated to the superposition of the normal and the axial components of the applied load. Figure 3.7 also shows the equivalent nodal forces for these two loads calculated using the formulas given in Appendix I.

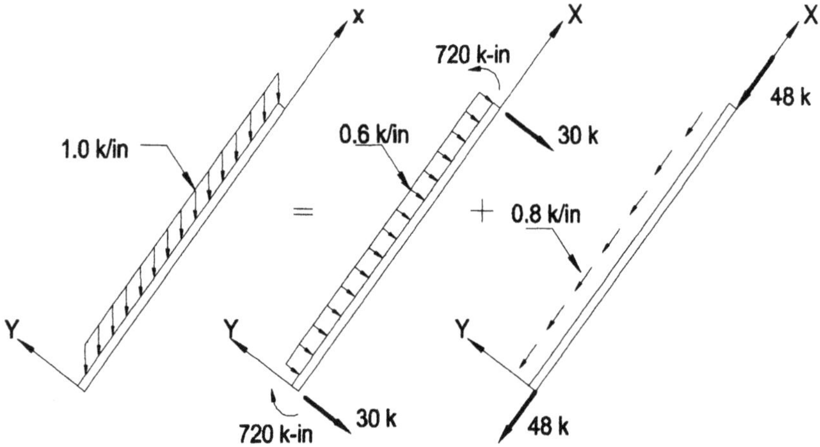

Fig. 3.7 Element 1 showing its load equated to the superposition of the normal and the axial components and also showing the equivalent forces at the normal coordinates.

The equivalent nodal forces for element 1 are calculated using the formulas in Appendix I as follows (L = 120 in):

$$Q_1 = \frac{-0.8L}{2} = -48(\text{kip}) \quad Q_2 = \frac{-0.6L}{2} = -36(\text{kip}) \quad Q_3 = \frac{-0.6L^2}{12} = -720(\text{kip.in})$$

$$Q_4 = \frac{-0.8L}{2} = -48(\text{kip}) \quad Q_5 = \frac{-0.6L}{2} = -36(\text{kip}) \quad Q_6 = \frac{0.6L^2}{12} = 720(\text{kip.in})$$

or in vector notation as

$$\{Q\}_1 = \begin{Bmatrix} -48 \\ -36 \\ -720 \\ -48 \\ -36 \\ 720 \end{Bmatrix} \tag{f}$$

ELEMENT 2

The equivalent nodal forces are calculated using the formulas for case (a) of Appendix I to obtain

$$Q_1 = 0 \qquad Q_2 = \frac{W}{2} = -50\,(\text{kip}) \qquad Q_3 = \frac{WL}{8} = -\frac{100\text{x}144}{8} = -1800(\text{kip.in})$$

$$Q_4 = 0 \qquad Q_5 = \frac{W}{2} = -50(\text{kip}) \qquad Q_6 = -\frac{WL}{8} = 1800(\text{kip.in})$$

Or in vector notation

$$\{Q\}_2 = \begin{Bmatrix} 0 \\ -50 \\ -1800 \\ 0 \\ -50 \\ 1800 \end{Bmatrix} \tag{g}$$

7. Element equivalent nodal forces (global coordinates).
   The equivalent nodal forces $\{\overline{Q}\}$ in global coordinates are calculated using the transformation of forces obtained from eq.(3.13):

$$\{\overline{Q}\} = [T]^T \{Q\} \tag{h}$$

ELEMENT 1

The substitution into eq.(h) of the transpose of matrix $[T]_1$ from eq.(c) and of $\{Q\}_1$ from eq.(f) results in

$$\{\overline{Q}\}_1 = \begin{bmatrix} 0.60 & -0.80 & 0 & 0 & 0 & 0 \\ 0.80 & 0.60 & 0 & 0 & 0 & 0 \\ 0 & 0 & 1 & 0 & 0 & 0 \\ 0 & 0 & 0 & 0.60 & -0.80 & 0 \\ 0 & 0 & 0 & 0.80 & 0.60 & 0 \\ 0 & 0 & 0 & 0 & 0 & 1 \end{bmatrix} \begin{Bmatrix} -48 \\ -36 \\ -720 \\ -48 \\ -36 \\ 720 \end{Bmatrix} = \begin{Bmatrix} 0 \\ -60 \\ -720 \\ 0 \\ 0 \\ 720 \end{Bmatrix}$$

or

$$\{\overline{Q}\}_1 = \begin{Bmatrix} 0 \\ -60 \\ -720 \\ 0 \\ -60 \\ 720 \end{Bmatrix} \begin{matrix} 4 \\ 5 \\ 6 \\ 1 \\ 2 \\ 3 \end{matrix} \tag{i}$$

ELEMENT 2 (no transformation is needed since $\theta = 0$) from eq.(g):

$$\{\overline{Q}\}_2 = \{Q\}_2 = \begin{Bmatrix} 0 \\ -50 \\ -1800 \\ 0 \\ -50 \\ 1800 \end{Bmatrix} \begin{matrix} 1 \\ 2 \\ 3 \\ 7 \\ 8 \\ 9 \end{matrix} \tag{j}$$

8. Assemblage of the reduced system force vector.

The reduced system force vector is assembled by transferring the components of the force vectors in eqs.(i) and (j) to the locations indicated on the right side of these vectors and adding the force applied directly to a nodal coordinate: namely,

$$\{F\}_r = \begin{Bmatrix} 0+0 \\ -60-50 \\ 720-1800 \end{Bmatrix} + \begin{Bmatrix} 0 \\ 0 \\ 108 \end{Bmatrix} = \begin{Bmatrix} 0 \\ -110 \\ -972 \end{Bmatrix} \tag{k}$$

9. System stiffness equation.

Substitution of the reduced stiffness matrix [eq.(e)] and of the reduced force vector [eq.(k)] into the reduced system stiffness equation, $\{F\}_R = [K]_R\{u\}$, results in:

$$\begin{Bmatrix} 0 \\ -110 \\ -972 \end{Bmatrix} = \begin{bmatrix} 0.7307E4 & 0.2804E4 & 0.8820E4 \\ 0.2804E4 & 0.4028E4 & 0.1041E4 \\ 0.8820E4 & 0.1041E4 & 0.1617E7 \end{bmatrix} \begin{Bmatrix} u_1 \\ u_2 \\ u_3 \end{Bmatrix} \tag{l}$$

10. Solution of the unknown displacements.

$$
\begin{aligned}
u_1 &= 0.0153 \text{ (in)} \\
u_2 &= -0.0378 \text{ (in)} \\
u_3 &= -\,6.602\text{E}{-4} \text{ (rad)}
\end{aligned}
\qquad \text{(m)}
$$

11. Element nodal displacement vectors (global coordinates).

The displacements at the nodal coordinates of elements 1 and 2 are identified from the global coordinates assigned in the analytical model shown in Fig. 3.6.

$$
\{\bar{\delta}\}_1 = \begin{Bmatrix} 0 \\ 0 \\ 0 \\ 0.0153 \\ -0.0378 \\ -6.602\text{E}-4 \end{Bmatrix} \qquad \{\bar{\delta}\}_2 = \begin{Bmatrix} 0.0153 \\ -0.0378 \\ -6.602\text{E}-4 \\ 0 \\ 0 \\ 0 \end{Bmatrix} \qquad \text{(n)}
$$

12. Element nodal displacement vectors (local coordinates).

The transformation of the element nodal coordinates is given by eq.(3.15) as

$$
\{\delta\} = [T]\{\bar{\delta}\} \qquad \text{(3.15) repeated}
$$

ELEMENT 1

The substitution into eq.(3.15) of $\{\bar{\delta}\}_1$, and $[T]_1$, respectively, from eqs.(n) and (c) results in

$$
\{\delta\}_1 = \begin{bmatrix} 0.6 & 0.8 & 0 & 0 & 0 & 0 \\ -0.8 & 0.6 & 0 & 0 & 0 & 0 \\ 0 & 0 & 1 & 0 & 0 & 0 \\ 0 & 0 & 0 & 0.6 & 0.8 & 0 \\ 0 & 0 & 0 & -0.8 & 0.6 & 0 \\ 0 & 0 & 0 & 0 & 0 & 1 \end{bmatrix} \begin{Bmatrix} 0 \\ 0 \\ 0 \\ 0.0153 \\ -0.0378 \\ -0.00066 \end{Bmatrix} = \begin{Bmatrix} 0 \\ 0 \\ 0 \\ -0.211 \\ -0.035 \\ -0.00066 \end{Bmatrix}
$$

ELEMENT 2

For element 2, $\{\delta\}_2 = \{\bar{\delta}\}_2$ since $[T]_2 = [I]$.

13. Element end forces (local coordinates).

Element end forces are calculated by eq.(3.20):

$$\{P\} = [k]\{\delta\} - \{Q\} \qquad \text{(3.20) repeated}$$

ELEMENT 1

$$\begin{Bmatrix} P_1 \\ P_2 \\ P_3 \\ P_4 \\ P_5 \\ P_6 \end{Bmatrix} = \begin{bmatrix} 6.025E3 & 0 & 0 & -6.025E6 & 0 & 0 \\ 0 & 1.838E2 & 1.103E4 & 0 & -1.838E2 & 1.103E4 \\ 0 & 1.103E4 & 8.820E5 & 0 & -1.103E4 & 4.410E5 \\ -6.025E6 & 0 & 0 & 6.025E3 & 0 & 0 \\ 0 & -1.838E2 & -1.103E4 & 0 & 1.838E2 & -1.103E4 \\ 0 & 1.103E4 & 4.410E5 & 0 & -1.103E4 & 8.820E5 \end{bmatrix} \begin{Bmatrix} 0 \\ 0 \\ 0 \\ -0.211 \\ -0.035 \\ -6.602E-4 \end{Bmatrix} - \begin{Bmatrix} -48 \\ -36 \\ -720 \\ -48 \\ -36 \\ 720 \end{Bmatrix} = \begin{Bmatrix} 174.81 \\ 35.13 \\ -813.63 \\ -78.81 \\ 36.87 \\ -917.52 \end{Bmatrix}$$

ELEMENT 2

$$\begin{Bmatrix} P_1 \\ P_2 \\ P_3 \\ P_4 \\ P_5 \\ P_6 \end{Bmatrix} = \begin{bmatrix} 5.02E3 & 0 & 0 & -5.02E3 & 0 & 0 \\ 0 & 1.063E2 & 7.656E3 & 0 & -1.063E2 & 7.656E3 \\ 0 & 7.656E3 & 7.350E5 & 0 & -7.656E3 & 3.675E5 \\ -5.02E3 & 0 & 0 & 5.02E2 & 0 & 0 \\ 0 & -1.063E2 & -7.656E3 & 0 & 1.063E2 & -7.656E3 \\ 0 & 7.656E3 & 3.675E5 & 0 & -7.656E3 & 7.350E5 \end{bmatrix} \begin{Bmatrix} 0.0153 \\ -0.0378 \\ -6.602E-4 \\ 0 \\ 0 \\ 0 \end{Bmatrix} - \begin{Bmatrix} 0 \\ -50 \\ -1800 \\ 0 \\ -50 \\ 1800 \end{Bmatrix} = \begin{Bmatrix} 76.78 \\ 40.93 \\ 1025 \\ -76.78 \\ 59.07 \\ -2331.0 \end{Bmatrix}$$

Figure 3.8 shows the end forces on the two elements on the plane frame of Illustrative Example 3.1.

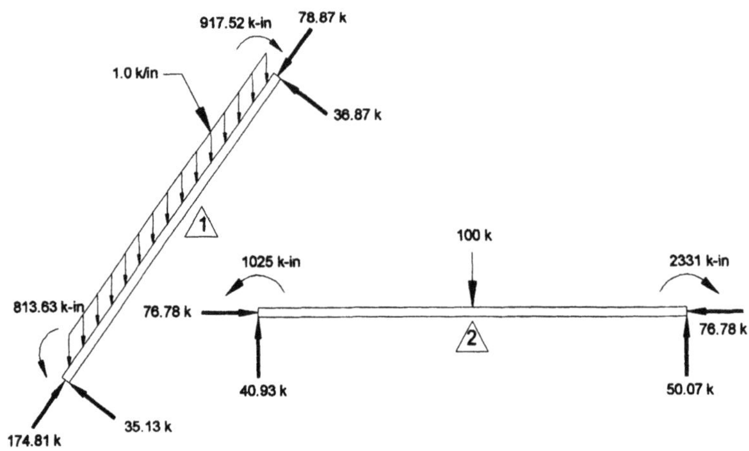

Fig. 3.8 End forces for elements 1 and 2.

14. Support reactions (global coordinates).

The support reactions in reference to the global coordinates are identified from the components of the calculated element end forces at the supports as follows:

Joint ①

$$\overline{R}_4 = 0.6\,P_{11} - 0.8\,P_{21}$$

$$\overline{R}_4 = 0.6\text{x}174.81 - 0.8\text{x}35.13 = 76.792\,(\text{kip})$$

$$\overline{R}_5 = 0.8\,P_{11} + 0.6\,P_{21}$$

$$\overline{R}_5 = 0.8\text{x}174.81 + 0.6\text{x}35.13 = 160.926\,(\text{kip})$$

$$\overline{R}_6 = P_{31} = 813.63\,(\text{kip.in})$$

Joint ③

$$\overline{R}_7 = P_{42} = -76.78\,(\text{kip})$$

$$\overline{R}_8 = P_{52} = 59.07\,(\text{kip})$$

$$\overline{R}_9 = P_{62} = -2331.86\,(\text{kip.in})$$

where P_{ij} is the force (in reference to local coordinates) at nodal coordinate i for element j.

3.8 Inclined Roller Supports

The structures considered thus far have been supported such that fixed joint displacements are in the direction of the global coordinate axes oriented in the horizontal and vertical directions. Occasionally, situations may exist in which prevented displacements are inclined with respect to direction of the global axes. Figure 3.9 shows an example of a plane frame having an inclined roller support. This frame is analyzed later in this chapter using SAP2000 (Illustrative Example 3.6). A simple example of a one span beam with an inclined roller support is presented now to illustrate the necessary calculations required for the analysis of structures having inclined roller supports.

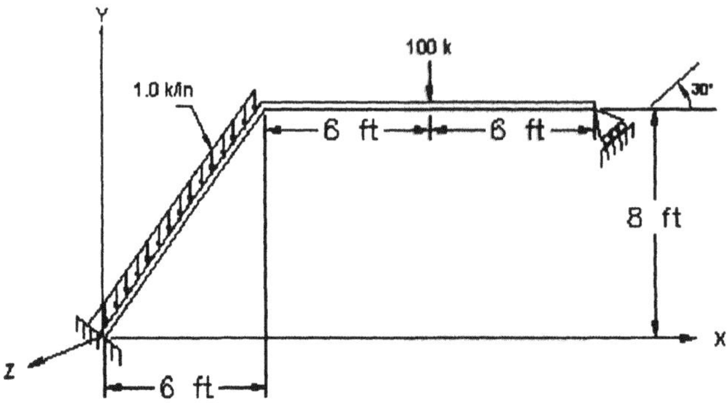

All Members $E = 30{,}000$ ksi $I = 822$ in^4 $A = 24.1$ in^2

Fig. 3.9 Example of a plane frame having an inclined roller support

Illustrative Example 3.2

Consider in Fig. 3.10(a) a beam having on the right end an inclined roller support. Use matrix structural analysis to determine:

(a) Displacements on the right end of the beam.
(b) End-forces on the beam element.
(c) Reactions at the supports.

$E = 10{,}000$ ksi $I = 100$ in^4 $A = 10$ in^2

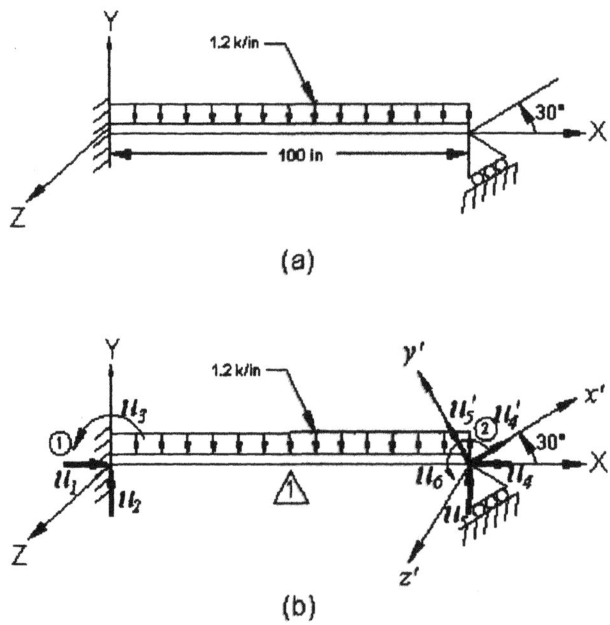

Fig. 3.10 (a) Beam having an inclined roller support for Illustrative Example 3.2. (b) Analytical model showing nodal coordinates in global axes (X, Y, Z) and in the auxiliary axes $\{x', y', z'\}$.

1. Analytical Model.

   Figure 3.10(b) shows the analytical model in which the nodal coordinates in reference to global axes, u_1 through u_6 are indicated. This figure also shows the nodal coordinates u_4' and u_5' oriented along the axialliary axis (x', y', z') with axes x' and y' oriented along the direction of the inclined support and normal to it, respectively.

2. Stiffness matrix (*X, Y, Z* axes).

   The element stiffness matrix in local coordinates is given by eq.(3.8). For this example, the global coordinates coincide with the element local coordinates. The substitution of numerical values in eq.(3.8) yields:

$$[k] = \begin{bmatrix} 1000 & 0 & 0 & -1000 & 0 & 0 \\ 0 & 12 & 600 & 0 & -12 & 600 \\ 0 & 600 & 40000 & 0 & -600 & 20000 \\ -1000 & 0 & 0 & 1000 & 0 & 0 \\ 0 & -12 & -600 & 0 & 12 & -600 \\ 0 & 600 & 20000 & 0 & -600 & 40000 \end{bmatrix} \tag{a}$$

3. Stiffness matrix in reference to the axialliary coordinate axes (x', y', z').

   The relationship at node ② [(Fig. 3.10(b)] between the nodal displacements u_4 , u_5 along the axes *X*, *Y*, and nodal displacements u_4' and u_5' along the auxiliary axes x', y' is given by

$$\begin{aligned} u_4' &= u_4 \cos\alpha + u_5 \sin\alpha \\ u_5' &= -u_4 \sin\alpha + u_5 \cos\alpha \end{aligned} \tag{b}$$

   Solving for u_4 and u_5 and using matrix notation yields

$$\begin{Bmatrix} u_4 \\ u_5 \end{Bmatrix} = \begin{bmatrix} \cos\alpha & -\sin\alpha \\ \sin\alpha & \cos\alpha \end{bmatrix} \begin{Bmatrix} u_4' \\ u_5' \end{Bmatrix} \tag{c}$$

   where $\alpha = 30^o$ for this Illustrative Example. Therefore, we may write the following relationship [eq.(d)] to include all the nodal displacements shown in Fig. 3.10(b):

$$\begin{Bmatrix} u_1 \\ u_2 \\ u_3 \\ u_4 \\ u_5 \\ u_6 \end{Bmatrix} = \begin{bmatrix} 1 & 0 & 0 & 0 & 0 & 0 \\ 0 & 1 & 0 & 0 & 0 & 0 \\ 0 & 0 & 1 & 0 & 0 & 0 \\ 0 & 0 & 0 & \cos\alpha & -\sin\alpha & 0 \\ 0 & 0 & 0 & \sin\alpha & \cos\alpha & 0 \\ 0 & 0 & 0 & 0 & 0 & 1 \end{bmatrix} \begin{Bmatrix} u_1 \\ u_2 \\ u_3 \\ u_4' \\ u_5' \\ u_6 \end{Bmatrix} \tag{d}$$

or in condensed notation

$$\{u\} = [T'] \{u'\} \tag{e}$$

in which the transformation matrix $[T']$ defined in eq.(d) relates the displacement vector {u} with the displacement vector {u'}. These displacement vectors are in reference to the global axes and to the axialliary coordinate axes, respectively. The same transformation matrix $[T']$ also relates force vector $\{F\}$ with the force vector $\{F'\}$ in reference to these two systems of coordinates, namely

$$\begin{Bmatrix} F_1 \\ F_2 \\ F_3 \\ F_4 \\ F_5 \\ F_6 \end{Bmatrix} = \begin{bmatrix} 1 & 0 & 0 & 0 & 0 & 0 \\ 0 & 1 & 0 & 0 & 0 & 0 \\ 0 & 0 & 1 & 0 & 0 & 0 \\ 0 & 0 & 0 & \cos\alpha & -\sin\alpha & \\ 0 & 0 & 0 & \sin\alpha & \cos\alpha & 0 \\ 0 & 0 & 0 & 0 & 0 & 1 \end{bmatrix} \begin{Bmatrix} F_1 \\ F_2 \\ F_3 \\ F_4' \\ F_5' \\ F_6 \end{Bmatrix} \tag{f}$$

or in condensed notation

$$\{F\} = [T']\{F'\} \quad \text{and} \quad \{F'\} = [T']^T \{F\} \tag{g}$$

The substitution of $\{u\}$ from eq.(e) and of $\{F\}$ from eq.(g) into the stiffness equation $\{F\} = [k]\{u\}$| results in

$$[T']\{F'\} = [k][T']\{u'\}$$

or, since $[T']$ is an orthogonal matrix $[T']^{-1} = [T']^T$:

$$\{F'\} = [k']\{u'\}$$

where

$$[k'] = [T']^T [k][T'] \tag{h}$$

in which $[T']$, for this example, is given by

$$[T'] = \begin{bmatrix} 1 & 0 & 0 & 0 & 0 & 0 \\ 0 & 1 & 0 & 0 & 0 & 0 \\ 0 & 0 & 1 & 0 & 0 & 0 \\ 0 & 0 & 0 & 0.866 & -0.5 & 0 \\ 0 & 0 & 0 & 0.5 & 0.866 & 0 \\ 0 & 0 & 0 & 0 & 0 & 1 \end{bmatrix} \tag{i}$$

Substituting $[T']$ and its transpose from eq.(i), $[k]$ from eq.(a) into eq.(h) and performing the multiplications indicated in eq.(h) results in

$$\begin{Bmatrix} F_1 \\ F_2 \\ F_3 \\ F_4' \\ F_5' \\ F_6 \end{Bmatrix} = \overset{\begin{matrix} 1 & \quad 2 & \quad 3 & \quad 4 & \quad 5 & \quad 6 \end{matrix}}{\begin{bmatrix} 1 & 0 & 0 & -866 & 500 & 0 \\ 0 & 12 & 600 & -6 & -10.39 & 600 \\ 0 & 600 & 40000 & -300 & -519.6 & 20000 \\ -866 & -6 & -300 & 753 & -428 & -300 \\ 500 & -10.39 & -519.6 & -428 & -259 & -519.6 \\ 0 & 600 & 20000 & -300 & -519.6 & 40000 \end{bmatrix}} \begin{Bmatrix} u_1 \\ u_2 \\ u_3 \\ u_4' \\ u_5' \\ u_6 \end{Bmatrix} \begin{matrix} 1 \\ 2 \\ 3 \\ 4 \\ 5 \\ 6 \end{matrix} \tag{j}$$

4. Equivalent nodal forces vector (X, Y, Z axes).

The equivalent nodal force vector for the distributed load on the span of this beam is obtained by substituting numerical values in Case (b) of Appendix I:

$$\{Q\} = \begin{Bmatrix} 0 \\ wL/2 \\ wL^2/12 \\ 0 \\ wL/2 \\ -wL^2/12 \end{Bmatrix} = \begin{Bmatrix} 0 \\ -60 \\ -1000 \\ 0 \\ -60 \\ 1000 \end{Bmatrix} \tag{k}$$

in which w = -1.2 k/in and L = 100 in.

5. Equivalent nodal force vector $\{Q'\}$ (axialliary coordinates).

The equivalent nodal force vector in the axialliary coordinates is given as in eq.(g) by

$$\{Q'\} = [T']^T \{Q\}$$

Substituting $[T']^T$ from eq.(i) and $\{Q\}$ from eq.(k) yields

$$\{Q'\} = \begin{Bmatrix} Q_1 \\ Q_2 \\ Q_3 \\ Q_4' \\ Q_5' \\ Q_6 \end{Bmatrix} = \begin{Bmatrix} 0 \\ -60 \\ -1000 \\ -30 \\ -52 \\ 1000 \end{Bmatrix} \begin{matrix} 1 \\ 2 \\ 3 \\ 4 \\ 5 \\ 6 \end{matrix} \qquad (l)$$

6. Reduced stiffness matrix equation (axialliary system).

The reduced stiffness matrix equation is assembled by transferring the coefficients corresponding to the free coordinates u_4' and u_6 from the matrix in eq.(j). Likewise, the reduced system force vector is obtained from eq.(l) by transferring the coefficients in the force vector $\{Q'\}$ corresponding to these two free coordinates. Namely,

$$\begin{Bmatrix} -30 \\ 1000 \end{Bmatrix} = \begin{Bmatrix} 753 & -300 \\ -300 & 40000 \end{Bmatrix} \begin{Bmatrix} u_4' \\ u_6 \end{Bmatrix} \qquad (m)$$

7. Nodal displacements (axialliary system).

The solution of eq.(m) gives the displacements at the free nodal coordinates as

$$u_4' = -0.030 \text{ in}$$
$$u_6 = 0.0248 \text{ rad}$$

8. Element end-forces.

The element end-forces are given from eq.(3.20) as

$$\{P\} = [k]\{\delta\} - \{Q\} \qquad (3.20) \text{ repeated}$$

in which the displacement vector $\{\delta\}$ in reference to the coordinate system (X, Y, Z) is identified from Fig. 3.10(b) as

$$\{\delta\} = \begin{Bmatrix} u_1 \\ u_2 \\ u_3 \\ u_4 \\ u_5 \\ u_6 \end{Bmatrix} = \begin{Bmatrix} 0 \\ 0 \\ 0 \\ u_4' \cos\alpha \\ u_4' \sin\alpha \\ u_6 \end{Bmatrix} = \begin{Bmatrix} 0 \\ 0 \\ 0 \\ -0.026 \\ -0.015 \\ 0.0248 \end{Bmatrix} \qquad \text{(n)}$$

in which $\alpha = 30^o$.
Then substituting $[k]$ from eq.(a), $\{\delta\}$ from eq.(n) and $\{Q\}$ from eq.(k) into eq.(3.20), we obtain

$$\begin{Bmatrix} P_1 \\ P_2 \\ P_3 \\ P_4 \\ P_5 \\ P_6 \end{Bmatrix} = \begin{bmatrix} 1000 & 0 & 0 & -1000 & 0 & 0 \\ 0 & 12 & 600 & 0 & -12 & 600 \\ 0 & 600 & 40000 & 0 & -600 & 20000 \\ -1000 & 0 & 0 & 1000 & 0 & 0 \\ 0 & -12 & -600 & 0 & 12 & -600 \\ 0 & 600 & 20000 & 0 & -600 & 40000 \end{bmatrix} \begin{Bmatrix} 0 \\ 0 \\ 0 \\ -0.026 \\ -0.015 \\ 0.0248 \end{Bmatrix} - \begin{Bmatrix} 0 \\ -60 \\ -1000 \\ 0 \\ -60 \\ 1000 \end{Bmatrix}$$

or

$$\begin{Bmatrix} P_1 \\ P_2 \\ P_3 \\ P_4 \\ P_5 \\ P_6 \end{Bmatrix} = \begin{Bmatrix} 26 \\ 75.04 \\ 1504.5 \\ -26 \\ 45.04 \\ 0 \end{Bmatrix} \qquad \text{(o)}$$

9. Reaction at the supports.

   The reactions at the supports are calculated using the end-forces given in eq.(o):

   At the left support:

$$R_1 = P_1 = 26 \text{ kip}$$
$$R_2 = P_2 = 75.04 \text{ kip}$$
$$R_3 = P_3 = 15014.5 \text{ kip.in}$$

   At the right support:

$$R_5' = -P_4 \sin 30^\circ + P_5 \cos 30^\circ$$
$$R_5' = (-26\}(0.5) + (45.04)(0.866)$$
$$R_5' = 52.00 \text{ kip}$$

Illustrative Example 3.3

Use SAP2000 to analyze the plane frame in Illustrative Example 3.1. For convenience, the plane frame for this example is reproduced in Fig. 3.11 using the global coordinate axes with axis Z along the vertical direction as used in SAP2000. The input data is given in Table 3.1.

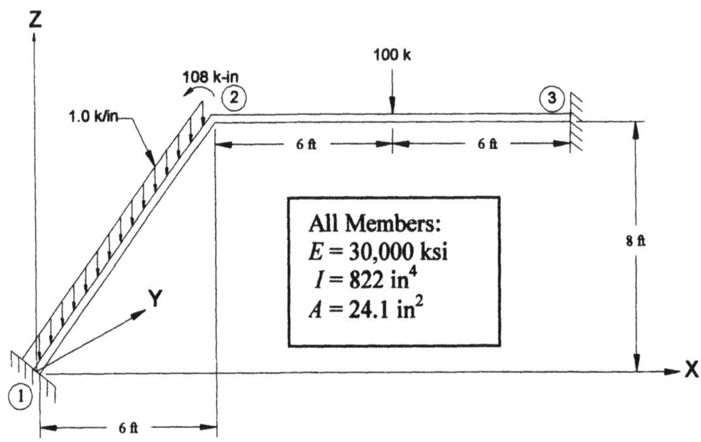

Fig 3.11 Plane Frame for Illustrative Example 3.3

Table 3. 1 Edited Input Data for Illustrative Example 3.3 (units: kips and inches)

JOINT DATA

JOINT	GLOBAL-X	GLOBAL-Y	GLOBAL-Z	RESTRAINTS
1	0.00000	0.00000	0.00000	1 1 1 1 1 1
2	72.00000	0.00000	96.00000	0 0 0 0 0 0
3	216.00000	0.00000	96.00000	1 1 1 1 1 1

FRAME ELEMENT DATA

FRAME	JNT-1	JNT-2	SECTION	RELEASES	SEGMENTS	LENGTH
1	1	2	W14X82	000000	2	120.000
2	2	3	W14X82	000000	4	144.000

JOINT FORCES Load Case LOAD1

JOINT	GLOBAL-X	GLOBAL-Y	GLOBAL-Z	GLOBAL-XX	GLOBAL-YY	GLOBAL-ZZ
2	0.000	0.000	0.000	0.000	-108.000	0.000

FRAME SPAN DISTRIBUTED LOADS Load Case LOAD1

FRAME	TYPE	DIRECTION	DISTANCE-A	VALUE-A	DISTANCE-B	VALUE-B
1	FORCE	GLOBAL-Z	0.0000	-1.0000	1.0000	-1.0000

FRAME SPAN POINT LOADS Load Case LOAD1

FRAME	TYPE	DIRECTION	DISTANCE	VALUE
2	FORCE	GLOBAL-Z	0.5000	-100.0000

Solution:

The following commands are implemented in SAP2000.

Begin: Open SAP2000.

Enter: "OK" to disable the "Tip of the Day" screen.

Hint: Maximize both screens for full views of all windows.

Select: In the lower right-hand corner of the screen use the drop-down menu to select "kip-in".

Select: From the Main Menu Bar enter
FILE>NEW MODEL

In the Coordinate System Definition: Set x, y, and z to 10 divisions.
Change Grid Spacing $x = 24$, $y = 24$, and $z = 24$.
Accept the other default values.

Select: Click on icon XZ on the toolbar to select this view on the screen

Edit: Minimize the screen showing the 3-D view and maximize the screen X-Z plane. Click on the icon on the toolbar with the symbol of a hand (PAN) and drag the plot to the center of the screen.

Draw: From the Main Menu enter:

DRAW>DRAW FRAME ELEMENT
Click in the lower left grid intersection point and drag the cursor to a point 3 grid lines to the right and 4 grid lines upward. Then click again at this location.

Drag again horizontally 6 grid lines and quickly double-click at this location. Then press enter to disable the command to draw frame elements.

Select: Translate the origin of the global coordinate system to left lower node of the frame by using the following commands:

SELECT>SELECT ALL This command will mark all the elements of the frame. Then enter:
EDIT>MOVE
In the screen "Move Selected Points" change $x = 240$. Then OK.

Define: Beam material. From the Main Menu enter:
DEFINE >MATERIALS Select STEEL Then OK.

In the Define Materials screen select
MODIFY / SHOW MATERIALS
Change the value of the Modulus of elasticity to 30000. Then OK.

Define: Cross-Sections: From the Main Menu enter:
DEFINE>FRAME SECTIONS
Select Import / Wide Flange. Then OK
In the next window open the file "section.prop"
Scroll on the Section labels screen and select "W 14 X 82". Then OK.

Click on Modification Factors
Change shear area in Z direction to 0 (in order not to include shear deformation in stiffness matrix of the elements). Then OK, OK, OK.

Assign: Assign Frame Sections: From the Main Menu enter:
SELECT>SELECT ALL

Then from the Main Menu enter:
ASSIGN>FRAME>SECTIONS
In the Define Frame Section screen select "W 14 X 82". Then OK.

Label: For viewing convenience, label joints and elements:
From the Main Menu enter:
VIEW>SET ELEMENTS
Click on the boxes labeled "joint labels" and "frame labels". Then OK.

Modify: Restraint Setting: Click on the dots that locate Joints ① and ③. This command will mark these joints with an "X".

From the Main Menu enter:
ASSIGN>JOINTS>RESTRAINTS
In the window "Joint Restraints" select restraints in all directions.
Then OK

Assign: Loads: From the Main Menu enter:
DEFINE>STATIC LOAD CASES

In the window "Define Static Load Case Names" change the label DEAD to LIVE using the drop-down menu. Also change self-weight multiplier to 0 (zero). Click on the button "Change Load". Then OK.

Assign: Select Joint ②. Then from the Main Menu enter:
ASSIGN>JOINT STATIC LOAD>FORCES
Change the value of the moment YY to –108. Then OK.

Assign: Select Frame 1 by clicking on this element on the screen. Frame 1 will change from a continuous line to dashes. On the Main Menu select:
ASSIGN>FRAME STATIC LOADS>POINT AND UNIFORM LOADS
Select Forces and Global Direction Z.
Then enter Uniform Load = -1.0

Assign: Select Frame 2. Then from the Main Menu enter:
ASSIGN>FRAME STATIC LOADS>POINT AND UNIFORM LOAD

Warning: Make certain to zero out any previous entries made on this screen!

Select Forces and Global Direction Z then enter:
Distance = 0.5 and Load = -100

Set Analysis: From the Main Menu enter:
ANALYZE>SET OPTIONS
In the box "Available DOF" check only UX, UZ and RY.
Alternatively, click on the picture labeled "Plane Frame". Then OK.

Final Analysis: On the Main Menu enter:
ANALYZE>RUN.
Name the model enter its name: "Example 3.3". Click on "SAVE". The calculations will appear on the screen as they take place. When the analysis is completed, a message will appear on the bottom of the pop-up window "The analysis is complete". Click OK.

Observe: On the Main Menu select DISPLAY>SHOW DEFORMED SHAPE. Click on the selection for "Wire Shadow". Also click on the "XZ" icon to see the plot on that plane. Then OK. To display the displacement values at any node, right-click on the node and a pop-up window will show the nodal displacements.

Print Input Tables: From the Main Menu enter:
FILE>PRINT INPUT TABLES.

(The previous Table 3.1 contains the edited input tables for Illustrative Example 3.3.)

Print Out put Tables: From the Main Menu enter:
FILE>PRINT OUTPUT TABLES

(Table 3.2 contains the edited output tables for Illustrative Example 3.3)

Table 3.2 Edited Output Tables for Illustrative Example 3.3 (Units: kips and inches)

JOINT DISPLACEMENTS

JOINT	LOAD	UX	UY	UZ	RX	RY	RZ
1	LOAD1	0.0000	0.0000	0.0000	0.0000	0.0000	0.0000
2	LOAD1	0.0153	0.0000	-0.0378	0.0000	6.602E-04	0.0000
3	LOAD1	0.0000	0.0000	0.0000	0.0000	0.0000	0.0000

JOINT REACTIONS

JOINT	LOAD	F1	F2	F3	M1	M2	M3
1	LOAD1	76.7783	0.0000	160.9282	0.0000	-813.6295	0.0000
3	LOAD1	-76.7783	0.0000	59.0718	0.0000	2331.8601	0.0000

FRAME ELEMENT FORCES

FRAME	LOAD	LOC	P	V2	V3	T	M2	M3
1	LOAD1							
		0.00	-174.81	-35.13	0.00	0.00	0.00	-813.63
		60.00	-126.81	8.657E-01	0.00	0.00	0.00	214.43
		120.00	-78.81	36.87	0.00	0.00	0.00	-917.52
2	LOAD1							
		0.00	-76.78	-40.93	0.00	0.00	0.00	-1025.52
		36.00	-76.78	-40.93	0.00	0.00	0.00	447.90
		72.00	-76.78	59.07	0.00	0.00	0.00	1921.31
		108.00	-76.78	59.07	0.00	0.00	0.00	-205.27
		144.00	-76.78	59.07	0.00	0.00	0.00	-2331.86

Plot Displacements: From the Main Menu enter:
DISPLAY>SHOW DEFORMED SHAPE, then enter
FILE>PRINT GRAPHICS
(The deformed shape shown on the screen is reproduced in Fig. 3.12)

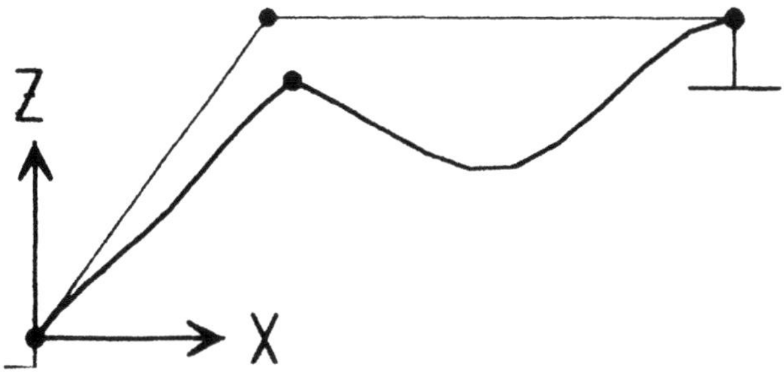

Fig. 3.12 Deformed shape for the Plane Frame in Illustrative Example 3.3.

Plot Shear Forces: From the Main Menu enter:
DISPLAY>SHOW ELEMENT FORCES / STRESSES>FRAMES
Select Shear -22.
Then
FILE>PRINT GRAPHICS
(The shear force diagram shown on the screen is depicted in Fig. 3.13)

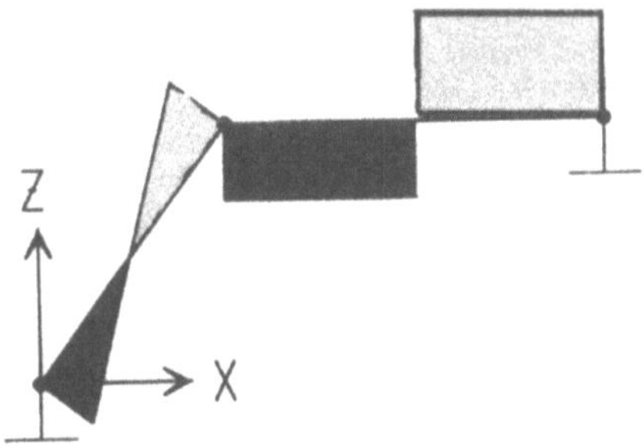

Fig. 3.13 Shear Force diagram for the Plane Frame in Illustrative Example 3.3.

Plot the Bending Moment: From the Main Menu enter:
DISPLAY>SHOW FORCES / STRESSES>FRAME
Select Moment 3-3 and OK.
Then
FILE>PRINT GRAPHICS

(The bending moment diagram shown on the screen is depicted in Fig. 3.14)

Note: To plot the Moment Diagram on the compression side, from the Main Menu enter:
OPTIONS then remove check on "Moment Diagram on the Tension Side"

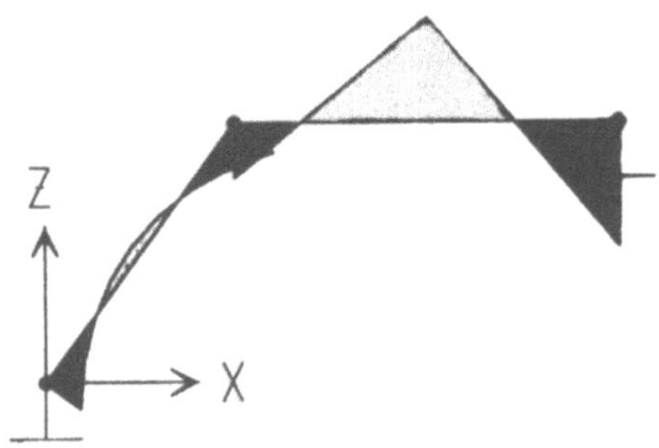

Fig 3.14 Bending Moment Diagram for the plane frame of Illustrative Example 3.3.

Note: To view plots of either the shear force or bending moment diagrams for any element of the frame, right-click on the element and a pop-up window will depict the diagram for the selected element. It should be observed that as the cursor is moved across the plot in this window, values of shear force or of the bending moment will be displayed.

Illustrative Example 3.4

Use SAP2000 to analyze the plane frame shown in Fig. 3.15. The cross bracings are to be modeled as rod elements resisting only axial force. These elements are modeled as beam elements to which a small value is assigned the cross-sectional moment of inertia, for example, I = 1.0 mm^4. Input data tables for this Illustrative Example are given in Table 3.3.

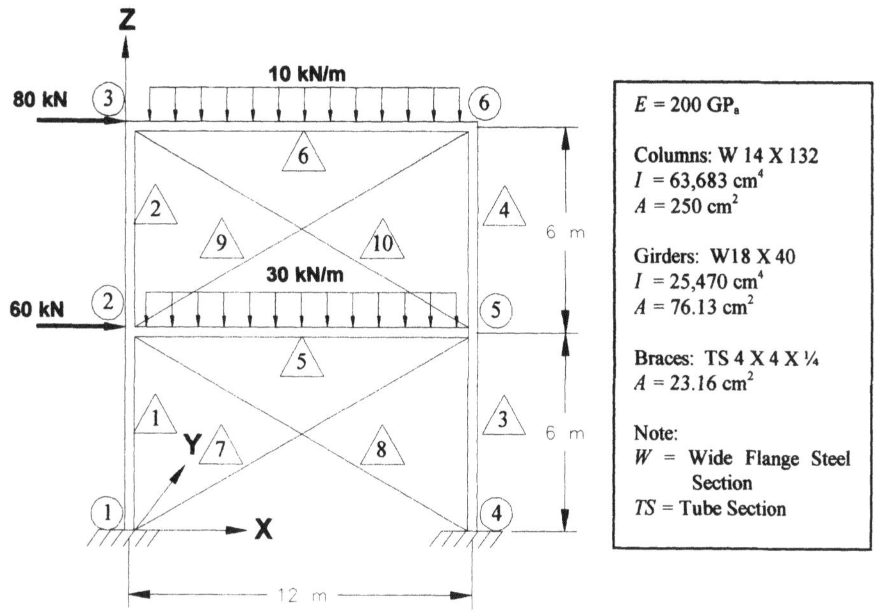

Fig. 3.15 Plane Frame of Illustrative Example 3.4

Table 3.3 Edited Input data tables for Illustrative Example 3.4 (Units: KN, m)

JOINT DATA

JOINT	GLOBAL-X	GLOBAL-Y	GLOBAL-Z	RESTRAINTS
1	0.00000	0.00000	0.00000	1 1 1 1 1 1
2	0.00000	0.00000	6.00000	0 0 0 0 0 0
3	0.00000	0.00000	12.00000	0 0 0 0 0 0
4	12.00000	0.00000	0.00000	1 1 1 1 1 1
5	12.00000	0.00000	6.00000	0 0 0 0 0 0
6	12.00000	0.00000	12.00000	0 0 0 0 0 0

Table 3.3 Continued

FRAME ELEMENT DATA

FRAME	JNT-1	JNT-2	SECTION	RELEASES	SEGMENTS	LENGTH
1	1	2	W14X132	000000	2	6.000
2	2	3	W14X132	000000	2	6.000.
4	5	6	W14X132	000000	2	6.000
3	4	5	W14X132	000000	2	6.000
5	2	5	W18X40	000000	4	12.000
6	3	6	W18X40	000000	4	12.000
7	2	6	TS4X4X1/	000000	2	13.416
8	3	5	TS4X4X1/	000000	2	13.416
9	1	5	TS4X4X1/	000000	2	13.416
10	2	4	TS4X4X1/	000000	2	13.416

JOINT FORCES Load Case LOAD1

JOINT	GLOBAL-X	GLOBAL-Y	GLOBAL-Z	GLOBAL-XX	GLOBAL-YY	GLOBAL-ZZ
2	60.000	0.000	0.000	0.000	0.000	0.000
3	80.000	0.000	0.000	0.000	0.000	0.000

FRAME SPAN DISTRIBUTED LOADS Load Case LOAD1

FRAME	TYPE	DIRECTION	DISTANCE-A	VALUE-A	DISTANCE-B	VALUE-B
6	FORCE	GLOBAL-Z	0.0000	-10.0000	1.0000	-10.0000
5	FORCE	GLOBAL-Z	0.0000	-30.0000	1.0000	-30.0000

Solution:

Table 3.4 contains the edited output tables for Illustrative Example 3.4. The deformed shape, the axial force, the shear force diagram and the bending moment diagrams are reproduced in Figs. 3.16(a) through (d).

Table 3.4 Edited Output tables for Illustrative Example 3.4 (Units: KN, m)

JOINT DISPLACEMENTS

JOINT	LOAD	UX	UY	UZ	RX	RY	RZ
1	LOAD1	0.0000	0.0000	0.0000	0.0000	0.0000	0.0000
2	LOAD1	2.317E-03	0.0000	-1.966E-04	0.0000	2.333E-03	0.0000
3	LOAD1	4.167E-03	0.0000	-2.374E-04	0.0000	6.219E-04	0.0000
4	LOAD1	0.0000	0.0000	0.0000	0.0000	0.0000	0.0000
5	LOAD1	2.262E-03	0.0000	-3.731E-04	0.0000	-1.558E-03	0.0000
6	LOAD1	3.545E-03	0.0000	-4.638E-04	0.0000	-3.081E-04	0.0000

Table 3.4 Continued

JOINT REACTIONS

JOINT	LOAD	F1	F2	F3	M1	M2	M3
1	LOAD1	-24.2079	0.0000	135.3580	0.0000	49.8529	0.0000
4	LOAD1	-115.7921	0.0000	344.6420	0.0000	-114.1489	0.0000

FRAME ELEMENT FORCES

FRAME	LOAD	LOC	P	V2	V3	T	M2	M3
1	LOAD1							
		0.00	-164.02	-33.12	0.00	0.00	0.00	-49.85
		3.00	-164.02	-33.12	0.00	0.00	0.00	49.52
		6.00	-164.02	-33.12	0.00	0.00	0.00	148.90
2	LOAD1							
		0.00	-34.06	-49.64	0.00	0.00	0.00	-185.23
		3.00	-34.06	-49.64	0.00	0.00	0.00	-36.32
		6.00	-34.06	-49.64	0.00	0.00	0.00	112.59
3	LOAD1							
		0.00	-311.28	49.07	0.00	0.00	0.00	114.15
		3.00	-311.28	49.07	0.00	0.00	0.00	-33.06
		6.00	-311.28	49.07	0.00	0.00	0.00	-180.28
4	LOAD1							
		0.00	-75.69	48.68	0.00	0.00	0.00	172.56
		3.00	-75.69	48.68	0.00	0.00	0.00	26.52
		6.00	-75.69	48.68	0.00	0.00	0.00	-119.52
5	LOAD1							
		0.00	-6.99	-178.44	0.00	0.00	0.00	-334.12
		3.00	-6.99	-88.44	0.00	0.00	0.00	66.20
		6.00	-6.99	1.56	0.00	0.00	0.00	196.52
		9.00	-6.99	91.56	0.00	0.00	0.00	56.84
		12.00	-6.99	181.56	0.00	0.00	0.00	-352.84
6	LOAD1							
		0.00	-78.90	-59.42	0.00	0.00	0.00	-112.59
		3.00	-78.90	-29.42	0.00	0.00	0.00	20.68
		6.00	-78.90	0.58	0.00	0.00	0.00	63.95
		9.00	-78.90	30.58	0.00	0.00	0.00	17.22
		12.00	-78.90	60.58	0.00	0.00	0.00	-119.52
7	LOAD1							
		0.00	33.79	0.00	0.00	0.00	0.00	0.00
		6.71	33.79	0.00	0.00	0.00	0.00	0.00
		13.42	33.79	0.00	0.00	0.00	0.00	0.00
8	LOAD1							
		0.00	-56.72	0.00	0.00	0.00	0.00	0.00
		6.71	-56.72	0.00	0.00	0.00	0.00	0.00
		13.42	-56.72	0.00	0.00	0.00	0.00	0.00
9	LOAD1							
		0.00	64.10	0.00	0.00	0.00	0.00	0.00
		6.71	64.10	0.00	0.00	0.00	0.00	0.00
		13.42	64.10	0.00	0.00	0.00	0.00	0.00
10	LOAD1							
		0.00	-74.60	0.00	0.00	0.00	0.00	0.00
		6.71	-74.60	0.00	0.00	0.00	0.00	0.00
		13.42	-74.60	0.00	0.00	0.00	0.00	0.00

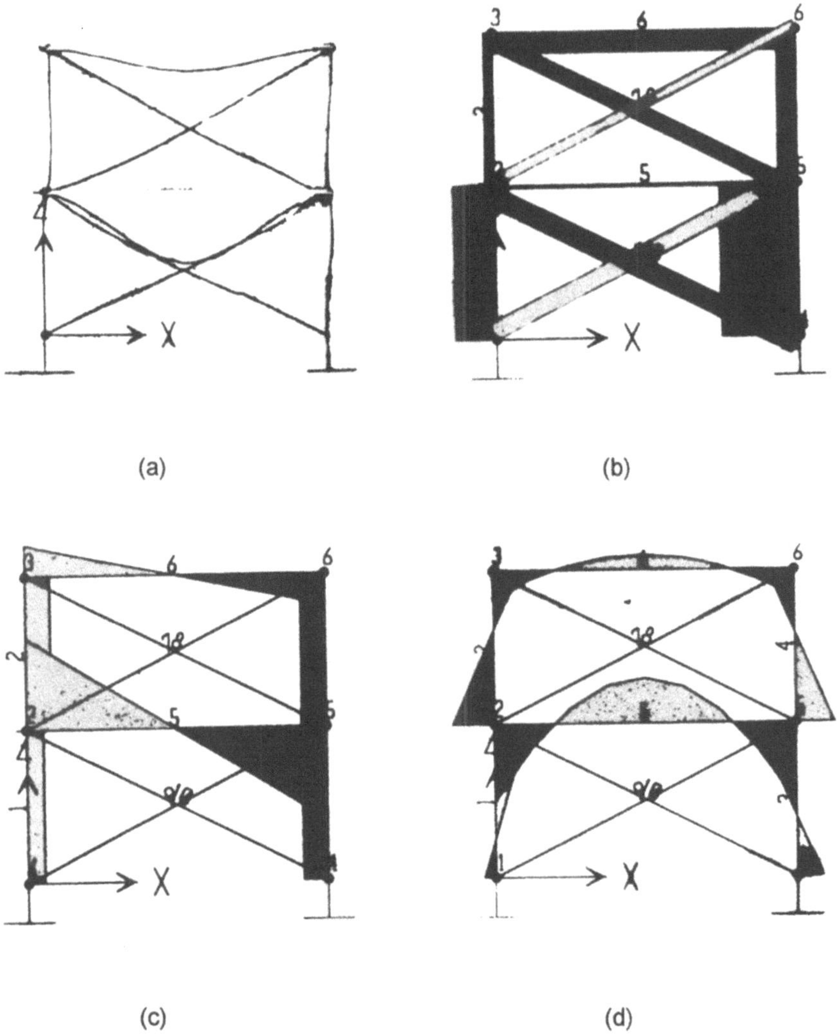

Fig. 3.16 Plots for the plane frame of Illustrative Example 3.4:
(a) Deformed shape, (b) Axial force diagram, (c) Shear Force Diagram and (d) Bending moment diagram.

Illustrative Example 3.5

Use SAP2000 to solve Illustrative Example 3.2 which consists of a beam having an inclined roller support as shown in Fig. 3.17.

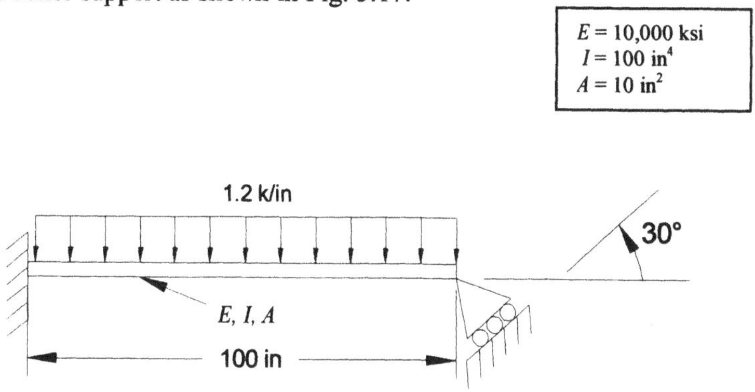

Fig. 3.17 Beam for Illustrative Example 3.5 having an inclined roller support

Solution:

The following commands are implemented in SAP2000:

Begin: Open SAP2000.

Hint: Maximize both screens for a full view of all windows.

Units: Select KN-m (Kilo-Newton, meters) in the drop-down menu located on the lower right hand corner of the screen.

Model: FILE>NEW MODEL
Set: Number x = 2
Distance x = 100
Then OK.

Edit: Minimize 3-D screen
Maximize 2-D screen (X-Z screen). Then drag the figure to the center of the screen using the PAN icon on the toolbar.

Draw: DRAW>DRAW FRAME MEMBER
Click at the origin of XZ coordinates and drag the cursor horizontally to the next grid line on the right. Then double-click and press the enter key.

Labels: VIEW>SET ELEMENTS
Check Joints-Labels and Frames-Labels. Then OK.

Boundaries: Click on joint ①

ASSIGN>JOINT>RESTRAINTS
Check restraints in all directions. Then OK.

Click on joint ②.
ASSIGN>JOINT>LOCAL AXES
Enter: Rotation in degrees about Y = -30

Click on joint ②.
ASSIGN>JOINTS>RESTRAINTS
Check (only) Joint restraint Translation 3. Then OK

Material: DEFINE >MATERIALS
Select OTHER and Modify/Show Materials

Set: Modulus of Elasticity = 10000. Then OK, OK.

Section: DEFINE>FRAME>SECTIONS
Click: Add/Wide Flange
Add General Section

Set: Enter: Cross-section area = 10.

Moment of Inertia about 3 axis = 100

Shear area in 2 direction = 0. Then OK.

Select material = OTHER. Then OK, OK.

Click on the beam element and enter
ASSIGN>FRAME>SECTIONS
Click on FSEC2. Then OK.

Loads: DEFINE>STATIC LOAD CASES
Change DEAD load to LIVE load.
Set self-weight multiplier = 0.
Then click on "Change Load". Then OK.

Click on beam element 1, then enter:
ASSIGN>FRAME STATIC LOADS>POINTS>AND UNIFORM
Enter: Uniform load = -1.2. Then OK.

Analyze: ANALYZE>SET OPTIONS
Click on XZ plane. Then OK.

ANALYZE>RUN
Enter: File Name = "Example 3.5". Then SAVE.

At the conclusion of the calculation click OK.

Plots: Click on XZ view in the toolbar.
Use the PAN icon to center the deformed structure in the screen.
DISPLAY>SHOW DEFORMED SHAPE
FILE>PRINT GRAPHICS

DISPLAY> SHOW ELEMENT FORCES/STRESES>FRAMES
Click: Axial force. Then OK.
FILE>PRINT GRAPHICS

DISPLAY>SHOW ELEMENT FORCES/STRESSES>FRAMES
Click: Moment 3-3. Then OK.
FILE>PRINT GRAPHICS

Tables: FILE>PRINT INPUT TABLES
Click: Print to File. Then OK.

FILE>PRINT OUTPUT TABLES
Click Print to File and click Append. Then OK.

Notes: Use a text editor (Notepad or WORD, for example) to edit and print the edited Input and Output files that have been stored as text (… *.txt*) files.

(The edited Input Tables and Output Tables for Illustrative Example 3.5 are shown in Tables 3.5 and 3.6, respectively.)

Table 3.5 Edited Input Tables for Illustrative Example 3.5 (Units: Kip-in).

JOINT DATA

JOINT	GLOBAL-X	GLOBAL-Y	GLOBAL-Z	RESTRAINTS	ANG-A	ANG-B	ANG-C
1	0.00000	48.00000	0.00000	1 1 1 1 1 1	0.000	0.000	0.000
2	100.00000	48.00000	0.00000	0 0 1 0 0 0	0.000	-30.000	0.000

FRAME ELEMENT DATA

FRAME	JNT-1	JNT-2	SCTN	ANG	RLS	SGMNTS	R1	R2	FCTR	LENGTH
1	1	2	FSEC2	0.000	000000	4	0.000	0.000	1.000	100.000

Table 3.5 Continued

FRAME SPAN DISTRIBUTED LOADS Load Case LOAD1

FRAME	TYPE	DIRECTION	DISTANCE-A	VALUE-A	DISTANCE-B	VALUE-B
1	FORCE	GLOBAL-Z	0.0000	-1.2000	1.0000	-1.2000

Table 3.6 Output tables for Illustrative Example 3.5 (Units: Kip-in).

JOINT DISPLACEMENTS

JOINT	LOAD	UX	UY	UZ	RX	RY	RZ
1	LOAD1	0.0000	0.0000	0.0000	0.0000	0.0000	0.0000
2	LOAD1	-0.0260	0.0000	- 0.0150	0.0000	-0.0248	0.0000

JOINT REACTIONS

JOINT	LOAD	F1	F2	F3	M1	M2	M3
1	LOAD1	25.9548	0.0000	75.0450	0.0000	-1504.4955	0.0000
2	LOAD1	0.0000	0.0000	51.9096	0.0000	0.0000	0.0000

FRAME ELEMENT FORCES

FRAME	LOAD	LOC	P	V2	V3	T	M2	M3
1	LOAD1							
		0.00	-25.95	-75.04	0.00	0.00	0.00	-1504.50
		25.00	-25.95	-45.04	0.00	0.00	0.00	-3.37
		50.00	-25.95	-15.04	0.00	0.00	0.00	747.75
		75.00	-25.95	14.96	0.00	0.00	0.00	748.88
		100.00	-25.95	44.96	0.00	0.00	0.00	0.00

The deformed plot and the axial force diagrams are reproduced in Fig. 3.18 (a) and (b). The shear force and the bending moment diagram are reproduced in Fig. 3.19 (a) and (b).

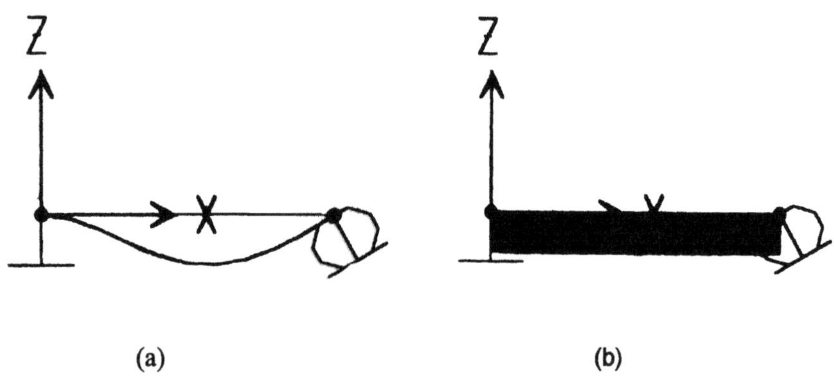

Fig. 3.18 (a) Deformed shape, and (b) Axial force diagram for the beam of Illustrative Example 3.5.

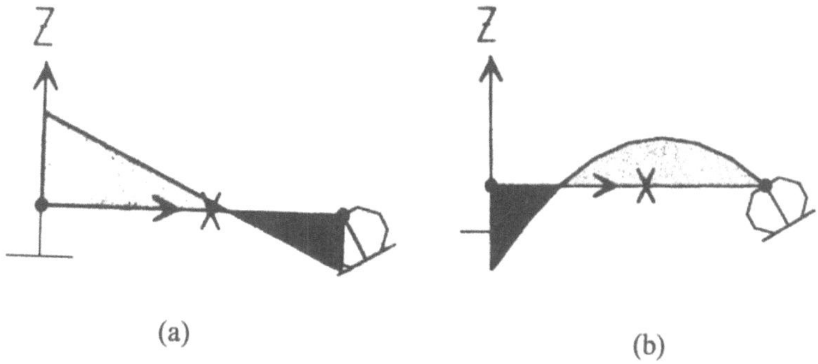

Fig. 3.19 (a) Shear force and (b) Bending moment diagrams for the beam of Illustrative Example 3.5.

Illustrative Example 3.6

Consider the plane frame in Fig. 3.20 having an inclined roller support. Use SAP2000 to determine:

(a) Displacements at the joints
(b) End-forces on the members
(c) Reactions at the supports

$E = 30{,}000$ ksi $I = 882$ in^4 $A = 24.1$ in^2

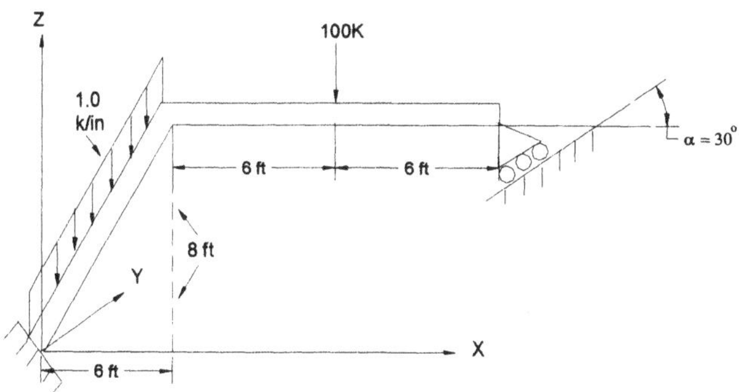

Fig. 3.20 Plane frame having an inclined roller support for Illustrative Example 3.6.

Solution:

The necessary commands for SAP2000 can easily be implemented by analogy with detailed solutions presented in Illustrative Example 3.3 and 3.5. In particular, Illustrative Example 3.5 provides the necessary commands in SAP2000 to implement the condition of an inclined roller support.

Tables containing the edited input data and the edited output results for Illustrative Example 3.6 are reproduced as Tables 3.7 and 3.8, respectively. The deformed shape for the frame of Illustrative Example 3.6 is reproduced in Fig. 3.20.

Table 3.7 Edited Input tables for Illustrative Example 3.6 (Units: kips, in)

JOINT DATA

JOINT	GLOBAL-X	GLOBAL-Y	GLOBAL-Z	RESTRAINTS	ANGLE-A	ANGLE-B	ANGLE-C
1	0.00000	30.00000	0.00000	1 1 1 1 1 1	0.000	0.000	0.000
2	80.00000	30.00000	60.00000	0 0 0 0 0 0	0.000	0.000	0.000
3	180.00000	30.00000	60.00000	0 0 1 0 0 0	0.000	-30.000	0.000

FRAME ELEMENT DATA

FRAME	J1	J2	SECTION	ANGLE	RELEASES	SEGS	R1	R2	FACTOR	LENGTH
1	1	2	FSEC2	0.000	000000	2	0.000	0.000	1.000	100.000
2	2	3	FSEC2	0.000	000000	4	0.000	0.000	1.000	100.000

JOINT FORCES Load Case LOAD1

JOINT	GLOBAL-X	GLOBAL-Y	GLOBAL-Z	GLOBAL-XX	GLOBAL-YY	GLOBAL-ZZ
2	0.000	0.000	-10.000	0.000	0.000	0.000

Table 3.8 Edited Output tables for Illustrative Example 3.6 (Units: kips, in)

JOINT DISPLACEMENTS

JOINT	LOAD	UX	UY	UZ	RX	RY	RZ
1	LOAD1	0.0000	0.0000	0.0000	0.0000	0.0000	0.0000
2	LOAD1	0.1473	0.0000	-0.1980	0.0000	9.029E-04	0.0000
3	LOAD1	0.1471	0.0000	0.0849	0.0000	-4.696E-03	0.0000

JOINT REACTIONS

JOINT	LOAD	F1	F2	F3	M1	M2	M3
1	LOAD1	1.3964	0.0000	7.5814	0.0000	-280.8675	0.0000
3	LOAD1	0.0000	0.0000	2.7928	0.0000	0.0000	0.0000

FRAME ELEMENT FORCES

FRAME	LOAD	LOC	P	V2	V3	T	M2	M3
1	LOAD1							
		0.00	-5.67	-5.23	0.00	0.00	0.00	-280.87
		50.00	-5.67	-5.23	0.00	0.00	0.00	-19.50
		100.00	-5.67	-5.23	0.00	0.00	0.00	241.86

Table 3.8 Continued

2	LOAD1							
		0.00	-1.40	2.42	0.00	0.00	0.00	241.86
		25.00	-1.40	2.42	0.00	0.00	0.00	181.40
		50.00	-1.40	2.42	0.00	0.00	0.00	120.93
		75.00	-1.40	2.42	0.00	0.00	0.00	60.47
		100.00	-1.40	2.42	0.00	0.00	0.00	0.00

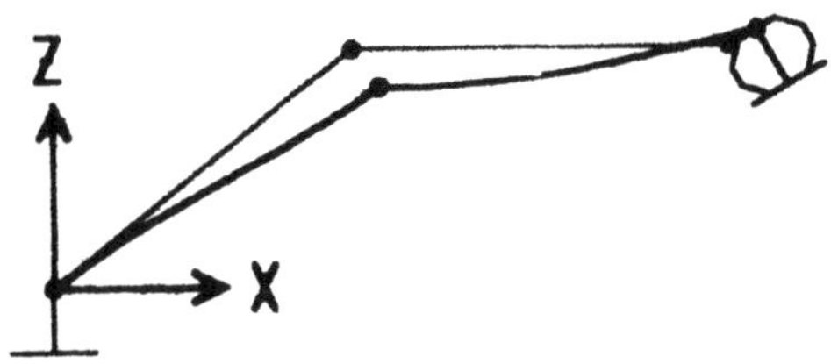

Fig. 3.21 The deformed shape for the plane frame having an inclined roller support of Illustrative Example 3.6.

3.9 Analytical Problems

Problem 3.1

Demonstrate that the stiffness coefficients k_{ij} for axial effect may be calculated by eq.(3.7), repeated here for convenience.

$$k_{ij} = \int_0^L AE\, u_i'(x)\, u_j' dx \qquad \text{(3.7) repeated}$$

in which

E is the modulus of elasticity
A is the cross-sectional area
L is the length of the beam element, and
u_i', u_j' are derivatives with respect to x of shape functions given by

$$u_1(x) = 1 - \frac{x}{L} \qquad \text{(3.5) repeated}$$

and

$$u_2(x) = \frac{x}{L} \qquad \text{(3.6) repeated}$$

Solution:

Consider in Fig. P3.1 a beam element to which a unit displacement $\delta_1 = 1$ has been applied at node 1.

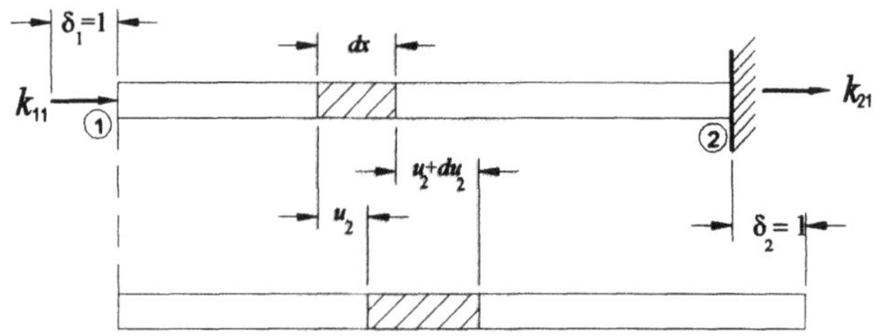

Fig. P3.1 Axial displacement given at node 2 ($\delta_2 = 1$) of a beam element subjected to a unitary axial displacement at node 1 ($\delta_1 = 1$)

The axial force required to produce this displacement is given by eq.(3.4) as

$$P = AE\frac{du}{dx} = AEu_1' \qquad \text{(a)}$$

To this displaced beam, we superimpose an axial deformation by giving a unit displacement ($\delta_2 = 1$) at node ② which results in deformation du_2 of a differential element dx of the beam which may be expressed as

$$du_2 = u_2'(x)dx \qquad \text{(b)}$$

The Principle of Virtual Work which states that during the virtual displacement the work of external forces is equal to the work performed by the internal force is applied. The external work W_E is calculated as the product of the external force k_{21} displaced in one unit $\delta_2 = 1$. That is

$$W_E = k_{21} \qquad \text{(c)}$$

The internal work dW_I on the differential element dx is then calculated as the product of the axial force P times the differential displacement du_2:

$$dW_I = Pdu_2$$

or using eqs.(a) and (b)

$$dW_I = AEu_1' \, u_2' dx$$

and for the entire beam element as

$$W_I = \int_0^L AE \, u_1' \, u_2' \, dx \qquad \text{(d)}$$

Equating the external virtual work, eq.(c), with the internal virtual work, eq.(d), results in

$$k_{21} = \int_0^L AE \, u_i' \, u_2' \, dx \qquad \text{(e)}$$

and in general

$$k_{ij} = \int_1^L AE \, u_i' \, u_j' \, dx \qquad \text{Q.E.D}$$

<u>Problem 3.2</u>

Demonstrate that the equivalent nodal forces for a beam element load axially by a force function $p(x)$ may be calculated by

$$Q_i = \int_0^L p(x) \, u_i(x) dx \qquad \text{(3.19) repeated}$$

in which $u_i(x)$ is the shape function [eq.(3.5) or (3.6)] resulting from a unit nodal displacement.

Solution:

Consider in Fig. P3.2(a) a beam element showing a general axial load $p(x)$ and in Fig. P3.2(b) the beam element showing the nodal equivalent axial forces Q_1 and Q_2. We give to both beams a unit virtual displacement at nodal coordinate ($\delta_1 = 1$) which will result in an axial displacement along the length of the beam given by eq.(3.5):

$$u_1(x) = 1 - \frac{x}{L} \qquad \text{(3.5) repeated}$$

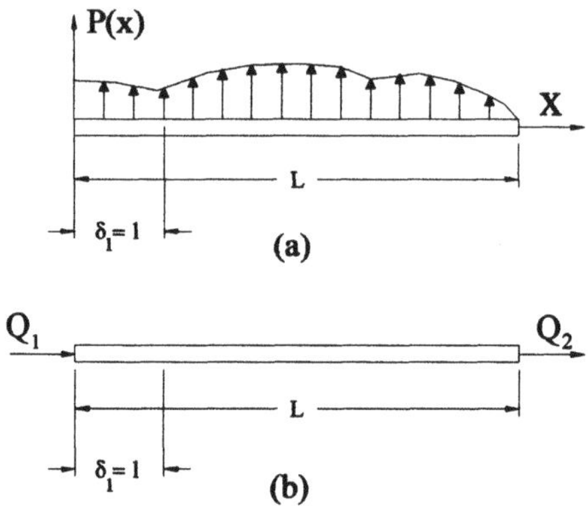

Fig. P3.2 (a) Beam element supporting a general axial force $p(x)$
(b) Beam element supporting the equivalent nodal force Q_1 and Q_2.

We require that the resulting virtual work of the forces Q_i be equal to the virtual work performed by the externally applied axial load $p(x)$. The work $W_E^{'}$ of the equivalent forces Q_i is simply

$$W_E^{'} = Q_1\delta_1 = Q_1 \tag{a}$$

since Q_1 is the only equivalent force undergoing displacement and $\delta_1 = 1.0$.

The work performed by the external force $p(x)\,dx$ is equal to $p(x)\,u_1(x)dx$ and the total work $W_E^{''}$ is then

$$W_E^{''} = \int_0^L p(x)\,u_1\,(x)dx \tag{b}$$

Equating these two calculations of the virtual work results in the following expression for determining the equivalent nodal force Q_1:

$$Q_1 = \int_0^L p(x)\,u_1\,(x)dx$$

In general, the expression for the equivalent nodal force Q_i is given by

$$Q_i = \int_0^L p(x)\,u_i(x)dx \qquad \text{Q.E.D.}$$

where u_i $(i = 1,2)$ is given by eq.(3.5) or by eq.(3.6).

3.10 Practice Problems

Problem 3.3

For the plane frame shown in Fig. P3.3 determine:

(a) Displacements at node ①, and
(b) Reaction at the supports

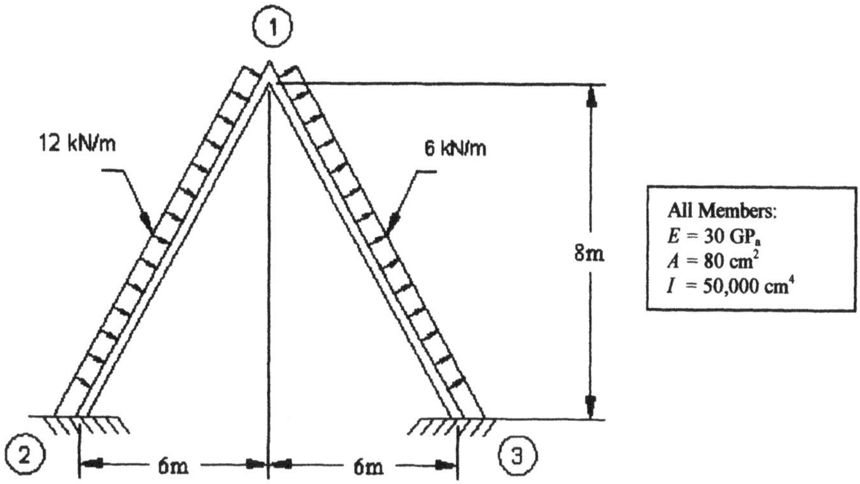

Fig. P3.3 Plane frame for Problem 3.3.

Problem 3.4

Use SAP2000 to solve the plane frame shown in Fig. P3.4 which has an inclined roller support.

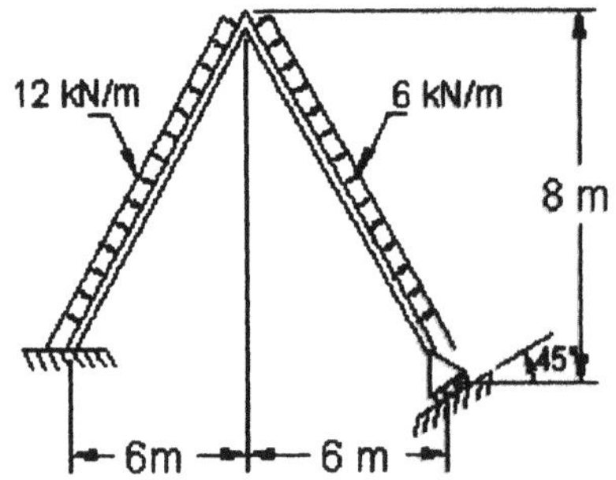

All Members
$E = 30$ GPa
$A = 80$ cm^2
$I = 50{,}000$ cm^4

Fig. P3.4

Problems 3.5, 3.6 and 3.7

Determine the joint displacements, element end forces and support reactions for the frames shown in Figs.P3.5, P3.6 and P3.7.

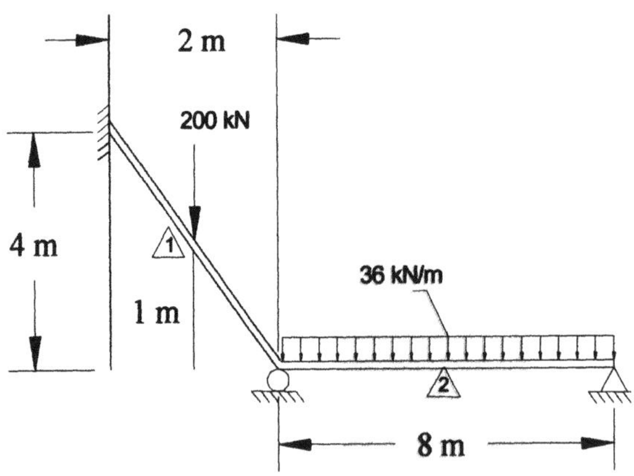

$E = 200$ GPa
$A = 150$ cm^2
$I = 84{,}000$ cm^4

Fig. P3.5

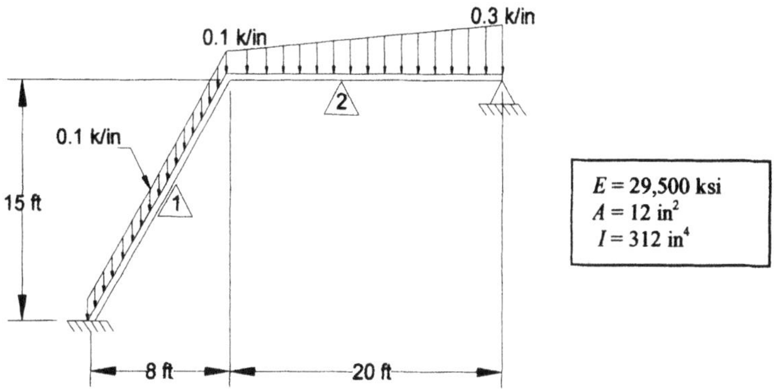

Fig. P3.6

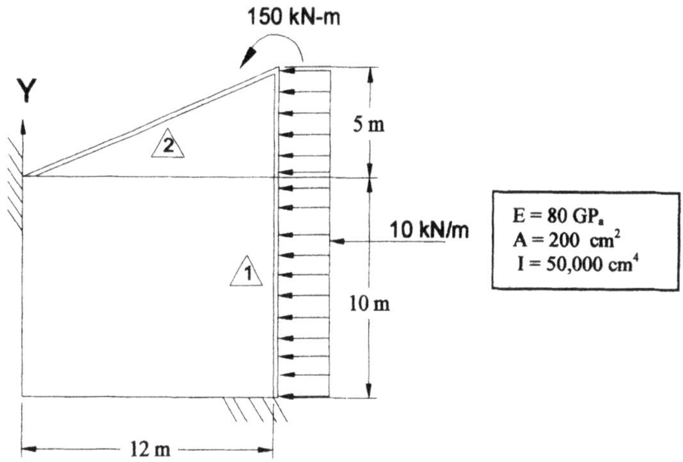

Fig. P3.7

Problem 3.8

For the concrete frame shown in Fig. P3.8 determine:

(a) Joint displacements
(b) Element forces
(c) Reactions at the supports

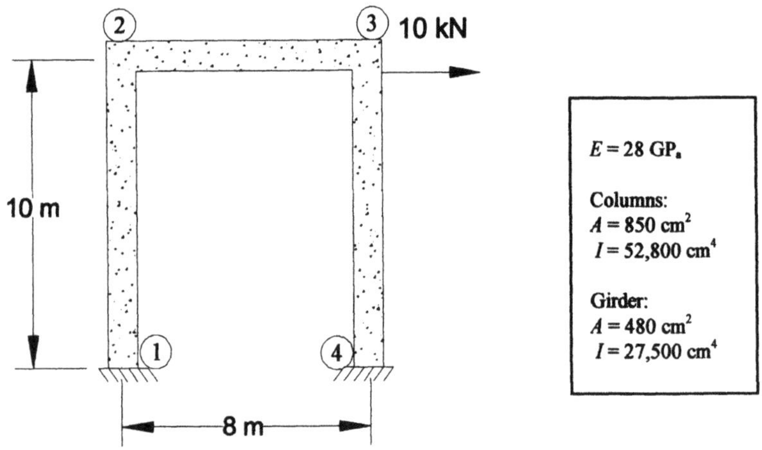

Fig. P3.8

Problem 3.9

For the frame shown in Fig. P3.9 determine:

(a) Joint displacements
(b) Member end-forces
(c) Support reactions

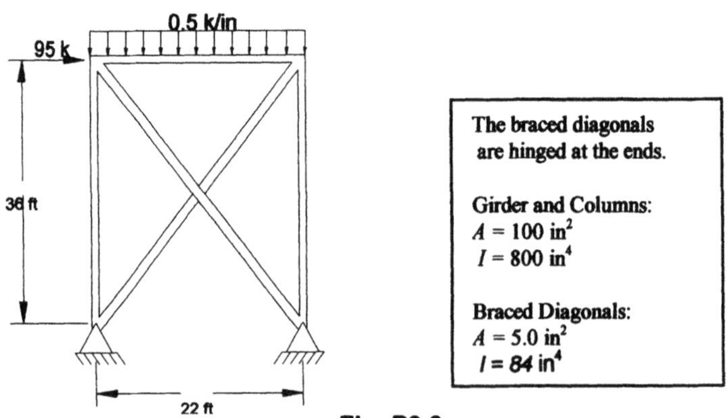

Fig. P3.9

Problem 3.10

For the plane frame shown in Fig. P3.10 determine:

(a) The displacements at the joints
(b) Member end-forces
(c) Reactions at the supports

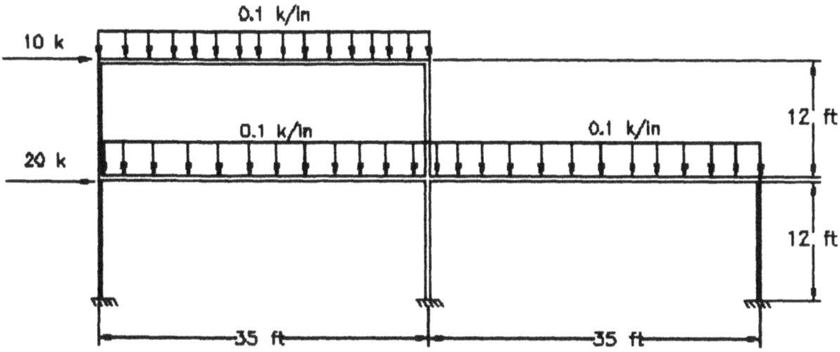

Lower story girders: $I = 280$ in^4; $A = 28$ in^2
Upper story girder: $I = 250$ in^4; $A = 16$ in^2
Lower story columns: $I = 188$ in^4; $A = 9.8$ in^2
Upper story columns: $I = 108$ in^4; $A = 8.6$ in^2

Fig. P3.10

Problem 3.11

For the plane frame shown in Fig. P3.11 determine:

(a) Joint displacements
(b) Member end-forces
(c) Reactions at the supports

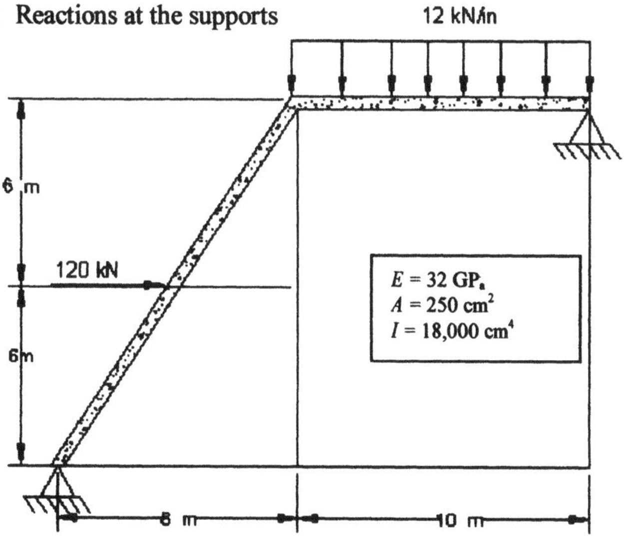

Fig. P3.11

Problem 3.12

For the plane frame shown in Fig. P3.12 determine: (a) Joint Displacements, (b) Member end-forces and (c) Support reactions.

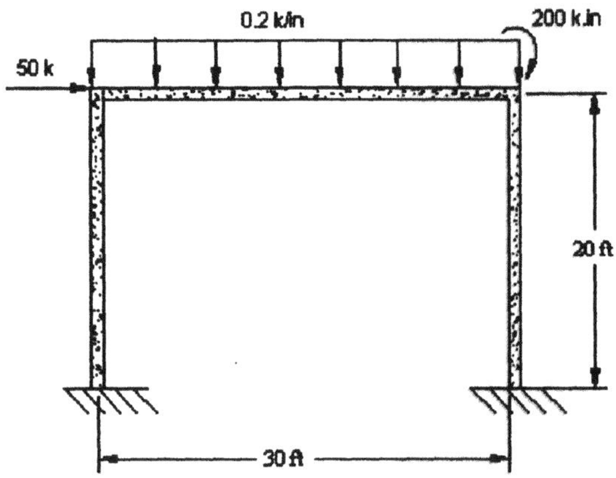

$E = 4500$ ksi
Columns:
$A = 100$ in^2
$I = 600$ in^4
Girder:
$A = 120$ in^2
$I = 1500$ in^4

Fig. P3.12

Problem 3.13

For the plane frame shown in Fig. P3.13 determine:

(a) Joint displacements
(b) Member end-forces
(c) Support reactions

All Members:
E = 29,500 ksi
A = 24 in^2
I = 1200 in^4

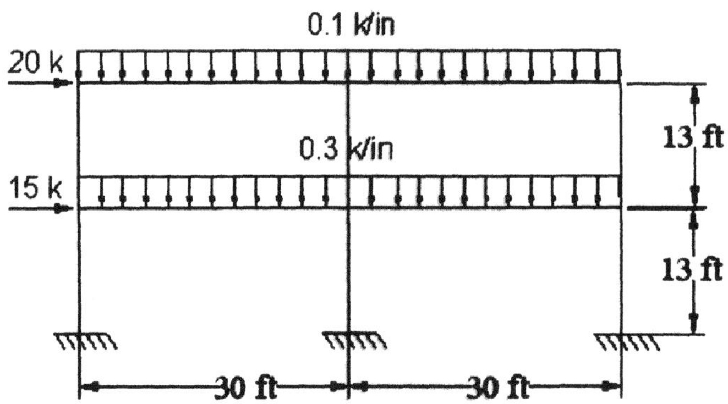

Fig. P3.13

Problem 3.14

For the plane frame shown in Fig. P3.14 determine:

(a) Joint displacements
(b) Member end-forces
(c) Support reactions

E = 29,500 ksi
I = 800 in^4
A = 22 in^2

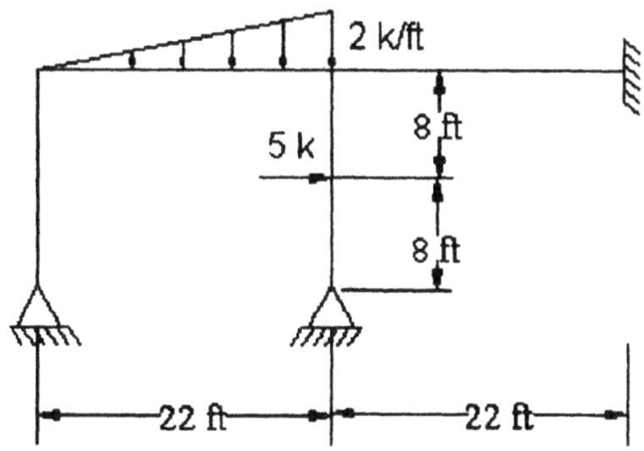

Fig. P3.14

4 Grid Frames

4.1 Introduction

In Chapters 1 and 2 consideration was given to the structural analysis of beams and in Chapter 3 to the analysis of plane frames with loads applied in the plane of the frame. When the plane structural frame is subjected to loads applied normally to its plane, the structure is referred to as a grid frame. This structure can also be treated as a special case of the three-dimensional frame that will be presented in Chapter 5. Plane frames and grid frames are treated as special cases because there is an immediate reduction of unknown nodal coordinates for an element of the frame.

When analyzing the plane frame under the action of loads in its plane, the possible components of nodal displacements that must be considered are translations in the X and Y directions and a rotation about the Z axis. However, if a frame on the plane XY is loaded normal to the plane of the structure, the components of the joint displacements at a node are a translation in the Z direction and rotations about the X and Y axes. Thus, treating the plane grid structure as a special case, it will be necessary to consider only three displacement components at each node of an element of a plane grid.

4.2 Torsional Effects

The analysis of plane grids using the stiffness method, that is, for plane frames subjected to normal loads, requires the determination of the torsional stiffness coefficients of an element of the grid frame. The derivation of these coefficients is essentially identical to the derivation for the axial stiffness coefficient of a beam element. Similarity between these two derivations occurs because the differential equation for both problems has the same mathematical form. For the axial problem, the differential equation for the displacement function is given by eq.(3.4) as

$$\frac{du}{dx} = \frac{P}{AE} \tag{4.1}$$

Likewise, the differential equation for torsional angular displacement is

$$\frac{d\theta}{dx} = \frac{T}{JG} \tag{4.2}$$

in which

θ = angular torsional displacement
T = torsional moment
G = modulus of elasticity in shear
J = torsional constant of the cross-section (polar moment of inertia for circular sections)

As a consequence of the analogy between eqs.(4.1) and (4.2) we can express the displacement functions for torsional effects as corresponding functions for displacements for axial effects; hence by analogy to eqs.(3.5) and (3.6) and in reference to the nodal coordinates shown in Fig. 4.1, we have

$$\theta_1(x) = \left(1 - \frac{x}{L}\right) \tag{4.3}$$

and

$$\theta_2(x) = \frac{x}{L} \tag{4.4}$$

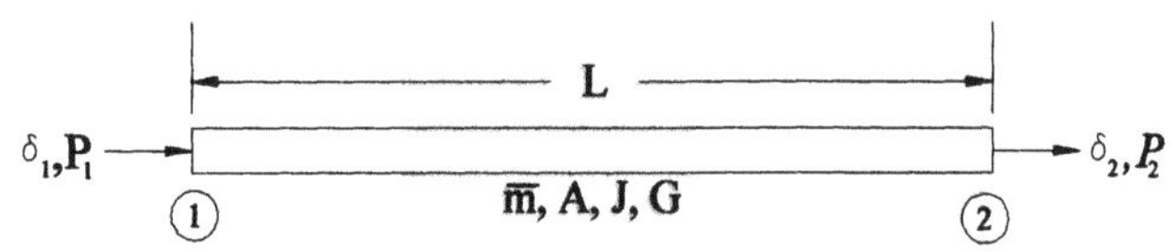

Fig. 4.1. Nodal torsional coordinates for a beam element.

in which the angular displacement function $\theta_1(x)$ corresponds to a unit angular displacement $\delta_1 = 1$ at nodal coordinate 1 and $\theta_2(x)$ corresponds to the displacement function resulting from a unit angular displacement $\delta_2 = 1$ at nodal coordinate 2. Analogously to eq.(3.7), the stiffness coefficient for torsional effects may be calculated from

$$k_{ij} = \int_0^L JG\ \theta_i'(x)\ \theta_j'(x)\ dx \tag{4.5}$$

in which $\theta_i'(x)$ and $\theta_j'(x)$ are the derivatives with respect to x of the displacement functions $\theta_1(x)$ and $\theta_2(x)$.

4.3 Stiffness Matrix for a Grid Element

The application of eq.(4.5) for a uniform element of a grid frame yields the stiffness coefficients $k_{11} = k_{22} = -k_{12} = -k_{21} = JG/L$. The equations relating the torsional moments T_1 and T_2 and the angular displacements δ_1 and δ_2 at the two nodes of the element are then given by

$$\begin{Bmatrix} T_1 \\ T_2 \end{Bmatrix} = \frac{JG}{L} \begin{bmatrix} 1 & -1 \\ -1 & 1 \end{bmatrix} \begin{Bmatrix} \delta_1 \\ \delta_2 \end{Bmatrix} \tag{4.6}$$

Finally, the torsional stiffness matrix in eq.(4.6) is combined with the flexural stiffness matrix in eq.(1.11) to obtain the stiffness matrix for an element of a grid frame. In reference to the local coordinate system indicated in Fig.4.2(a), the stiffness equation for a uniform grid element is

$$\begin{Bmatrix} P_1 \\ P_2 \\ P_3 \\ P_4 \\ P_5 \\ P_6 \end{Bmatrix} = \frac{EI}{L^3} \begin{bmatrix} JGL^2/EI & 0 & 0 & JGL^2/EI & 0 & 0 \\ 0 & 4L^2 & -6L & 0 & 3L^2 & 6L \\ 0 & -6L & 12 & 0 & -6L & -12 \\ -JGL^2/EI & 0 & 0 & JGL^2/EI & 0 & 0 \\ 0 & 2L^2 & -6L & 0 & 4L^2 & 6L \\ 0 & 6L & -12 & 0 & 6L & 12 \end{bmatrix} \begin{Bmatrix} \delta_1 \\ \delta_2 \\ \delta_3 \\ \delta_4 \\ \delta_5 \\ \delta_6 \end{Bmatrix} \tag{4.7}$$

or in condensed notation

$$\{P\} = [k]\{\delta\} \tag{4.8}$$

in which $\{P\}$ and $\{\delta\}$ are, respectively, the force and the displacement vectors at the nodes of the grid element and $[k]$ is the element stiffness matrix defined in eq.(4.7) in reference to the local system of coordinates.

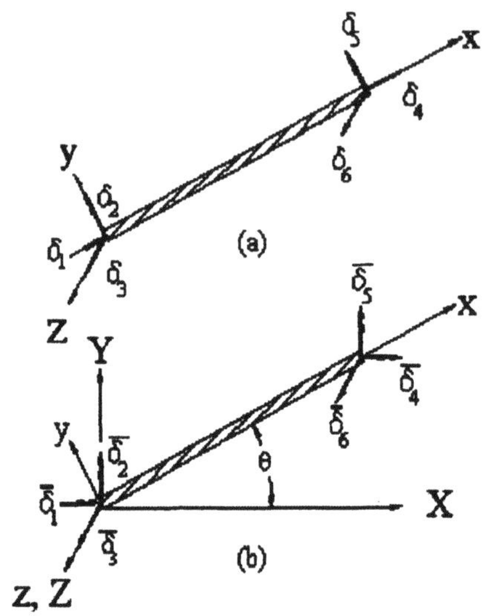

Fig. 4.2. Components of nodal displacements for a grid element.
(a) Local coordinate system, (b) Global coordinate system

4.4 Transformation of Coordinates

The stiffness matrix [eq.(4.7)], is in reference to the local system of coordinates as shown in Fig. 4.2 (a). Therefore, before the assemblage of the stiffness matrix for the structure, it is necessary to transform the reference of the element stiffness matrix to the global system of coordinates as shown in Fig. 4.2(b). Since the *z* axis in the local coordinate system coincides with the *Z* axis for the global system, it is only necessary to perform a transformation of coordinates in the *x-y* plane. The corresponding matrix for this transformation may be obtained by establishing the relationship between components in these two systems of coordinates of the moments at the nodes. In reference to Fig. 4.3, these relations when written for node ① are

$$
\begin{aligned}
P_1 &= \overline{P}_1 \cos\theta + \overline{P}_2 \sin\theta \\
P_2 &= -\overline{P}_1 \sin\theta + \overline{P} \cos\theta \\
P_3 &= \overline{P}_3
\end{aligned}
\tag{4.9}
$$

and for node ②

$$
\begin{aligned}
P_4 &= \overline{P}_4 \cos\theta + \overline{P}_5 \sin\theta \\
P_5 &= -\overline{P}_4 \sin\theta + \overline{P}_5 \cos\theta \\
P_6 &= \overline{P}_6
\end{aligned}
\tag{4.10}
$$

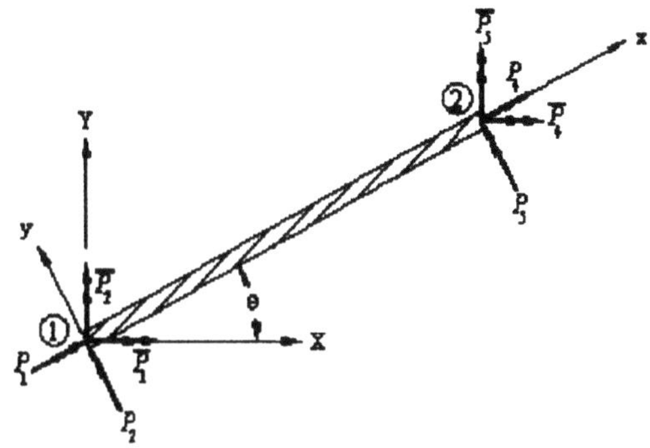

Fig. 4.3 Components of the nodal moments in a grid element; local and global coordinates.

It should be noted that eqs. (4.9) and (4.10) are identical to those derived for the transformation of coordinates for the nodal forces of an element of a plane frame. Equations (4.9) and (4.10) may be written in matrix notation as

$$\begin{Bmatrix} P_1 \\ P_2 \\ P_3 \\ P_4 \\ P_5 \\ P_6 \end{Bmatrix} = \begin{bmatrix} \cos\theta & \sin\theta & 0 & 0 & 0 & 0 \\ -\sin\theta & \cos\theta & 0 & 0 & 0 & 0 \\ 0 & 0 & 1 & 0 & 0 & 0 \\ 0 & 0 & 0 & \cos\theta & \sin\theta & 0 \\ 0 & 0 & 0 & -\sin\theta & \cos\theta & 0 \\ 0 & 0 & 0 & 0 & 0 & 1 \end{bmatrix} \begin{Bmatrix} \overline{P}_1 \\ \overline{P}_2 \\ \overline{P}_3 \\ \overline{P}_4 \\ \overline{P}_5 \\ \overline{P}_6 \end{Bmatrix} \tag{4.11}$$

or in short notation

$$\{P\} = [T]\{\overline{P}\} \tag{4.12}$$

in which $\{P\}$ and $\{\overline{P}\}$ are, respectively, the vectors of the nodal forces of a grid element in local and global coordinates and $[T]$ the transformation matrix defined in eq.(4.11). The same transformation matrix $[T]$ also serves to transform the nodal components of the displacements from a global to a local system of coordinates. In condensed notation, this transformation is given by

$$\{\delta\} = [T]\{\overline{\delta}\} \tag{4.13}$$

where $\{\delta\}$ and $\{\bar{\delta}\}$ are, respectively, the components of nodal displacements in local and global coordinates. The substitution of eqs.(4.12) and (4.13) in the stiffness equation, eq.(4.8), yields the element stiffness matrix in reference to the global coordinate systems:

$$[T]\{\bar{P}\} = [k][T]\{\bar{\delta}\}$$

or since $[T]$ is an orthogonal matrix, it follows that

$$\{\bar{P}\} = [T]^T [k][T]\{\bar{\delta}\}$$

or

$$\{\bar{P}\} = [\bar{k}]\{\bar{\delta}\} \qquad (4.14)$$

in which

$$[\bar{k}] = [T]^T [k] [T] \qquad (4.15)$$

is the element stiffness matrix in reference to the global coordinate system.

4.5 Analysis of Grid Frames

The structural analysis of grid frames is identical to the analysis of beams or plane frames presented in Chapters 1 and 3, respectively. These analyses differ only in the selection of nodal coordinates and the expressions corresponding to the stiffness matrix for the elements in each structure. The following numerical example provides a detailed analysis of a simple grid frame.

Illustrative Example 4.1

For the grid frame shown in Fig. 4.4 perform the structural analysis to determine the following:

(a) Displacements at the node between elements.
(b) End-forces on the elements.
(c) Reactions of the supports.

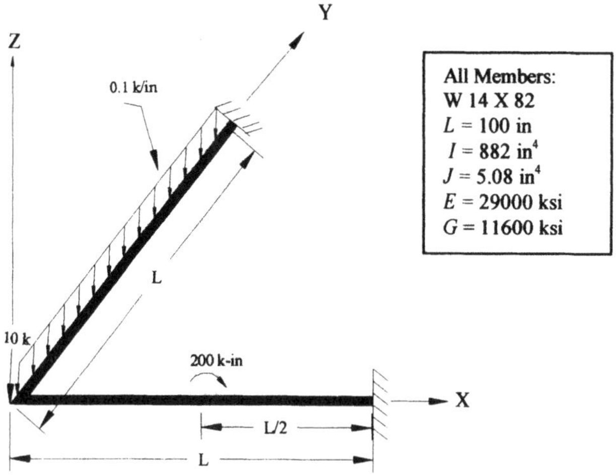

Fig. 4.4 Grid frame for Illustrative Example 4.1

Solution:

1. Model the grid frame.

The grid frame of Illustrative Example 4.1 has been modeled with two elements, three nodes and nine system nodal coordinates. The first three system nodal coordinates correspond to the free nodal coordinates and the last six to the fixed nodal coordinates as shown in Fig, 4.5.

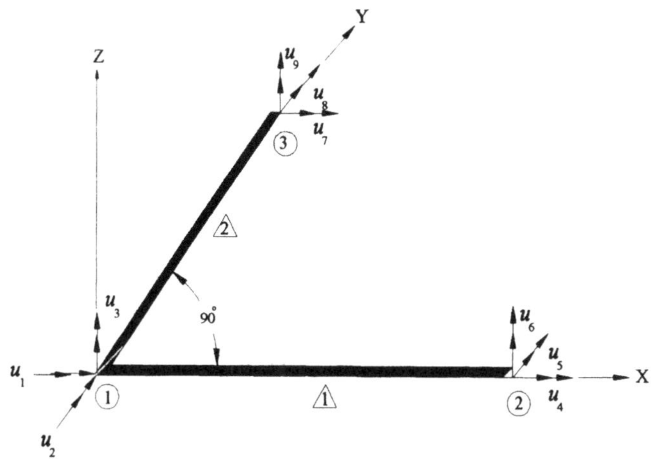

Fig. 4.5 Modeled grid frame of Illustrative Example 4.1 showing the nodal coordinates u_1 through u_9 .

2. Element stiffness matrices (local coordinates).

ELEMENT 1 or 2

Substituting numerical values into eq.(4.7) yields

$$[k]_1 = [k]_2 = \begin{array}{c} \begin{matrix} 1 & 2 & 3 & 4 & 5 & 6 \end{matrix} \\ \left[\begin{matrix} 589.3 & 0 & 0 & -589.3 & 0 & 0 \\ 0 & 1.023E6 & -1.535E4 & 0 & 5.116E5 & 1.535E4 \\ 0 & -1.535E4 & 306.9 & 0 & -1.535E4 & -3.069E2 \\ -589.3 & 0 & 0 & 589.3 & 0 & 0 \\ 0 & 5.116E5 & -1.535E4 & 0 & 1.023E6 & 1.535E4 \\ 0 & 1.535E4 & -3.069E2 & 0 & 1.535E4 & 3.069E2 \end{matrix}\right] \begin{matrix} 1 \\ 2 \\ 3 \\ 4 \\ 5 \\ 6 \end{matrix} \end{array} \quad \text{(a)}$$

3. Element transformation matrices.

ELEMENT 1: Since $\theta = 0$

$$[T]_1 = [I]$$

ELEMENT 2

From eq.(4.11) with $\theta = 90^o$:

$$[T]_2 = \begin{bmatrix} 0 & 1 & 0 & 0 & 0 & 0 \\ -1 & 0 & 0 & 0 & 0 & 0 \\ 0 & 0 & 1 & 0 & 0 & 0 \\ 0 & 0 & 0 & 0 & 1 & 0 \\ 0 & 0 & 0 & -1 & 0 & 0 \\ 0 & 0 & 0 & 0 & 0 & 1 \end{bmatrix} \quad \text{(b)}$$

4. Element stiffness matrices (global coordinates).

ELEMENT 1: Since $\theta = 0$:

$$[\bar{k}]_1 = [k]_1$$

ELEMENT 2: with $\theta = 90^o$

Substitution of the stiffness matrix $[k]_2$ from eq.(a) and the transformation matrix $[T]_2$ from eq.(b) into the expression for the element stiffness matrix in global coordinates,

$$[\bar{k}]_2 = [T]_2^T [k]_2 [T]_2 \qquad \text{(4.15) repeated}$$

yields:

$$[\bar{k}]_2 = \begin{array}{c} \begin{matrix} 1 & 2 & 3 & 7 & 8 & 9 \end{matrix} \\ \begin{bmatrix} 1.023E6 & 0 & 1.535E4 & 5.116E5 & 0 & -1.535E4 \\ 0 & 5.893E2 & 0 & 0 & -5.839E2 & 0 \\ 1.535E4 & 0 & 3.069E2 & 1.535E4 & 0 & -3.069E2 \\ 5.116E5 & 0 & 1.535E4 & 1.023E6 & 0 & -1.535E4 \\ 0 & -5.839E2 & 0 & 0 & 5.893E2 & 0 \\ -1.535E4 & 0 & -3.069E2 & -1.535E4 & 0 & 3.069E2 \end{bmatrix} \end{array} \begin{matrix} 1 \\ 2 \\ 3 \\ 7 \\ 8 \\ 9 \end{matrix} \qquad \text{(c)}$$

5. Assemblage of the reduced system stiffness matrix.

The transfer to the reduced system stiffness matrix of coefficients corresponding to the free coordinates, from element stiffness matrices, eqs.(a) and (c), to the locations indicated at the top and on the right of these matrices yields

$$[K]_R = \begin{bmatrix} 5.893E2 + 1.023E6 & 0+0 & 0+1.535E4 \\ 0+0 & 1.023E6 + 5.893E2 & -1.535E4 + 0 \\ 0+1.535E4 & -1.535E4 & 3.069E2 + 3.069E2 \end{bmatrix}$$

or

$$[K]_R = \begin{bmatrix} 1.024E6 & 0 & 1.535E4 \\ 0 & 1.024E6 & -1.535E4 \\ 1.535E4 & -1.535E4 & 6.139E2 \end{bmatrix} \qquad \text{(d)}$$

6. Equivalent nodal forces.

ELEMENT 1

From Appendix I, Case (b) (concentrated moment)

$$Q_1 = -\frac{6ML_1}{L^3}L_2 = \frac{6x200\text{x}50^2}{100^3} = 3.0\text{kip}$$

$$Q_2 = \frac{ML_2}{L^2}(L_2 - 2L_1) = \frac{-200\text{x}50}{100^2}(50 - 100) = 50\,\text{kip·in}$$

$$Q_3 = \frac{6ML_1L_2}{L^3} = -3.0\,\text{kip}$$

$$Q_4 = \frac{ML_1}{L^2}(L_1 - 2L_2) = 50\,\text{kip·in}$$

Arranging these values of the equivalent nodal forces in a vector, and in accordance with the global nodal coordinates shown in Fig. 4.5, we obtain

$$\{Q\}_1 = \begin{Bmatrix} 0 \\ 50 \\ 3.0 \\ 0 \\ 50 \\ -3.0 \end{Bmatrix} \begin{matrix} 1 \\ 2 \\ 3 \\ 4 \\ 5 \\ 6 \end{matrix} \qquad \text{(e)}$$

ELEMENT 2

From Appendix I, Case (c) (uniform distributed force).

$$Q_1 = \frac{wL}{2} = -\frac{0.1\text{x}100}{2} = -5\,\text{kip}$$

$$Q_2 = \frac{wL^2}{12} = -\frac{0.1\text{x}100^2}{12} = -83.33\,\text{kip}$$

$$Q_3 = \frac{wL}{L} = -5$$

$$Q_4 = \frac{wL^2}{12} = +83.33$$

or in vector form in accordance with the global nodal coordinate for element 2 indicated in Fig. 4.5, we obtain

$$\{\bar{Q}_2\} = \begin{Bmatrix} -83.33 \\ 0 \\ -5 \\ 0 \\ 83.33 \\ -5 \end{Bmatrix} \begin{matrix} 1 \\ 2 \\ 3 \\ 7 \\ 8 \\ 9 \end{matrix} \tag{f}$$

7. Reduced system force vector.

The transfer to the system force vector of the coefficients corresponding to the free nodal coordinates 1, 2, and 3, [from eqs.(e) and (f)] yields:

$$\{F\}_R = \begin{Bmatrix} 0-83.33 \\ 50+0 \\ 3.0-5 \end{Bmatrix} + \begin{Bmatrix} 0 \\ 0 \\ -10 \end{Bmatrix} = \begin{Bmatrix} -83.33 \\ 50 \\ -12.0 \end{Bmatrix} \tag{g}$$

Equation (g) includes the force −10 kips directly applied at nodal coordinate 3.

8. Reduced system stiffness equation.

Substitution of the reduced system stiffness matrix [eq.(d)] and of the reduced system force vector, [eq.(g)] into the stiffness equation $\{F\}_R=[K]_R\{u\}$ results in

$$\begin{bmatrix} -83.33 \\ 50 \\ -12.0 \end{bmatrix} = \begin{bmatrix} 1.024E6 & 0 & 1.535E4 \\ 0 & 1.024E6 & -1.535E4 \\ 1.535E4 & -1.535E4 & 6.139E2 \end{bmatrix} \begin{bmatrix} u_1 \\ u_2 \\ u_3 \end{bmatrix} \tag{h}$$

9. Solution of the unknown displacements.

Solution of eq.(h) yields:

u_1 = 1.040 E-3 rad
u_2 = -1.170 E-3 rad
u_3 = -0.0748 in

10. Element nodal displacements (local coordinates).

The nodal displacements for elements 1 and 2 identified from Fig. 4.5 results in

$$\{\delta\}_1 = \begin{Bmatrix} 1.040E-3 \\ -1.170E-3 \\ -0.0748 \\ 0 \\ 0 \\ 0 \end{Bmatrix} \quad \text{and} \quad \{\delta\}_2 = \begin{Bmatrix} -1.170E-3 \\ -1.040E3 \\ -0.0748 \\ 0 \\ 0 \\ 0 \end{Bmatrix} \qquad \text{(i)}$$

11. Element end forces.

Element end forces are calculated as in eq.(3.20) for plane frames by

$$\{P\} = \{k]\ \{\delta\} - \{Q\} \qquad \text{(3.20) repeated}$$

ELEMENT 1:

$$\begin{Bmatrix} P_1 \\ P_2 \\ P_3 \\ P_4 \\ P_5 \\ P_6 \end{Bmatrix}_1 = \begin{bmatrix} 589.3 & 0 & 0 & -589.3 & 0 & 0 \\ 0 & 1.023E6 & -1.535E4 & 0 & 5.116E5 & 1.535E4 \\ 0 & -1.535E4 & 306.9 & 0 & -1.535E4 & -3.069E2 \\ -589.3 & 0 & 0 & 589.2 & 0 & 0 \\ 0 & 5.116E5 & -1.535E4 & 0 & 1.023E6 & 1.535E4 \\ 0 & 1.535E4 & -3.069E2 & 0 & 1.535E4 & 3.069E2 \end{bmatrix} \begin{Bmatrix} 1.040E-3 \\ -1.170E-3 \\ -0.0748 \\ 0 \\ 0 \\ 0 \end{Bmatrix}$$

$$- \begin{Bmatrix} 0 \\ 50 \\ 3.0 \\ 0 \\ 50 \\ -3.0 \end{Bmatrix} = \begin{Bmatrix} -0.589 \\ 0.663 \\ 8.000 \\ -0.589 \\ -599 \\ 8.000 \end{Bmatrix}$$

ELEMENT 2

Analogously, substituting numerical values for element 2 into eq.(3.20), yields

$$\begin{Bmatrix} P_1 \\ P_2 \\ P_3 \\ P_4 \\ P_5 \\ P_6 \end{Bmatrix}_2 = \begin{Bmatrix} 0.663 \\ 0.589 \\ 2.000 \\ 0.663 \\ -699 \\ 12.000 \end{Bmatrix}$$

12. Reactions at supports

Support ②

$R_4 = P_{41} = -0.589$ (kip.in)
$R_5 = P_{51} = 599$ (kip.in)
$R_6 = P_{61} = 8.00$ (kip)

Support ③

$R_7 = P_{42} = -699$ (kip.in)
$R_8 = P_{52} = 0.663$ (kip.in)
$R_9 = P_{62} = 12.00$ (kip)

Notation: P_{ij} = Force at nodal coordinate i of element j.

Illustrative Example 4.2

Use SAP2000 to solve Illustrative Example 4.1.

Solution:

The following commands are implemented in SAP2000:

Begin: Open SAP2000.

Enter: "OK" to disable the "Tip of the Day" screen.

Hint: Maximize both screens for full view of all windows.

Select: In the lower right-hand corner of the screen use the drop-down menu to select "kip-in".

Select: From the Main menu enter:
FILE>NEW MODEL
In the Coordinate System Definition: Set x, y and z to 2 divisions.
Change Grid Spacing $x = 100$, $y = 100$ and $z = 100$.

Edit: Maximize the screen showing the 3-D view and minimize the 2-D screen. Click on the PAN icon and drag the plot to the center of the screen.

Draw: From the Main Menu enter:
DRAW>DRAW FRAME ELEMENT

Click at the origin of the coordinates and drag the cursor along the y axis to the next grid line ($y = 100$), click twice and press "Enter". Click at the origin of the coordinates and drag the cursor along the x axis to the next grid line ($x = 100$), click twice and press "Enter".

Click on the pointed arrow located on the toolbar. Click on element 2.
ASSIGN>FRAME STATIC LOADS>POINT AND UNIFORM
Select: FORCES. Direction Global Z.
Enter: Uniform Load = -0.1. Then OK
Click on Joint 1
ASSIGN>JOINT STATIC LOADS>FORCES
Enter: Force Global Z = -10.0. Then OK.

Label: For viewing convenience, label joints and elements:
From the Main Menu enter:
VIEW>SET ELEMENTS
Then Click on the boxes labeled "Joint labels" and "Frame labels".
Then OK.

Boundaries: Click on Joints ② and ③ and enter:
ASSIGN>JOINT>RESTRAINTS
Select restraints in all directions. Then OK.

Material: DEFINE>MATERIALS
Click on STEEL and on Modify/Show Material
Set Modulus of Elasticity = 29000. Then OK.

Define Section: DEFINE >FRAME SECTION
Click: Import/Wide Flange. Then OK.
Select ; "Sections.pro". Then OPEN.
Select: W14 X 82, Then OK.
Click: Modification Factors
Set: Shear area in 2 directions = 0. Then OK, OK, OK.

Assign Section: Click on beam elements 1 and 2.
ASSIGN>FRAMES>SECTIONS
Select W14 X 82. Then OK.

Define Loads: DEFINE>STATIC LOAD CASES
Select: LIVE LOAD and set self weight multiplier = 0
Click: Change Load. Then OK.

Assign Loads: Click on Beam element 1
ASSIGN>FRAME STATIC LOADS>POINT AND UNIFORM
Select: Moments, Direction: Global *Y*
Enter: Distance = 0.5
Load = 200. Then OK.

Analyze: ANALYZE>SET OPTIONS
Check: Available degrees of freedom *UZ*, *RX* and *RY*, then OK.
(or simply click on plane grid diagram)
ANALYZE>RUN
Enter: File Name: "Example 4.2". Then SAVE.
At the conclusion of the calculations, click OK.

Print Input Tables: FILE>PRINT INPUT TABLES

(Table 4.1 contains the edited Input Tables for Illustrative Example 4.2)

Table 4.1 Edited Input Tables for Illustrative Example 4.2 (Units: Kips, inches)

JOINT DATA

JOINT	GLOBAL-X	GLOBAL-Y	GLOBAL-Z	RESTRAINTS
1	0.00000	0.00000	0.00000	0 0 0 0 0 0
2	100.00000	0.00000	0.00000	1 1 1 1 1 1
3	0.00000	100.00000	0.00000	1 1 1 1 1 1

FRAME ELEMENT DATA

FRAME	JNT-1	JNT-2	SECTION	RELEASES	SEGMENTS	LENGTH
1	1	2	W14X82	000000	4	100.000
2	1	3	W14X82	000000	4	100.000

JOINT FORCES Load Case LOAD1

JOINT	GLOBAL-X	GLOBAL-Y	GLOBAL-Z	GLOBAL-XX	GLOBAL-YY	GLOBAL-ZZ
1	0.000	0.000	-10.000	0.000	0.000	0.000

FRAME SPAN DISTRIBUTED LOADS Load Case LOAD1

FRAME	TYPE	DIRECTION	DISTANCE-A	VALUE-A	DISTANCE-B	VALUE-B
2	FORCE	GLOBAL-Z	0.0000	-0.1000	1.0000	-0.1000

FRAME SPAN POINT LOADS Load Case LOAD1

FRAME	TYPE	DIRECTION	DISTANCE	VALUE
1	MOMENT	GLOBAL-Y	0.5000	200.0000

Print Output Tables: FILE>PRINT OUTPUT TABLES

(Table 4.2 contains the edited Output Tables for Illustrative Example 4.2)

Table 4.2 Edited Output Tables for Illustrative Example 4.2 (Units: Kips-inches)

JOINT DISPLACEMENTS

JOINT	LOAD	UX	UY	UZ	RX	RY	RZ
1	LOAD1	0.0000	0.0000	-0.0748	1.040E-03	-1.170E-03	0.0000
2	LOAD1	0.0000	0.0000	0.0000	0.0000	0.0000	0.0000
3	LOAD1	0.0000	0.0000	0.0000	0.0000	0.0000	0.0000

JOINT REACTIONS

JOINT	LOAD	F1	F2	F3	M1	M2	M3
2	LOAD1	0.0000	0.0000	8.0006	-0.5894	599.3922	0.0000
3	LOAD1	0.0000	0.0000	11.9994	-699.3553	0.6632	0.0000

FRAME ELEMENT FORCES

FRAME	LOAD	LOC	P	V2	V3	T	M2	M3
1	LOAD1							
		0.00	0.00	8.00	0.00	-5.894E-01	0.00	6.632E-01
		25.00	0.00	8.00	0.00	-5.894E-01	0.00	-199.35
		50.00	0.00	8.00	0.00	-5.894E-01	0.00	-199.36
		75.00	0.00	8.00	0.00	-5.894E-01	0.00	-399.38
		100.00	0.00	8.00	0.00	-5.894E-01	0.00	-599.39
2	LOAD1							
		0.00	0.00	2.00	0.00	6.632E-01	0.00	5.894E-01
		25.00	0.00	4.50	0.00	6.632E-01	0.00	-80.65
		50.00	0.00	7.00	0.00	6.632E-01	0.00	-224.38
		75.00	0.00	9.50	0.00	6.632E-01	0.00	-430.62
		100.00	0.00	12.00	0.00	6.632E-01	0.00	-699.36

Plot Deformed Shape: DISPLAY>SHOW DEFORMED SHAPE
FILE>PRINT GRAPHICS

(The deformed shape shown on the screen is reproduced in Fig. 4.6)

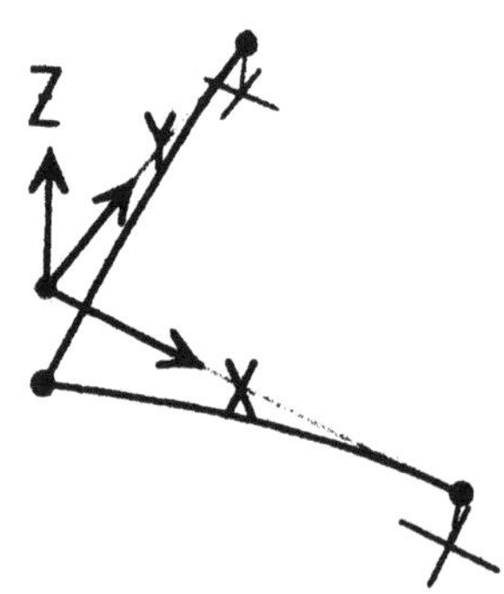

Fig. 4.6 The deformed shape for the plane grid in Illustrative Example 4.2.

Illustrative Example 4.3

Consider the grid frame shown in Fig. 4.7, having fixed supports at joints ①, ⑦ and ⑨, and a simple support at joint ③. Use SAP2000 to analyze this grid frame supporting the load shown in the figure.

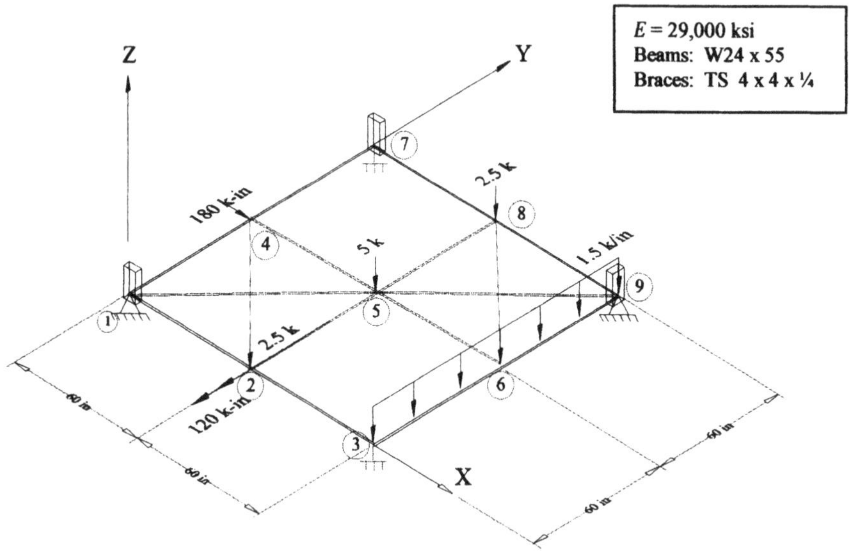

Fig. 4.7 Grid frame for Illustrative Example 4.3.

Solution:

The following commands are implemented in SAP2000:

Begin: Open SAP2000.

Enter: "OK" to disable "Tip of the Day" screen.

Hint: Maximize both screens for full view of all windows.

Units: In the lower right-hand corner of the screen use the drop-down menu to select "kip-in".

Model: FILE>NEW MODEL
Set divisions: x = 2, y = 2, z = 0.
Change grid spacing x = 60, y = 60, z = 60.

Edit: Maximize the 3-D screen and minimize the 2-D screen. Click on PAN icon and drag the plot to the center of the screen.

View: VIEW>SET 3-D VIEW
Set view angles: Plan = 290°; Elevation = 16° and Aperture = 0°, then OK.

Draw: DRAW>QUICK DRAW ELEMENT
Beam element will be drawn by clicking successively on every grid line parallel to the axis *X* and *Y*. Then enter:
DRAW >FRAME ELEMENT
Click successively on two joints to draw the beam elements along the diagonals as shown in Fig. 4.7.

Move Axes: SELECT>SELECT ALL
EDIT>MOVE
Set x = 60
y = 60. Then OK.

Cancel Grid: VIEW
Uncheck: Show grid

Material: DEFINE>MATERIALS>STEEL
Click on Modify/Show Material
Set: Modulus of Elasticity = 29,500. Then OK, OK.

Define Sections: DEFINE>FRAME SECTIONS>IMPORT/WIDE FLANGE, OK
Click on Sections.pro, then Open
Click on W24 X 55. Then OK, OK, OK.

DEFINE>FRAME SECTIONS>IMPORT BOX/TUBE
(Use drop-down menu)
Click on Sections.pro, then OPEN.
Select TS4 X 4 X 1/4, Then OK.

Assign Sections: Mark (click) on all the beams parallel to the axes X and Y.
ASSIGN>FRAMES>SECTIONS
Click on W24 X 55. Then OK.

Mark all the braces (diagonal members)
ASSIGN>FRAMES>SECTIONS
Click on ST4 X 4 X 1/4. Then OK.

Labels: VIEW>SET ELEMENTS
Check Joint Labels and Frame Labels. Then OK.

Boundaries: Mark (click) on joints ③ and ⑦.
ASSIGN>JOINT RESTRAINTS
Check restraints in all local directions. Then OK.
Mark (click on) joints ① and ⑨
ASSIGN>JOINT RESTRAINTS
Check restraint Translation 1, 2, and 3. Then OK.

Mark (click) on Joints ②, ④, ⑤, ⑥, ⑧.
ASSIGN>JOINT RESTRAINTS
Uncheck in all directions or just click on the dot. Then OK.

Define Loads: DEFINE>STATIC LOAD CASES
Change Load Type "DEAD" to "LIVE".
Set: Self-Weight Multiplier = 0.
Click on Change Load. Then OK.

Assign Loads: Mark joints ② and ⑧ and enter
ASSIGN>JOINT STATIC LOADS>FORCES
Set: Force Global Z = -2.5. Then OK.
Mark joint ⑤ and enter:
ASSIGN>JOINT STATIC LOADS>FORCES
Set Force Global Z = -5.0 Then OK
Mark joint ② and enter:
ASSIGN>JOINT STATIC LOADS>FORCES
Set Moment Global YY = -120. Then OK.
Mark joint ④ and enter:
ASSIGN>JOINT STATIC LOADS>FORCES
Set Moment Global X = 180. Then OK.
Mark beam elements 9 and 12.

Enter: ASSIGN>FRAME STATIC LOADS>POINT AND UNIFORM FORCE
Direction: Global Z
Uniform Load = -1.5. Then OK.

Set Options: ANALYZE>SET OPTIONS
Available DOF's: Check UZ, RX and RY. Then OK.

Analyze: ANALYZE>RUN
At the conclusion of calculation click OK.

Plots: DISPLAY>UNDEFORMED SHAPE
FILE>PRINT GRAPHICS

(The plot of the Grid Frame of Illustrative Example 4.2 is reproduced as Fig. 4.8.)

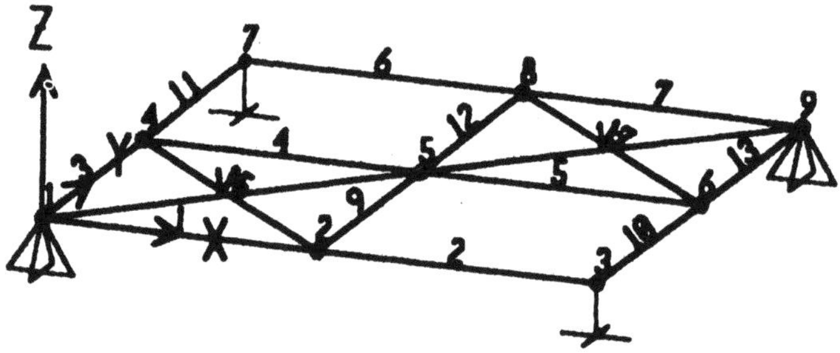

Fig. 4.8 Undeformed plot of grid frame for Illustrative Example 4.3.

Print Tables: FILE>PRINT INPUT TABLES
Check: Print to File. Then OK.

FILE>PRINT OUTPUT TABLES
Check: Displacements, Reactions and Frame Forces
Check: Print to File and Append. Then OK.

(The edited Input and Output Tables for Illustrative Example 4.3 are shown in Tables 4.3 and 4.4, respectively.)

Table 4.3 Edited Input Tables for Illustrative Example 4.3 (Units: Kips, inches)

JOINT DATA

JOINT	GLOBAL-X	GLOBAL-Y	GLOBAL-Z	RESTRAINTS	ANG-A	ANG-B	ANG-C
1	0.00000	0.00000	0.00000	1 1 1 0 0 0	0.000	0.000	0.000
2	60.00000	0.00000	0.00000	0 0 0 0 0 0	0.000	0.000	0.000
3	120.00000	0.00000	0.00000	1 1 1 1 1 1	0.000	0.000	0.000
4	0.00000	60.00000	0.00000	0 0 0 0 0 0	0.000	0.000	0.000
5	60.00000	60.00000	0.00000	0 0 0 0 0 0	0.000	0.000	0.000
6	120.00000	60.00000	0.00000	0 0 0 0 0 0	0.000	0.000	0.000
7	0.00000	120.00000	0.00000	1 1 1 1 1 1	0.000	0.000	0.000
8	60.00000	120.00000	0.00000	0 0 0 0 0 0	0.000	0.000	0.000
9	120.00000	120.00000	0.00000	1 1 1 0 0 0	0.000	0.000	0.000

Table 4.3 Continued

FRAME ELEMENT DATA

FRM	JNT-1	JNT-2	SECTN	RELEASES	SGMNT	R1	R2	FACTOR	LENGTH
1	1	2	W24X55	000000	4	0.000	0.000	1.000	60.000
2	2	3	W24X55	000000	4	0.000	0.000	1.000	60.000
3	1	4	W24X55	000000	4	0.000	0.000	1.000	60.000
4	4	5	W24X55	000000	4	0.000	0.000	1.000	60.000
5	5	6	W24X55	000000	4	0.000	0.000	1.000	60.000
6	7	8	W24X55	000000	4	0.000	0.000	1.000	60.000
7	8	9	W24X55	000000	4	0.000	0.000	1.000	60.000
9	2	5	W24X55	000000	4	0.000	0.000	1.000	60.000
10	3	6	W24X55	000000	4	0.000	0.000	1.000	60.000
11	4	7	W24X55	000000	4	0.000	0.000	1.000	60.000
12	5	8	W24X55	000000	4	0.000	0.000	1.000	60.000
13	6	9	W24X55	000000	4	0.000	0.000	1.000	60.000
14	1	5	TS4X4X1/4	000000	4	0.000	0.000	1.000	84.853
15	2	4	TS4X4X1/4	000000	4	0.000	0.000	1.000	84.853
16	5	9	TS4X4X1/4	000000	4	0.000	0.000	1.000	84.853
17	6	8	TS4X4X1/4	000000	4	0.000	0.000	1.000	84.853

JOINT FORCES Load Case LOAD1

JOINT	GLOBAL-X	GLOBAL-Y	GLOBAL-Z	GLOBAL-XX	GLOBAL-YY	GLOBAL-ZZ
2	0.000	0.000	-2.500	0.000	-120.000	0.000
8	0.000	0.000	-2.500	0.000	0.000	0.000
5	0.000	0.000	-5.000	0.000	0.000	0.000
4	0.000	0.000	0.000	180.000	0.000	0.000

FRAME SPAN DISTRIBUTED LOADS Load Case LOAD1

FRAME	TYPE	DIRECTION	DISTANCE-A	VALUE-A	DISTANCE-B	VALUE-B
10	FORCE	GLOBAL-Z	0.0000	-1.5000	1.0000	-1.5000
13	FORCE	GLOBAL-Z	0.0000	-1.5000	1.0000	-1.5000

Table 4.4 Output Tables for Illustrative Example 4.3 (Units: kips-inches)

JOINT DISPLACEMENTS

JOINT	LOAD	UX	UY	UZ	RX	RY	RZ
1	LOAD1	0.0000	0.0000	0.0000	4.417E-05	1.356E-04	0.0000
2	LOAD1	0.0000	0.0000	-6.724E-03	-2.990E-04	-4.767E-05	0.0000
3	LOAD1	0.0000	0.0000	0.0000	0.0000	0.0000	0.0000
4	LOAD1	0.0000	0.0000	2.756E-03	5.503E-05	3.974E-04	0.0000
5	LOAD1	0.0000	0.0000	-0.0225	1.897E-06	5.850E-04	0.0000
6	LOAD1	0.0000	0.0000	-0.0675	-1.106E-04	7.755E-04	0.0000
7	LOAD1	0.0000	0.0000	0.0000	0.0000	0.0000	0.0000
8	LOAD1	0.0000	0.0000	-6.560E-03	3.056E-04	1.071E-05	0.0000
9	LOAD1	0.0000	0.0000	0.0000	1.587E-03	-1.236E-04	0.0000

Table 4.4 Continued

JOINT REACTIONS

JOINT	LOAD	F1	F2	F3	M1	M2	M3
1	LOAD1	0.0000	0.0000	4.2328	0.0000	0.0000	0.0000
3	LOAD1	0.0000	0.0000	113.1427	2412.0447	187.1865	0.0000
7	LOAD1	0.0000	0.0000	1.7965	93.0201	-190.7070	0.0000
9	LOAD1	0.0000	0.0000	70.8280	0.0000	0.0000	0.0000

FRAME ELEMENT FORCES

FRAME	LOAD	LOC	P	V2	V3	T	M2	M3
1	LOAD1							
		0.00	0.00	-4.01	0.00	-7.657E-02	0.00	1.45
		15.00	0.00	-4.01	0.00	-7.657E-02	0.00	61.54
		30.00	0.00	-4.01	0.00	-7.657E-02	0.00	121.64
		45.00	0.00	-4.01	0.00	-7.657E-02	0.00	181.74
		60.00	0.00	-4.01	0.00	-7.657E-02	0.00	241.83
2	LOAD1							
		0.00	0.00	5.19	0.00	6.672E-02	0.00	124.08
		15.00	0.00	5.19	0.00	6.672E-02	0.00	46.22
		30.00	0.00	5.19	0.00	6.672E-02	0.00	-31.64
		45.00	0.00	5.19	0.00	6.672E-02	0.00	-109.50
		60.00	0.00	5.19	0.00	6.672E-02	0.00	-187.36
3	LOAD1							
		0.00	0.00	-2.158E-01	0.00	5.842E-02	0.00	7.294E-01
		15.00	0.00	-2.158E-01	0.00	5.842E-02	0.00	3.97
		30.00	0.00	-2.158E-01	0.00	5.842E-02	0.00	7.20
		45.00	0.00	-2.158E-01	0.00	5.842E-02	0.00	10.44
		60.00	0.00	-2.158E-01	0.00	5.842E-02	0.00	3.68

Bending Moment:

DISPLAY>SHOW ELEMENT FORCES/STRESSES>FRAMES
Click on Moment 3-3 and Fill Diagram. Then OK.
FILE>PRINT GRAPHICS

(Fig. 4.9 reproduces the bending moment diagram for the grid frame of Illustrative Example 4.3)

Note: Analogously, other plots such as torsional moment or shear force could readily be obtained by checking the desired plot.

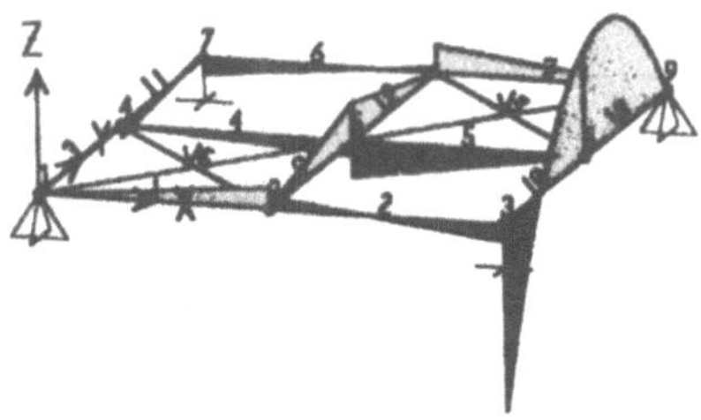

Fig. 4.9 Bending moment diagram for the grid frame of Illustrative Example 4.3.

4.6 Problems

Problems 4.1 through 4.6

For the grid frames shown in Figs. 4.1 through 4.6 determine:

(a) The joint displacements
(b) Member end-forces.
(c) Support reactions.

Neglect self-weight of the members and shear deformation.

Problem 4.1:

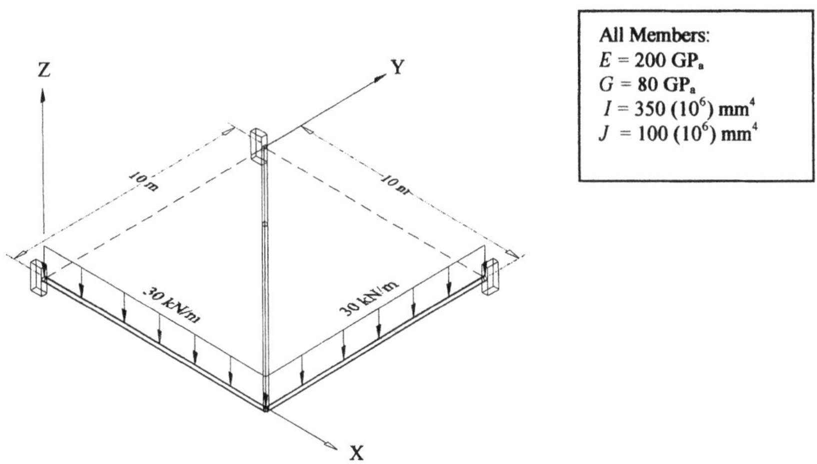

Fig. P4.1

Problem 4.2:

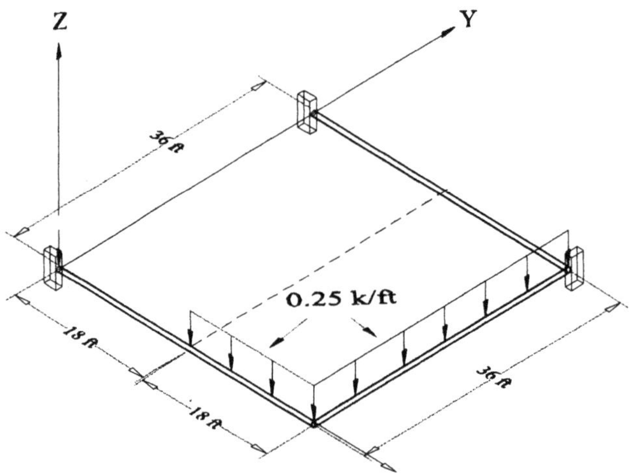

All Members:
E = 29,000 ksi
G = 11,500 ksi
I = 1350 in^4
J = 250 in^4

Fig. P4.2

Problem 4.3

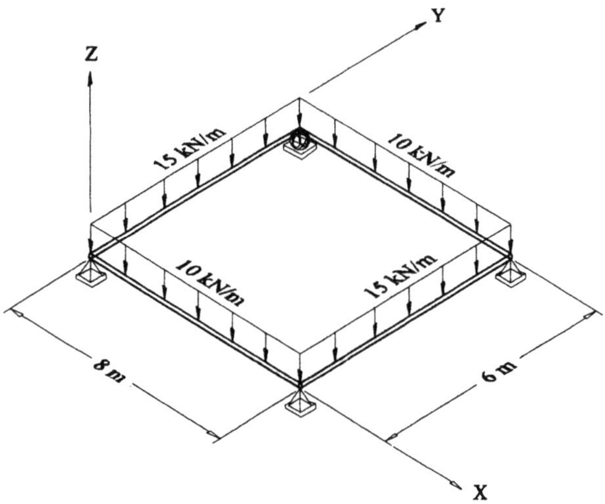

All Members:
E = 200 GP$_a$
G = 80 GP$_a$
I = 26,000 cm^4
J = 1,000 cm^4

Fig. P4.3

Problem 4.4:

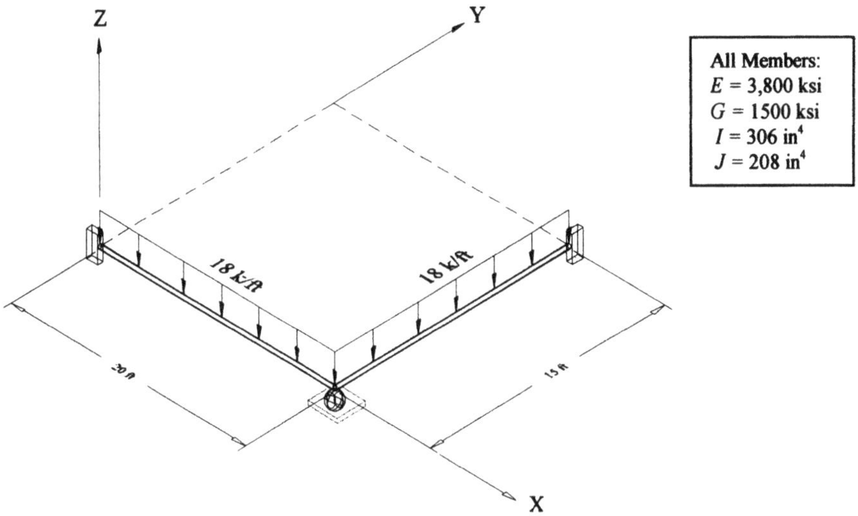

Fig. P4.4

Problem 4.5:

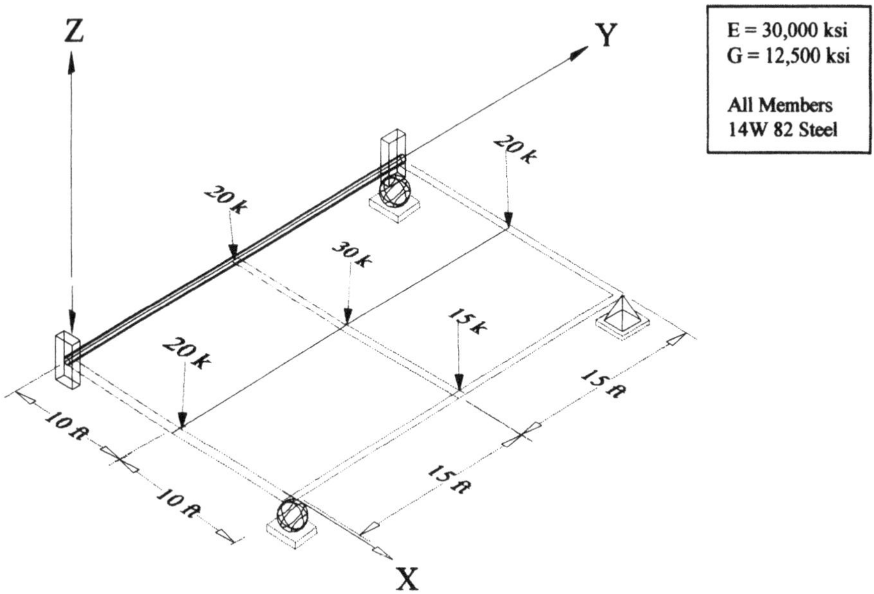

Fig. P4.5

Problem 4.6:

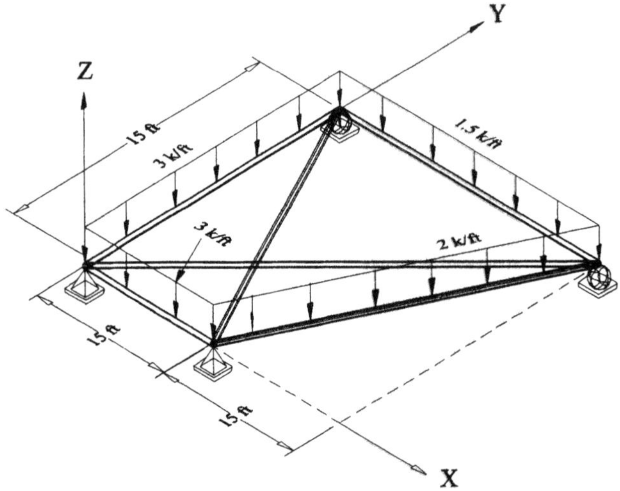

Fig. P4.6

E = 29,500 ksi
G = 12,000 ksi

All members
Except diagonals
W 24 x 55

Diagonal Members
ST 4 x 4 x ¼

5 Space Frames

5.1 Introduction

The stiffness method for analysis presented in Chapter 3 for plane frames and in Chapter 4 for grid frames can readily be extended to the analysis of space frames. The plane frame or grid frame has three nodal coordinates at each joint. The space frame, however, has a total of six possible nodal displacements at each unconstrained joint; three translations along the X, Y, and Z axes and three rotations about these axes. Consequently, a beam element of a space frame has for its two joints a total of 12 nodal coordinates; hence, the resulting element stiffness matrix will be of dimension 12 x 12.

The analysis of space frames results in a comparatively longer computer analysis, requiring more input data as well as substantially more computational time. Except for size, the analysis of space frames by the stiffness method is identical to the analysis of plane frames or grid frames.

5.2 Element Stiffness Matrix

Figure 5.1 shows the beam element of a space frame with its 12 nodal coordinates numbered consecutively. The convention adopted is to first label the three translatory displacements of the first joint, followed by the three rotational displacements of the same joint; then continuing with the three translatory displacements of the second joint and finally the three rotational displacements of the second joint. The double arrows used in Fig. 5.1 serve to indicate rotational nodal coordinates; hence, these are distinguished from translational nodal coordinates for which single arrows are used.

The stiffness matrix for a three-dimensional uniform beam segment is readily written by the superposition of the axial stiffness matrix from eq.(3.3), the torsional stiffness matrix from eq.(4.6) and the flexural stiffness matrix from eq.(1.11). The flexural stiffness matrix is used twice in forming the stiffness matrix of an element space frame to account for the flexural effects in the two principal planes of the cross-section. Proceeding to combine these matrices in an appropriate manner, we obtain in eq.(5.1) the stiffness equation for a uniform beam element of a space frame, namely

$$\begin{Bmatrix} P_1 \\ P_2 \\ P_3 \\ P_4 \\ P_5 \\ P_6 \\ P_7 \\ P_8 \\ P_9 \\ P_{10} \\ P_{11} \\ P_{12} \end{Bmatrix} = \begin{bmatrix} \frac{EA}{L} & & & & & & & & & \text{symmetric} & & \\ 0 & \frac{12EI_z}{L_3} & & & & & & & & & & \\ 0 & 0 & \frac{12EI_y}{L^3} & & & & & & & & & \\ 0 & 0 & 0 & \frac{GJ}{L} & & & & & & & & \\ 0 & 0 & \frac{-6EI_y}{L^2} & 0 & \frac{4EI_y}{L} & & & & & & & \\ 0 & \frac{6EI_z}{L^2} & 0 & 0 & 0 & \frac{4EI_z}{L} & & & & & & \\ \frac{-EA}{L} & 0 & 0 & 0 & 0 & 0 & \frac{EA}{L} & & & & & \\ 0 & \frac{-12EI_z}{L^3} & 0 & 0 & 0 & \frac{-6EI_z}{l^2} & 0 & \frac{12EI_z}{L^3} & & & & \\ 0 & 0 & \frac{-12EI_y}{L^3} & 0 & \frac{6EI_y}{L^2} & 0 & 0 & 0 & \frac{-12EI_y}{L^3} & & & \\ 0 & 0 & 0 & \frac{-GJ}{L} & 0 & 0 & 0 & 0 & 0 & \frac{GJ}{L} & & \\ 0 & 0 & \frac{-6EI_y}{L^2} & 0 & \frac{2EI_y}{L} & 0 & 0 & 0 & \frac{6EI_y}{L^2} & 0 & \frac{4EI_y}{L} & \\ 0 & \frac{6EI_z}{L^2} & 0 & 0 & 0 & \frac{2EI_z}{L} & 0 & \frac{-6EI_z}{L^2} & 0 & 0 & 0 & \frac{4EI_z}{L} \end{bmatrix} \begin{bmatrix} \delta_1 \\ \delta_2 \\ \delta_3 \\ \delta_4 \\ \delta_5 \\ \delta_6 \\ \delta_7 \\ \delta_8 \\ \chi_9 \\ \delta_{10} \\ \delta_{11} \\ \delta_{12} \end{bmatrix} \tag{5.1}$$

or in condensed notation

$$\{P\} = [K]\{\delta\} \tag{5.2}$$

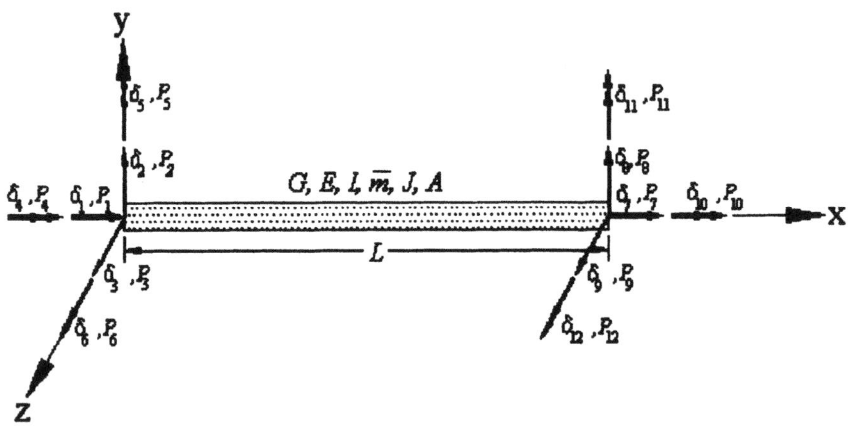

Fig. 5.1 Beam segment of a space frame showing forces and displacements at the nodal coordinates.

5.3 Transformation of Coordinates

The stiffness matrix given in eq.(5.1) refers to local coordinate axes fixed on the beam element. Inasmuch as the coefficients of these matrices (corresponding to the same nodal coordinates of the structure) should be added to obtain the system stiffness matrix, it is necessary to first transform these matrices to the same reference system, the global system of coordinates. Figure 5.2 shows these two reference systems, the (x, y, z) axes representing the local system of coordinates and the (X, Y, Z) axes representing the global system of coordinates. Also shown in this figure is a general vector **A** with its components X, Y, and Z along the global coordinate. The vector **A** may represent any force or displacement at the nodal coordinates of one of the joints of the structure. To obtain the components of vector **A** along one of the local axes x, y or z, it is necessary to add the projections along that axis of the components X, Y, Z. For example, the component x of vector **A** along the x coordinate is given by

$$x = X\cos xX + Y\cos xY + Z\cos xZ \tag{5.3a}$$

in which $\cos xY$ is the cosine of the angle between axes x and Y and corresponding definitions for other cosines. Similarly, the y and z components of **A** are

$$y = X\cos yX + Y\cos yY + Z\cos yZ \tag{5.3b}$$

$$z = X\cos zX + Y\cos zY + Z\cos zZ \tag{5.3c}$$

These equations are conveniently written in matrix notation as

$$\begin{Bmatrix} x \\ y \\ z \end{Bmatrix} = \begin{bmatrix} \cos xX & \cos xY & \cos xZ \\ \cos yX & \cos yY & \cos yZ \\ \cos zX & \cos zY & \cos zZ \end{bmatrix} \begin{Bmatrix} X \\ Y \\ Z \end{Bmatrix} \tag{5.4}$$

or in short notation

$$\{A\} = [T_1]\{\overline{A}\} \tag{5.5}$$

in which $\{\mathbf{A}\}$ and $\{\overline{\mathbf{A}}\}$ are, respectively, the components in the local and global systems of the general vector **A** and $[T_1]$ the transformation matrix given by

$$[T_1] = \begin{bmatrix} \cos xX & \cos xY & \cos xZ \\ \cos yX & \cos yY & \cos yZ \\ \cos zX & \cos zY & \cos zZ \end{bmatrix} \tag{5.6}$$

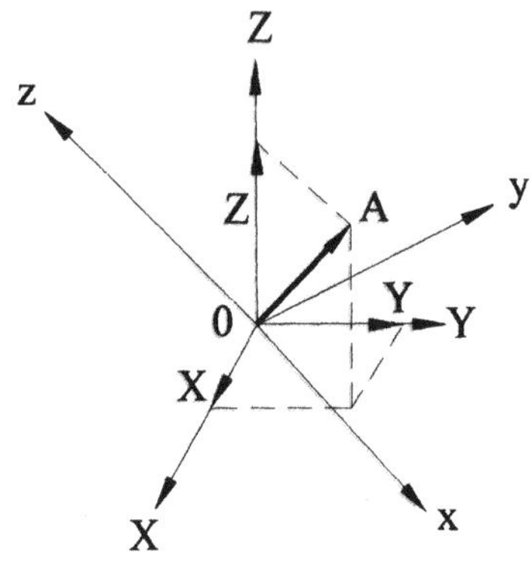

Fig. 5.2 Components of a general vector **A** in global coordinates.

The cosines required in eq.(5.6) to determine the transformation matrix $[T]_1$ are usually calculated from the global coordinates of the two points at the nodes of the beam element in addition to one more piece of information. This additional information could be either the angle of rolling or the global coordinates of a third point located in the local plane *x-y* of the beam element. The plane *x*-y is defined by the local axis *x* which is an axis along the centroid of the beam element and the axis *y* which is located in a plane perpendicular to *z* and directed along the minor principal axis *y* of the cross-sectional area of the beam element. The local axes *x* and *y*, together with a third axis *z* forming a right hand system of coordinates, constitute the so called element or local coordinate system. The positive direction of axis *x* is dictated by the order in which two nodes of the of the beam element are specified.

The angle of rolling is the angle by which the local axis *y* has been rotated from the "standard" orientation. This standard orientation exists when the local plane, formed by the local axes *x'* and *y'*, is vertical; that is, parallel to the global axis *Z* as shown in Fig. 5.3.

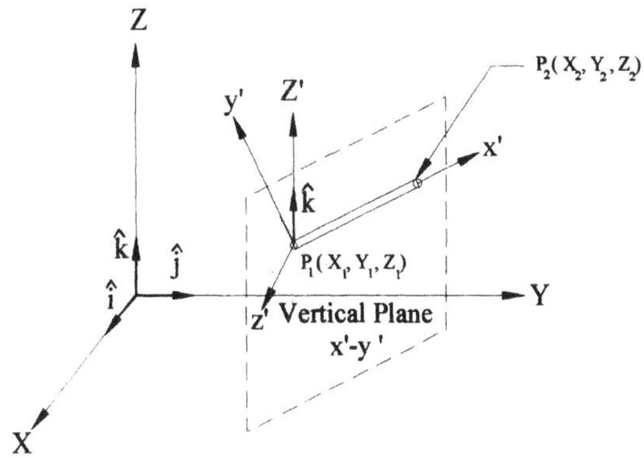

Fig. 5.3 Global system of coordinates (X, Y, Z) and local coordinate system (x', y', z') with the beam element in the "standard" orientation (plane x'-y' vertical).

The angle of roll, ϕ, is positive for the case in which the local axis y appears rotated around the local axis x counter-clockwise from the standard direction when observed from the positive end of axis x. The direction cosines for the local axis x, conveniently designated $c_1 = \cos xX$, $c_2 = \cos xY$, and $c_3 = \cos xZ$, are readily calculated from the coordinates (X_1, Y_1, Z_1) and (X_2, Y_2, Z_2) of the two nodes of the beam element as

$$c_1 = \frac{X_2 - X_1}{L} \qquad c_2 = \frac{Y_2 - Y_1}{L} \qquad c_3 = \frac{Z_2 - Z_1}{L} \tag{5.7}$$

in which

$$L = \sqrt{(X_2 - X_1)^2 = (Y_2 - Y_1)^2 + (Z_2 - Z_1)^2} \tag{5.8}$$

As already stated, the transformation matrix $[T]_1$ from the global coordinate system (X, Y, Z) to the local system (x, y, z) may be expressed in terms of the angle of roll ϕ and the direction cosines c_1, c_2, and c_3 of the local axis x' (see Problem 5.1). The final expression for the transformation matrix $[T]_1$ is

$$[T]_1 = \begin{bmatrix} c_1 & c_2 & c_3 \\ \frac{-c_1 c_3}{d}\cos\phi - \frac{c_2}{d}\sin\phi & \frac{c_1}{d}\sin\phi - \frac{c_2 c_3}{d}\cos\phi & d\cos\phi \\ \frac{c_1 c_3}{d}\sin\phi + \frac{c_2}{d}\cos\phi & \frac{c_1}{d}\cos\phi + \frac{c_2 c_3}{d}\sin\phi & -d\sin\phi \end{bmatrix} \tag{5.9}$$

in which

$$d = \sqrt{c_1^2 + c_2^2} \tag{5.10}$$

It should be noted that the transformation matrix $[T]_1$ is not defined if the local axis x is parallel to the global axis Z. In this case, eqs.(5.7) and (5.10) result in $c_1 = 0$, $c_2 = 0$, and $d = 0$. If the centroidal axis of the beam element is vertical, that is, the local axis x and the global axis Z are parallel, the angle of roll is then defined as the angle that the local axis y has been rotated about the axis x from the "standard" direction defined for a vertical beam element. The "standard" direction in this case exists when the local axis y is parallel to the global axis X as shown in Fig. 5.4.

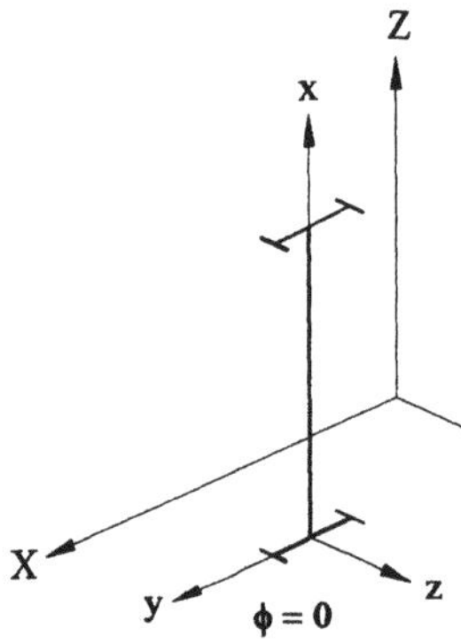

Fig. 5.4 Standard orientation of the cross-section for a vertical beam element

For a vertical beam element, the angle of roll ϕ is positive if the local axis y appears rotated from the "standard" direction counter-clockwise when observed from the positive end of the local axis x.

The final expression for the transformation matrix $[T]_1$ for the case of a vertical beam element is given by

$$[T]_1 = \begin{bmatrix} 0 & 0 & c \\ c\sin\phi & -c\cos\phi & 0 \\ \cos\phi & c\sin\phi & 0 \end{bmatrix} \tag{5.11}$$

where ϕ is the angle of roll between the global axis X and the local axis y for the case in which the centroidal axis of the beam has a vertical orientation. The coefficient $c = +1$ when the local axis x and the global axis Z have the same sense and $c = -1$ for the opposite sense of the axes. Figures 5.5(a) through 5.5(d) illustrate the measurement of the angle of roll for several cases. We have, therefore, shown that the knowledge of the coordinates of the two ends of the beam element and of the angle of roll, suffices to calculate the direction cosines of the transformation matrix $[T]_1$ in eq.(5.6) (See Problem 5.2).

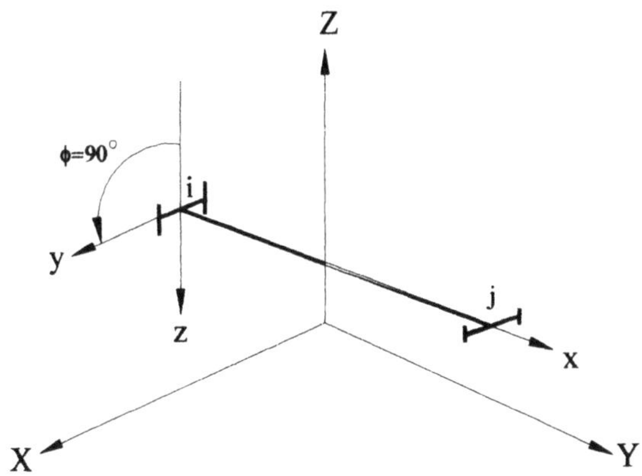

Fig. 5.5 (a) Local axis *x* is parallel to +Y Axis
Local axis *y* is rotated 90° from Z-x Plane

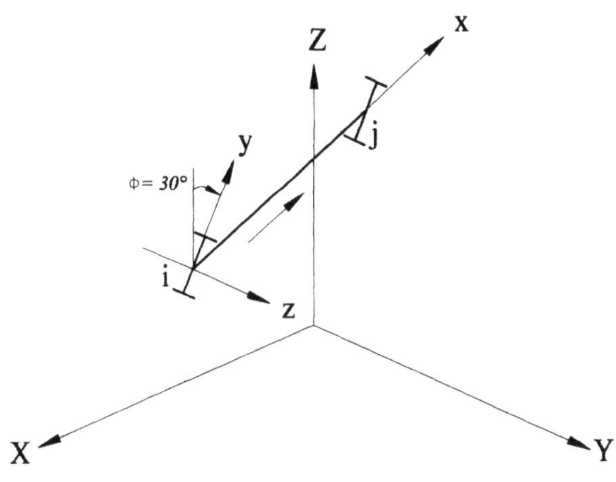

Fig. 5.5(b) Local x Axis is not parallel to X, Y or Z Axes
Local y Axis is rotated 30° from Z-x Plane

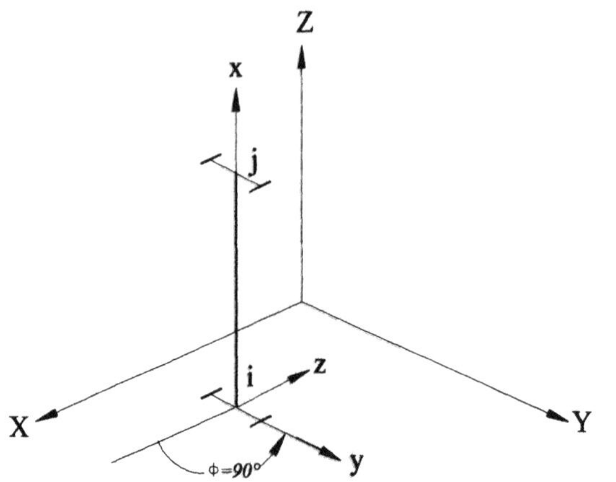

Fig. 5.5(c) Local axis *x* is parallel to +Z Axis
Local Axis *y* is rotated 90° from X-y Plane

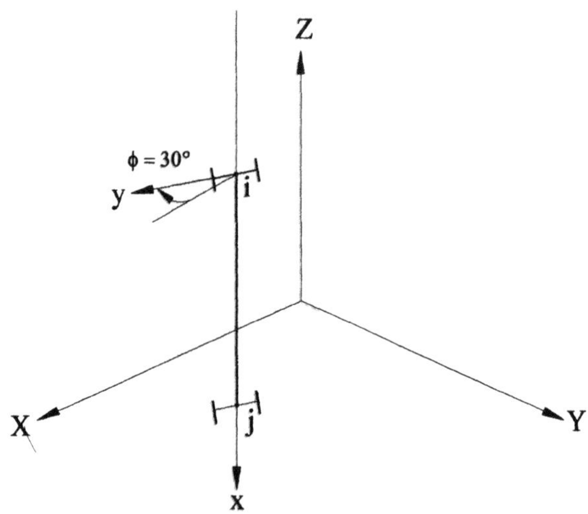

Fig. 5.5(d) Local Axis *x* is parallel to –Z Axis
Local Axis *y* is rotated 30° from X-y Plane

Alternatively, the direction cosines in eq.(5.6) may be calculated from the coordinates of two points at the two nodes of a beam element and the coordinates of a third point located in the x-y local plane, in which y is the minor principal axis of the cross-section and x is the axis along the beam element. Problem 5.3 contains the development of the relationship to calculate the transformation matrix $[T]_1$ from the coordinates of these three points.

For the beam element of a space frame, the transformations of the nodal displacement vectors involve the transformation of linear and angular displacement vectors at each node of the segment. Therefore, a beam element of a space frame requires the transformation of a total of four displacement vectors for the two nodes. The transformation of the 12 nodal displacements $\{\bar{\delta}\}$ in global coordinates to the displacements $\{\delta\}$ in the local coordinates may be written in abbreviated form as

$$\{\delta\} = [T]\{\hat{\delta}\} \tag{5.12}$$

in which

$$[T] = \begin{bmatrix} [T]_1 & & & \\ & [T]_1 & & \\ & & [T]_1 & \\ & & & [T]_1 \end{bmatrix} \tag{5.13}$$

Analogously, the transformation of the nodal forces $\{\bar{P}\}$ in global coordinates to the nodal forces $\{P\}$ in local coordinates is given by

$$\{P\} = [T]\{\bar{P}\} \tag{5.14}$$

Finally, to obtain the stiffness matrix $[\bar{K}]$ in reference to the global system of coordinates, we simply substitute $\{\delta\}$ from eq.(5.13) and $[P]$ from eq.(5.14) into eq.(5.2), to obtain

$$[T]\{\bar{P}\} = [K][T]\{\bar{\delta}\}$$

or since $[T]$ is an <u>orthogonal matrix</u>

$$\{\bar{P}\} = [T]^T[K][T]\{\bar{\delta}\} \tag{5.15}$$

From eq.(5.15), we may write

$$\{\bar{P}\} = [\bar{K}]\{\bar{\delta}\} \tag{5.16}$$

in which the element stiffness matrix $[\overline{K}]$ in reference to the global system of coordinates is given by

$$[\overline{K}] = [T]^T[K][T] \tag{5.17}$$

5.4 Reduced System Stiffness Equation

The reduced system stiffness matrix $[K]_R$ is assembled by transferring the coefficients of the element stiffness matrices to the appropriate locations in the system stiffness matrix, as presented in the preceding chapters for beams, plane frames and grid frames. Thus, the relationship between displacements $\{u\}$ and the forces $\{F\}_R$ at the free nodal coordinates is established through the reduced system stiffness matrix:

$$\{F\}_R = [K]_R\{u\}$$

in which the reduced force vector $\{F\}_R$ contains the equivalent nodal forces for forces applied on the beam elements of the structure and the forces applied directly to the free nodal coordinates.

Illustrative Example 5.1

Use SAP2000 to analyze the space frame shown in Fig. 5.6. The corresponding input data is given in Table 5.1.

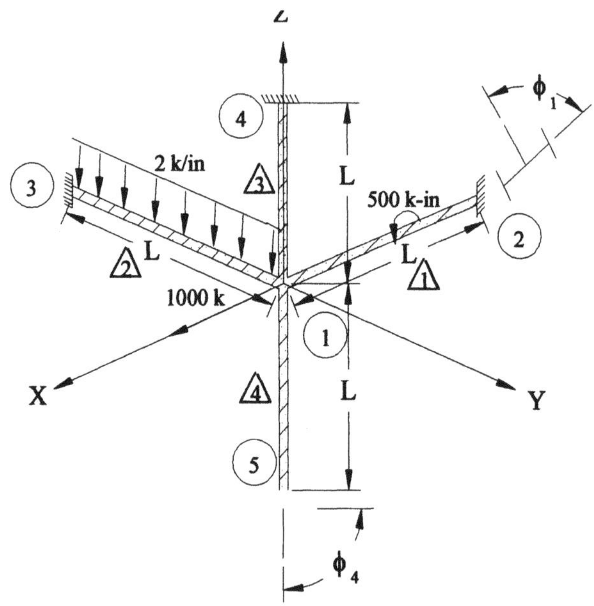

Fig. 5.6 Space frame for Illustrative Example 5.1

E = 29,500 ksi	Member 1: W 18 X 130	Angles of Roll:
$\mu = 0.3$	Member 2: W 18 X 130	$\phi_1 = 30^\circ$
L = 200 in.	Member 3: W 14 X 82	$\phi_4 = 90^\circ$
	Member 4: W 14 X 82	

Table 5.1 Edited Input Data for Illustrative Example 5.1 (Units: Kips, inches)

JOINT DATA

JOINT	GLOBAL-X	GLOBAL-Y	GLOBAL-Z	RESTRAINTS
1	0.00000	0.00000	0.00000	0 0 0 0 0 0
2	-200.00000	0.00000	0.00000	1 1 1 1 1 1
3	0.00000	-200.00000	0.00000	1 1 1 1 1 1
4	0.00000	0.00000	200.00000	1 1 1 1 1 1
5	0.00000	0.00000	-200.00000	1 1 1 1 1 1

FRAME ELEMENT DATA

FRAME	JNT-1	JNT-2	SECTION	ANGLE	RELEASES	SEGMENTS	LENGTH
1	1	2	W18X130	30.000	000000	4	200.000
2	1	3	W18X130	0.000	000000	4	200.000
3	1	4	W14X82	0.000	000000	2	200.000
4	1	5	W14X82	90.000	000000	2	200.000

JOINT FORCES Load Case LOAD1

JOINT	GLOBAL-X	GLOBAL-Y	GLOBAL-Z	GLOBAL-XX	GLOBAL-YY	GLOBAL-ZZ
1	1000.000	0.000	0.000	0.000	0.000	0.000

FRAME SPAN DISTRIBUTED LOADS

FRAME	TYPE	DIRECTION	ISTANCE-A	VALUE-A	DISTANCE-B	VALUE-B
2	FORCE	GLOBAL-Z	0.0000	-2.0000	1.0000	-2.0000

FRAME SPAN POINT LOADS Load Case LOAD1

FRAME	TYPE	DIRECTION	DISTANCE	VALUE
1	FORCE	GLOBAL-Y	0.5000	500.0000

Solution:

Begin: Open SAP2000

Enter: "OK" to disable the Tip of the Day on the screen.

Hint: Maximize both screens for a full view of both windows.

Select: In the lower right-hand corner of the screen, use the drop-down menu to select "kip-in".

Select: From the Main Menu:
FILE>NEW MODEL

Number of Grid Spaces
x direction = 4
y direction = 4
z direction = 4
Grid Spaces
x direction = 200
y direction = 200
z direction = 200
Then OK.

Select: Default screen to 3-D screen view
Then enter from the Main Menu:
VIEW>SET 3D VIEW
Rotate plane view to approximately 30^0 to have a frontal view of X axis.

Enter: VIEW>2-D VIEW
Set Plane Y-Z to $X = 0$
Plane X-Z to $Y = 0$
Plane X-Y to $Z = 0$
Then OK.

Select: Default to screen 2-D view
Set Plane Y-Z to $X = 0$
Plane X-Z to $Y = 0$
Plane X-Y to $Z = 0$ Then OK.

Add Grid: From the Main Menu enter:
DRAW>EDIT>GRID
Click on Z
Enter –200, Add grid. Then OK.

Select: Default to screen X-Z plane at $Y = 0$
From the Main Menu enter:
DRAW>DRAW FRAME ELEMENT
Start with the cursor at the origin and drag to the negative of axis X until next grid line. Then double click to stop drawing elements.

Draw: Default to view *Y-Z* at $X = 0$

Draw element from the origin along axis *Y* in the negative direction until the next grid line, then double click. Next, start at the origin in the positive direction of axis *Z* until the next grid line and double click. Finally, start at the origin and drag the cursor in the negative direction of axis *Z* until the next grid line and double click.

Labels: Default to 3-D view
From the Main menu enter:
VIEW>SET ELEMENTS
Click on Joint labels and Frame labels. Then OK

Restraints: Select joints ②, ③, ④ and ⑤ with a click of the mouse.
Enter from the Main Menu:
ASSIGN>JOINTS>RESTRAINTS
Select: "Restraints all". Then OK.

Materials: From the Main Menu enter:
DEFINE>MATERIALS
Select STEEL
Modify/Show Material
Set $E = 29{,}500$ and Poisson's Ratio = 0.3. Then OK.

Sections: From the Main Menu enter:
DEFINE>FRAME SECTIOINS
IMPORT I/WIDE FLANGE. Then OK.

Select Sections.prop., then OPEN.
Select W18 X 130
Holding down the Ctrl key
Select W14 X 82. Then OK.

Click on W18 X 130 and on Modification Factors
Set Shear area in *Z* direction = 0
Shear area in *Y* direction 0. Then OK's.

Click on W14 X 82 and on Modification Factors
Set Shear area *Z* direction = 0
Shear area *Y* direction = 0. Then OK's.

Select: Click on elements 1 and 2,
Then enter:
ASSIGN>FRAME SECTIONS
Select W18 X 130. Then OK.

Click on elements 3 and 4
ASSIGN>FRAME SECTIONS
Select W14 X 82. Then OK.

Loads: DEFINE>STATIC LOAD CASES
Change DEAD to LIVE LOAD
Change Self-weight multiplier to 0
Click on Change Load. Then OK.

Select: Select joint ① and enter
ASSIGN>JOINT STATIC LOADS>FORCES
Enter Force Global $X = 1000$. Then OK.

Select: Element 2 and enter
ASSIGN>FRAME STATIC LOADS>POINT AND UNIFORM
Check: Force
Select Global Z
Enter Uniform load = -2.0

Select: Select element 1 and enter
ASSIGN>FRAME STATIC LOADS>POINT AND UNIFORM

Check: Moments
Select: Direction Global Y
Distance 0.5 Load = 500

Warning: Make certain that Uniform Load value is set to zero.

Select: Element 1 and enter:
ASSIGN>FRAME LOCAL AXES, and enter:
Angle in degree = 30. Then OK.

Select: Select element 4
ASSIGN>FRAME>LOCAL AXES, AND ENTER
Angle in degree = 90

Execution: From the Main Menu enter:
ANALYZE>SET OPTIONS
Select all six degrees of freedom. Then OK

Enter ANALYZE>RUN

Enter: File name: "Example 5.1".
Then SAVE. When the analysis is completed, click OK.

Use icon "Rubber Band Zoom" and icon "Pan" to enlarge and center the deformed frame shown on the screen.

Results: From the Main Menu enter:
FILE>PRINT INPUT TABLES. Then OK.
(Edited Input Tables are shown as Table 5.1)

From the Main Menu enter:
FILE>PRINT OUTPUT TABLES, then OK
(Edited Output tables are shown in Table 5.2)

Table 5.2 Edited output Tables for Illustrative Example 5.1 (Units: kips and inches)

JOINT DISPLACEMENTS

JOINT	LOAD	UX	UY	UZ	RX	RY	RZ
1	LOAD1	0.1865	0.0316	-0.0280	3.095E-03	-9.256E-03	-0.0275
2	LOAD1	0.0000	0.0000	0.0000	0.0000	0.0000	0.0000
3	LOAD1	0.0000	0.0000	0.0000	0.0000	0.0000	0.0000
4	LOAD1	0.0000	0.0000	0.0000	0.0000	0.0000	0.0000
5	LOAD1	0.0000	0.0000	0.0000	0.0000	0.0000	0.0000

JOINT REACTIONS

JOINT	LOAD	F1	F2	F3	M1	M2	M3
2	LOAD1	-1050.6785	-311.5120	-35.5336	-2.5811	2334.0000	-16600.2305
3	LOAD1	31.5414	-178.1268	236.7357	9217.1768	7.7191	-2026.2944
4	LOAD1	26.3624	1.8055	99.3989	112.9852	-1432.0647	7.9267
5	LOAD1	-7.2252	-12.1667	99.3989	814.0115	-520.4639	7.9267

Table 5.2 Continued

FRAME ELEMENT FORCES

FRAME LOAD	LOC	P	V2	V3	T	M2	M3
1 LOAD1							
	0.00	1050.68	63.47	181.00	2.58	6108.39	-1984.38
	50.00	1050.68	63.47	181.00	2.58	-2941.72	-5157.93
	100.00	1050.68	-186.53	-252.01	2.58	-11991.83	-8331.48
	150.00	1050.68	-186.53	-252.01	2.58	608.70	994.97
	200.00	1050.68	-186.53	-252.01	2.58	13209.20	10321.42
2 LOAD1							
	0.00	178.13	-163.26	-31.54	-7.72	-4281.98	-1870.03
	50.00	178.13	-63.26	-31.54	-7.72	-2704.91	3793.18
	100.00	178.13	36.74	-31.54	-7.72	-1127.84	4456.40
	150.00	178.13	136.74	-31.54	-7.72	449.23	119.61
	200.00	178.13	236.74	-31.54	-7.72	2026.20	-9217.18
3 LOAD1							
	0.00	99.40	26.36	1.81	7.93	248.12	3840.41
	100.00	99.40	26.36	1.81	7.93	67.57	1204.17
	200.00	99.40	26.36	1.81	7.93	-112.99	-1432.06
4 LOAD1							
	0.00	-99.40	12.17	7.23	-7.93	924.59	1619.33
	100.00	-99.40	12.17	7.23	-7.93	202.06	402.66
	200.00	-99.40	12.17	7.23	-7.93	-520.46	-814.01

Plot Deformed Frame: From the Main Menu enter:
DISPLAY>SHOW DEFORMED SHAPE, then enter
FILE>PRINT GRAPHICS
(The deformed shape shown in the screen is reproduced in Fig. 5.7)

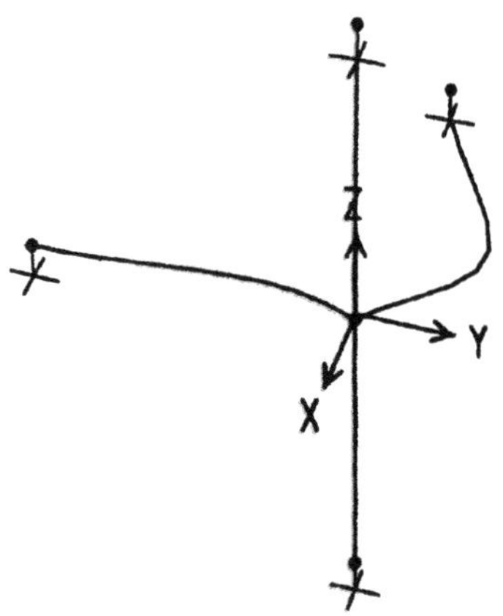

Fig. 5.7 Deformed frame of Illustrative Example 5.1

Plot the Bending Moment: From the Main Menu enter:

DISPLAY>SHOW FORCES/STRESSES>FRAMES
Select Moment 3-3 and OK. Then enter:
FILE>PRINT GRAPHICS
(The bending moment diagram shown on the screen is depicted in Fig. 5.8)

Fig. 5.8 Bending moment diagram for the space frame of Illustrative Example 5.1.

Illustrative Example 5.2

For the space frame shown in Fig. 5.9 determine:

(a) Joint displacements.
(b) End-force members.
(c) Support reactions.

All the beams of this space frame support uniformly distributed vertical loads of magnitude $w = -0.25$ k/in. In addition, the frame is subjected to horizontal loads of magnitude 50k parallel to the Y axis applied to joints ⑤, ⑥, ⑪ and ⑫ as shown in Fig. 5.9.

Member Orientations
Girders: Local axis y parallel to global axis Z
Columns: Local axis y parallel to global axis X
Material: $E = 29{,}500$ ksi
$V = 0.3$
Sections: All sections are
W 24 X 146
Column Supports: Fixed (in all directions)

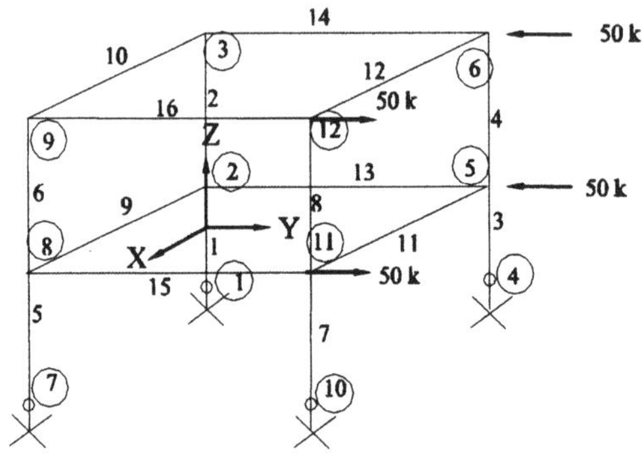

Fig. 5.9 Space frame for Illustrative Example 5.2

Solution:

The implementation of the pertinent data for this Illustrative example and the execution of SAP2000 provides the following results:

Edited tables for input data and output results are reproduced as Tables 5.3 and 5.4, respectively. The deformed shape and the bending moment diagrams for Illustrative Example 5.2 are reproduced in Figs. 5.10 and 5.11, respectively.

Table 5.3 Edited input tables for Illustrative Example 5.2 (Units: kips-inches)

JOINT DATA

JOINT	GLOBAL-X	GLOBAL-Y	GLOBAL-Z	RESTRAINTS
1	0.00000	0.00000	0.00000	1 1 1 1 1 1
2	0.00000	0.00000	144.00000	0 0 0 0 0 0
3	0.00000	0.00000	288.00000	0 0 0 0 0 0
4	0.00000	288.00000	0.00000	1 1 1 1 1 1
5	0.00000	288.00000	144.00000	0 0 0 0 0 0
6	0.00000	288.00000	288.00000	0 0 0 0 0 0
7	288.00000	0.00000	0.00000	1 1 1 1 1 1
8	288.00000	0.00000	144.00000	0 0 0 0 0 0
9	288.00000	0.00000	288.00000	0 0 0 0 0 0
10	288.00000	288.00000	0.00000	1 1 1 1 1 1
11	288.00000	288.00000	144.00000	0 0 0 0 0 0
12	288.00000	288.00000	288.00000	0 0 0 0 0 0

Table 5.3 Continued

FRAME ELEMENT DATA

FRAME	JNT-1	JNT-2	SECTION	ANGLE	RELEASES	SEGMENTS	LENGTH
1	1	2	W24X146	0.000	000000	2	144.000
2	2	3	W24X146	0.000	000000	2	144.000
3	4	5	W24X146	0.000	000000	2	144.000
4	5	6	W24X146	0.000	000000	2	144.000
5	7	8	W24X146	0.000	000000	2	144.000
6	8	9	W24X146	0.000	000000	2	144.000
7	10	11	W24X146	0.000	000000	2	144.000
8	11	12	W24X146	0.000	000000	2	144.000
9	2	8	W24X146	0.000	000000	4	288.000
10	3	9	W24X146	0.000	000000	4	288.000
11	5	11	W24X146	0.000	000000	4	288.000
12	6	12	W24X146	0.000	000000	4	288.000
13	2	5	W24X146	0.000	000000	4	288.000
14	3	6	W24X146	0.000	000000	4	288.000
15	8	11	W24X146	0.000	000000	4	288.000
16	9	12	W24X146	0.000	000000	4	288.000

JOINT FORCES Load Case LOAD1

JOINT	GLOBAL-X	GLOBAL-Y	GLOBAL-Z	GLOBAL-XX	GLOBAL-YY	GLOBAL-ZZ
11	0.000	50.000	0.000	0.000	0.000	0.000
12	0.000	50.000	0.000	0.000	0.000	0.000
5	0.000	-50.000	0.000	0.000	0.000	0.000
6	0.000	-50.000	0.000	0.000	0.000	0.000

FRAME SPAN DISTRIBUTED LOADS Load Case LOAD1

FRAME	TYPE	DIRECTION	DIST-A	VALUE-A	DIST-B	VALUE-B
9	FORCE	GLOBAL-Z	0.0000	-0.2500	1.0000	-0.2500
10	FORCE	GLOBAL-Z	0.0000	-0.2500	1.0000	-0.2500
11	FORCE	GLOBAL-Z	0.0000	-0.2500	1.0000	-0.2500
12	FORCE	GLOBAL-Z	0.0000	-0.2500	1.0000	-0.2500
13	FORCE	GLOBAL-Z	0.0000	-0.2500	1.0000	-0.2500
14	FORCE	GLOBAL-Z	0.0000	-0.2500	1.0000	-0.2500
15	FORCE	GLOBAL-Z	0.0000	-0.2500	1.0000	-0.2500
16	FORCE	GLOBAL-Z	0.0000	-0.2500	1.0000	-0.2500

Table 5.4 Edited output tables for Illustrative Example 5.2 (Units: Kips-inches)

JOINT DISPLACEMENTS

JOINT	LOAD	UX	UY	UZ	RX	RY	RZ
1	LOAD1	0.0000	0.0000	0.0000	0.0000	0.0000	0.0000
2	LOAD1	0.0591	-0.9291	-0.0190	5.011E-04	5.723E-04	3.426E-03
3	LOAD1	0.1266	-1.4566	-0.0277	-7.196E-04	6.596E-04	5.484E-03
4	LOAD1	0.0000	0.0000	0.0000	0.0000	0.0000	0.0000
5	LOAD1	-0.0614	-0.9339	-0.0137	2.473E-03	-1.735E-04	3.455E-03
6	LOAD1	-0.1230	-1.4639	-0.0213	1.803E-03	1.731E-04	5.514E-03
7	LOAD1	0.0000	0.0000	0.0000	0.0000	0.0000	0.0000
8	LOAD1	0.0614	0.9282	-0.0137	-2.461E-03	1.735E-04	3.426E-03

Table 5.4 Continued

JOINT	LOAD	UX	UY	UZ	RX	RY	RZ
9	LOAD1	0.1230	1.4582	-0.0213	-1.804E-03	-1.731E-04	5.484E-03
10	LOAD1	0.0000	0.0000	0.0000	0.0000	0.0000	0.0000
11	LOAD1	-0.0591	0.9348	-0.0190	-5.130E-04	-5.723E-04	3.455E-03
12	LOAD1	-0.1266	1.4622	-0.0277	7.210E-04	-6.596E-04	5.514E-03

JOINT REACTIONS

JOINT	LOAD	F1	F2	F3	M1	M2	M3
1	LOAD1	-6.8109	40.3840	167.5066	-2947.7888	-1027.3541	-3.6172
4	LOAD1	18.5768	34.1778	120.4934	-2658.8940	1500.2927	-3.6481
7	LOAD1	-18.5768	-33.9613	120.4934	2642.3523	-1500.2927	-3.6172
10	LOAD1	6.8109	-40.6005	167.5066	2964.3303	1027.3541	-3.6481

FRAME ELEMENT FORCES

FRAME	LOAD	LOC	P	V2	V3	T	M2	M3
1	LOAD1							
		0.00	-167.51	6.81	-40.38	3.62	-2947.79	1027.35
		72.00	-167.51	6.81	-40.38	3.62	-40.14	536.97
		144.00	-167.51	6.81	-40.38	3.62	2867.51	46.59
2	LOAD1							
		0.00	-76.49	-8.06	-24.56	2.17	-1670.59	-498.77
		72.00	-76.49	-8.06	-24.56	2.17	97.78	81.91
		144.00	-76.49	-8.06	-24.56	2.17	1866.16	662.58
3	LOAD1							
		0.00	-120.49	-18.58	-34.18	3.65	-2658.89	-1500.29
		72.00	-120.49	-18.58	-34.18	3.65	-198.09	-162.76
		144.00	-120.49	-18.58	-34.18	3.65	2262.71	1174.77
4	LOAD1							
		0.00	-67.51	-23.43	-10.05	2.17	-669.57	-1361.67
		72.00	-67.51	-23.43	-10.05	2.17	53.68	325.21
		144.00	-67.51	-23.43	-10.05	2.17	776.92	2012.09
5	LOAD1							
		0.00	-120.49	18.58	33.96	3.62	2642.35	1500.29
		72.00	-120.49	18.58	33.96	3.62	197.14	162.76
		144.00	-120.49	18.58	33.96	3.62	-2248.08	-1174.77
6	LOAD1							
		0.00	-67.51	23.43	10.08	2.17	673.00	1361.67
		72.00	-67.51	23.43	10.08	2.17	-52.61	-325.21
		144.00	-67.51	23.43	10.08	2.17	-778.22	-2012.09
7	LOAD1							
		0.00	-167.51	-6.81	40.60	3.65	2964.33	-1027.35
		72.00	-167.51	-6.81	40.60	3.65	41.09	-536.97
		144.00	-167.51	-6.81	40.60	3.65	-2882.14	-46.59
8	LOAD1							
		0.00	-76.49	8.06	24.53	2.17	1667.16	498.77
		72.00	-76.49	8.06	24.53	2.17	-98.85	-81.91
		144.00	-76.49	8.06	24.53	2.17	-1864.86	-662.58
9	LOAD1							
		0.00	9.86	-29.09	-5.01	-1.56	-721.98	-545.75
		72.00	9.86	-11.09	-5.01	-1.56	-360.99	900.68
		144.00	9.86	6.91	-5.01	-1.56	0.00	1051.10
		216.00	9.86	24.91	-5.01	-1.56	360.99	-94.47
		288.00	9.86	42.91	-5.01	-1.56	721.98	-2536.04

Table 5.4 Continued

FRAME	LOAD	LOC	P	V2	V3	T	M2	M3
10	LOAD1							
		0.00	-15.75	-31.32	-7.69	-5.726E-01	-1107.20	-662.84
		72.00	-15.75	-13.32	-7.69	-5.726E-01	-553.60	943.91
		144.00	-15.75	4.68	-7.69	-5.726E-01	0.00	1254.66
		216.00	-15.75	22.68	-7.69	-5.726E-01	553.60	269.41
		288.00	-15.75	40.68	-7.69	-5.726E-01	1107.20	-2011.84
11	LOAD1							
		0.00	9.86	-42.91	-5.03	-1.58	-724.35	-2536.04
		72.00	9.86	-24.91	-5.03	-1.58	-362.18	-94.47
		144.00	9.86	-6.91	-5.03	-1.58	0.00	1051.10
		216.00	9.86	11.09	-5.03	-1.58	362.18	900.68
		288.00	9.86	29.09	-5.03	-1.58	724.35	-545.75
12	LOAD1							
		0.00	-15.75	-40.68	-7.71	-5.711E-01	-1109.56	-2011.84
		72.00	-15.75	-22.68	-7.71	-5.711E-01	-554.78	269.41
		144.00	-15.75	-4.68	-7.71	-5.711E-01	0.00	1254.66
		216.00	-15.75	13.32	-7.71	-5.711E-01	554.78	943.91
		288.00	-15.75	31.32	-7.71	-5.711E-01	1109.56	-662.84
13	LOAD1							
		0.00	-20.84	-61.93	5.01	-3.937E-01	20.54	-4536.54
		72.00	-20.84	-43.93	5.01	-3.937E-01	359.68	-725.73
		144.00	-20.84	-25.93	5.01	-3.937E-01	-1.17	1789.08
		216.00	-20.84	-7.93	5.01	-3.937E-01	-362.03	3007.89
		288.00	-20.84	10.07	5.01	-3.937E-01	-722.88	2930.71
14	LOAD1							
		0.00	-32.25	-45.17	7.68	-2.568E-01	1105.02	-1865.58
		72.00	-32.25	-27.17	7.68	-2.568E-01	551.92	738.90
		144.00	-32.25	-9.17	7.68	-2.568E-01	-1.18	2047.38
		216.00	-32.25	8.83	7.68	-2.568E-01	-554.29	2059.87
		288.00	-32.25	26.83	7.68	-2.568E-01	1107.39	776.35
15	LOAD1							
		0.00	28.90	-10.07	5.01	-3.937E-01	720.54	2919.52
		72.00	28.90	7.93	5.01	-3.937E-01	359.68	2996.71
		144.00	28.90	25.93	5.01	-3.937E-01	-1.17	1777.90
		216.00	28.90	43.93	5.01	-3.937E-01	-362.03	-736.92
		288.00	28.90	61.93	5.01	-3.937E-01	-722.88	-4547.73
16	LOAD1							
		0.00	17.77	-26.83	7.68	-2.568E-01	1105.02	777.65
		72.00	17.77	-8.83	7.68	-2.568E-01	551.92	2061.17
		144.00	17.77	9.17	7.68	-2.568E-01	-1.18	2048.68
		216.00	17.77	27.17	7.68	-2.568E-01	-554.29	740.20
		288.00	17.77	45.17	7.68	-2.568E-01	-1107.39	1864.28

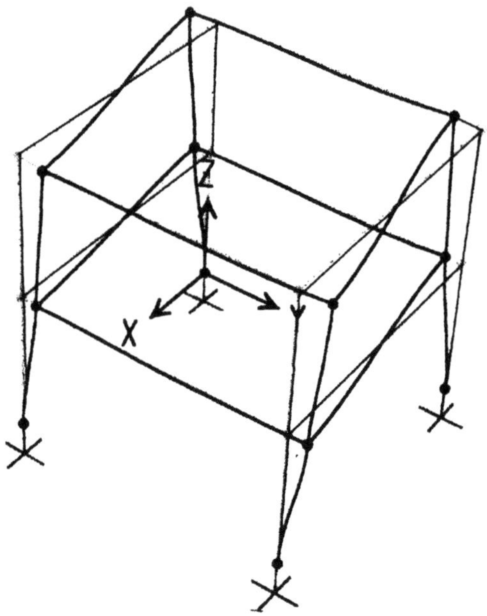

Fig . 5.10 Deformed shape for the space frame of Illustrative Example 5.2.

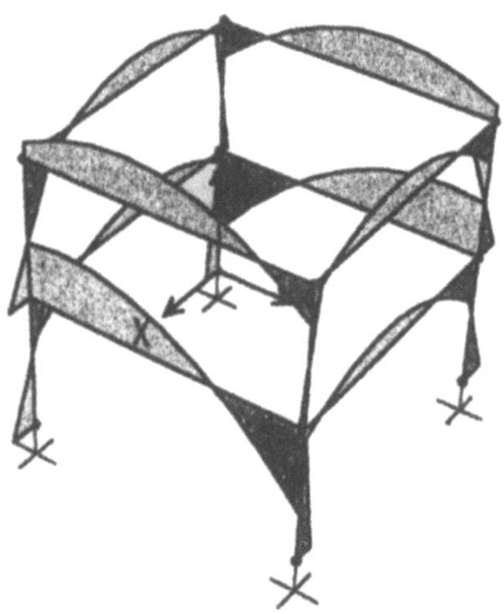

Fig. 5.11 Bending moment diagram for the space frame in Illustrative Example 5.2.

5.5 Analytical Problems

Problem 5.1

Show that the transformation matrix $[T]_1$ in eq.(5.6) between global axes (X,Y,Z) and local axes (x,y,z) may be expressed as given by eq.(5.11) in terms of the coordinates of the two points at the two nodes of the beam element and of the angle of roll (ϕ).

Solution:

The direction cosines of the local axis x along the beam element are given by eq.(5.7) as

$$c_1 = \frac{X_2 - X_1}{L} \qquad c_2 = \frac{Y_2 - Y_1}{L} \qquad c_3 = \frac{Z_2 - Z_1}{L} \tag{a}$$

in which (X_1, Y_1, Z_1) and (X_2, Y_2, Z_2) are the coordinates of the two points at the ends of the beam element and

$$L = \sqrt{(X_2 - X_1)^2 + (Y_2 - Y_1)^2 + (Z_2 - Z_1)^2} \tag{b}$$

we first consider the special case in which the local axes y' and x' forms a vertical plane parallel to the global axis Z (Fig. P5.1).

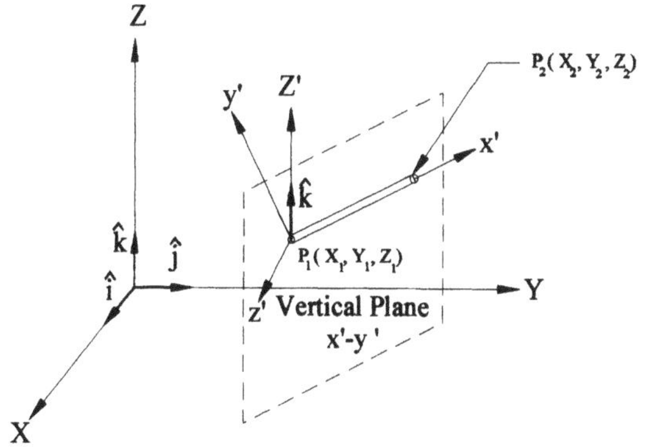

Fig. P5.1 Global system of coordinates (X,Y,Z) and local system of coordinates (x',y',z') with $x'y'$ plane vertical.

For the particular case in which the plane defined by the local axes $x'y'$ is vertical, the direction cosines for the local axes, y' and z', can also be expressed in terms of the coordinates of the two points at the two nodes of the beam element. Since in this particular case the local axis z' is perpendicular to the vertical plane defined by the local axis x', and an axis Z' parallel to the global axis Z as shown in Fig. P5.1, a vector along the local axis z may be found as

$$\mathbf{z} = \hat{x} \times \hat{k} = \begin{bmatrix} \hat{i} & \hat{j} & \hat{k} \\ c_1 & c_2 & c_3 \\ 0 & 0 & 1 \end{bmatrix} = c_2\hat{i} - c_1\hat{j} \tag{c}$$

in which

$$\hat{x} = c_1\hat{i} + c_2\hat{j} + c_3\hat{k} \tag{d}$$

is a unit vector along the local axis x, and $\hat{i}, \hat{j}, \hat{k}$, are respectively the unit vectors along the axes X, Y and Z of the global system.

A unit vector $\hat{z}$ along the local axis z' is then calculated from eq.(c) as

$$\hat{z} = \frac{\mathbf{z}}{d} = \frac{c_2\hat{i}}{d} - \frac{c_1\hat{j}}{d} \tag{e}$$

where

$$d = \sqrt{c_1^2 + c_2^2} \tag{f}$$

Finally, a unit vector $\hat{y}$ along the local axis y' is given by

$$\hat{y} = \hat{z} \times \hat{x} = \begin{bmatrix} i & j & k \\ \frac{c_2}{d} & -\frac{c_1}{d} & 0 \\ c_1 & c_2 & c_3 \end{bmatrix} \tag{g}$$

or

$$\hat{y} = -\frac{c_1c_3}{d}\hat{i} - \frac{c_2c_3}{d}\hat{j} + d\hat{k} \tag{h}$$

Therefore, using the unit vectors ($\hat{x}, \hat{y}, \hat{z}$) from eqs. (d), (e), and (h), in this case, the transformation in eq.(5.9) is given by

$$\begin{Bmatrix} x' \\ y' \\ z' \end{Bmatrix} = \begin{bmatrix} c_1 & c_2 & c_3 \\ \dfrac{-c_1c_3}{d} & \dfrac{-c_2c_3}{d} & d \\ \dfrac{c_2}{d} & -\dfrac{c_1}{d} & 0 \end{bmatrix} \begin{Bmatrix} X \\ Y \\ Z \end{Bmatrix} \tag{i}$$

where d is given by eq.(f). Equation (i) is the transformation matrix between local coordinates (x', y', z') and global coordinates (XYZ) for the particular case in which the local plane $x'-y'$ is vertical. If this plane is not vertical, it is necessary to rotate the local plane $x'-y'$ in an angle ϕ (angle of roll) until the axis y' reaches the actual direction of the local axis y. Denoting by (x', y', z') the auxiliary coordinate system in which the plane $x'-y'$ is vertical and (x, y, z) the element local coordinate system. The transformation matrix, due to rotation between these two local systems, is given by

$$\begin{Bmatrix} x \\ y \\ z \end{Bmatrix} = \begin{Bmatrix} 1 & 0 & 0 \\ 0 & \cos\phi & \sin\phi \\ 0 & -\sin\phi & \cos\phi \end{Bmatrix} \begin{Bmatrix} x' \\ y' \\ z' \end{Bmatrix} \tag{j}$$

in which ϕ is the angle of roll from the axis y' to the axis y. This angle is positive for a counter-clockwise rotation around axis x observing the rotation from the second end joint of the beam element to the first end joint. The transformation of the global system (X, Y, Z) to the local system (x, y, z) is then obtained by substituting eq.(i) into eq.(j), namely

$$\begin{Bmatrix} x \\ y \\ z \end{Bmatrix} = \begin{Bmatrix} 1 & 0 & 0 \\ 0 & \cos\phi & \sin\phi \\ 0 & -\sin\phi & \cos\phi \end{Bmatrix} \begin{bmatrix} c_1 & c_2 & c_3 \\ \dfrac{-c_1c_3}{d} & \dfrac{-c_2c_3}{d} & d \\ \dfrac{c_2}{d} & -\dfrac{c_1}{d} & 0 \end{bmatrix} \begin{bmatrix} X \\ Y \\ Z \end{bmatrix} \tag{k}$$

The final expression for the transformation matrix $[T]_1$ is then given by the product of the two matrices in eq.(k):

$$[T]_1 = \begin{bmatrix} c_1 & c_2 & c_3 \\ -\dfrac{c_1c_3}{d}\cos\phi - \dfrac{c_2}{d}\sin\phi & -\dfrac{c_1}{d}\sin\phi - \dfrac{c_2c_3}{d}\cos\phi & d\cos\phi \\ \dfrac{c_1c_3}{d}\sin\phi + \dfrac{c_2}{d}\cos\phi & -\dfrac{c_1}{d}\cos\phi + \dfrac{c_2c_3}{d}\sin\phi & -d\sin\phi \end{bmatrix} \tag{l}$$

in which

$$d = \sqrt{c_1^2 + c_2^2}$$

It should be noted that the transformation matrix $[T]_1$ is not defined if the local axis x is parallel to the global axis Z. In this case, eq.(a) results in $c_1 = 0$, $c_2 = 0$, and $d = 0$. The transformation matrix between the global system of coordinates (X, Y, Z) and the local system (x, y, z) is then obtained as a special case as it is developed in Problem 5.2.

Problem 5.2

Develop the transformation matrix in eq.(5.11) between the global coordinate system (X, Y, Z) and the local (x, y, z) for the special case in which the local axis is parallel to the global axis Z.

Solution:

If the centroidal axis of the beam element is vertical, that is, the local axis x and the global axis Z are parallel, the angle of roll is then defined as the angle that the local axis y has been rotated about the axis x from the "standard" direction defined for a vertical beam element. The "standard" direction in this case exists when the local axis y is parallel to the global axis X as shown in Fig. P5.2.

Let us designate by (x', y', z') an auxiliary system of coordinates in which the local axis y' is parallel to the global axis X as shown in Fig. P5.2. The transformation of coordinates between this auxiliary system (x', y', z') and the global system (X, Y, Z) obtained from Fig. P5.2 is given by

$$\begin{Bmatrix} x' \\ y' \\ z' \end{Bmatrix} = \begin{bmatrix} 0 & 0 & \lambda \\ 1 & 0 & 0 \\ 0 & \lambda & 0 \end{bmatrix} \begin{Bmatrix} X \\ Y \\ Z \end{Bmatrix} \tag{a}$$

in which $\lambda = 1$ when the local x has the sense of the global axis Z; otherwise, $\lambda = -1$.

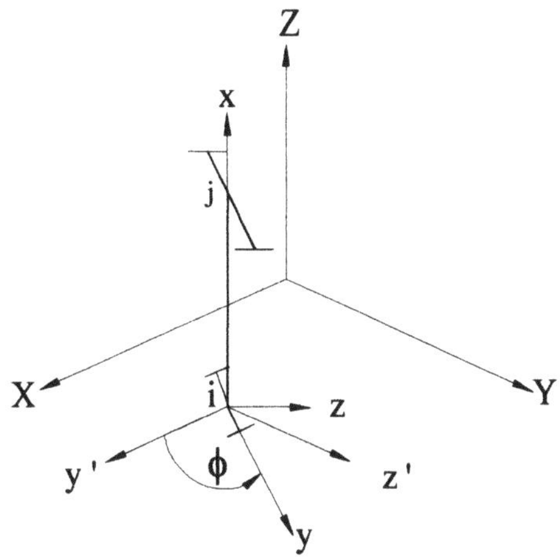

Fig. P5.2 Vertical beam element showing the local axes (*x*, *y*, *z*) and the auxiliary coordinate system (x', y', z') in which the local axis y' is parallel to the global axis *X*.

The simple transformation of coordinates from the (x', y', z') system to the (*x*, *y*, *z*) is

$$\begin{Bmatrix} x \\ y \\ z \end{Bmatrix} = \begin{bmatrix} 1 & 0 & 0 \\ 0 & \cos\phi & \sin\phi \\ 0 & -\sin\phi & \cos\phi \end{bmatrix} \begin{Bmatrix} x' \\ y' \\ z' \end{Bmatrix} \tag{b}$$

where the angle ϕ is the angle of roll around the local axis x measured from local axis y to the auxiliary axis y'. This angle of roll is positive for a counter-clockwise rotation observing the x axis from the second end toward the first end of the beam element.

The substitution of eq.(a) into eq.(b) yields

$$\begin{Bmatrix} x \\ y \\ z \end{Bmatrix} \begin{bmatrix} 1 & 0 & 0 \\ 0 & \cos\phi & \sin\phi \\ 0 & -\sin\phi & \cos\phi \end{bmatrix} \begin{bmatrix} 0 & 0 & \lambda \\ 1 & 0 & 0 \\ 0 & \lambda & 0 \end{bmatrix} \begin{Bmatrix} X \\ Y \\ Z \end{Bmatrix} \tag{c}$$

The product of two matrices in eq.(c) is the transformation matrix between the global coordinate system (*X*, *Y*, *Z*) and the local system (*x*, *y*, *z*) for the particular case in which the local axis x is parallel to the global axis *Z*. Thus from eq.(c) we obtain the transformation matrix $[T]_1$ as

$$[T]_1 = \begin{bmatrix} 0 & 0 & \lambda \\ \cos\phi & \lambda\sin\phi & 0 \\ -\sin\phi & \lambda\cos\phi & 0 \end{bmatrix} \tag{d}$$

in which $\lambda = 1$ when the local axis x has the same sense as the global axis Z; otherwise $\lambda = -1$, and ϕ is the angle of roll.

Problem 5.3

Demonstrate that direction cosines in eq.(5.6) for the transformation of global to local coordinates may be determined from the global coordinates of three points. The two points defining the ends of the beam element along the local axis x and any third point located in local plane x-y in which y is the minor principal axis of the cross sectional area of the member

$$[T_1] = \begin{bmatrix} \cos xX & \cos xY & \cos xZ \\ \cos yX & \cos yY & \cos yZ \\ \cos zX & \cos zY & \cos zZ \end{bmatrix} \qquad \text{(5.6) repeated}$$

Solution:

Designate the coordinates of the points at the ends of a beam element as P_i (X_i, Y_i, Z_i), $P_j(X_j, Y_j, Z_j)$ and of a third point P (X_P, Y_P, Z_P) located on the plane x-y of the cross-sectional area of the member. The direction cosines of the local axis x along the beam element are given by eq.(5.7) as

$$\cos xX = \frac{X_j - X_i}{L} \qquad \cos xY = \frac{Y_j - Y_i}{L} \qquad \cos xZ = \frac{Z_j - Z_i}{L} \qquad \text{eq.(5.7) repeated}$$

where

$$L = \sqrt{(X_j - X_i)^2 + (Y_j - Y_i)^2 + (Z_j - Z_i)^2} \qquad \text{(5.8) repeated}$$

The direction cosines of the z axis can be calculated from the condition that any vector **Z** along the z axis must be perpendicular to the plane formed by any two vectors in the local x-y plane. These two vectors simply could be the vector **X** from point i to point j along the x axis and the vector **P** from point i to point P. The orthogonality condition is expressed by the cross product between vectors **X** and **P** as

$$\mathbf{Z} = \mathbf{X} \times \mathbf{P} \tag{a}$$

or substituting the components of these vectors as

$$z_x\hat{i} + z_y\hat{j} + z_z\hat{k} = \begin{vmatrix} \hat{i} & \hat{j} & \hat{k} \\ x_j - x_i & y_j - y_i & z_j - z_i \\ x_P - x_i & y_P - y_i & z_P - z_i \end{vmatrix} \tag{b}$$

where $\hat{i}$, $\hat{j}$, and $\hat{z}$ are the unit vectors along the global coordinate axes X, Y and Z, respectively. Consequently, the direction cosines of axis z are given by

$$\cos zX = \frac{z_x}{|Z|} \qquad \cos zY = \frac{z_y}{|Z|} \qquad \cos zZ = \frac{z_z}{|Z|} \tag{c}$$

where

$$\begin{aligned} z_x &= (y_j - y_i\}(z_P - z_i) - (z_j - z_i)(y_P - y_i) \\ z_y &= (z_j - z_i\}(x_P - x_i) - (x_j - x_i)(z_P - z_i) \\ z_z &= (x_j - x_i\}(y_P - y_i) - (y_j - y_i)(x_P - x_i) \end{aligned} \tag{d}$$

and

$$|Z| = \sqrt{z_x^2 + z_y^2 + z_z^2} \tag{e}$$

Analogously, the direction cosines of the local axis y are calculated from the condition of orthogonality between a vector **Y** along the y axis and the unit vectors $\mathbf{X}_1$ and $\mathbf{Z}_1$ along the x and z axes, respectively. Hence,

$$\mathbf{Y} = \mathbf{X}_1 \times \mathbf{Z}_1$$

or in expanded notation

$$y_x\hat{i} + y_y\hat{j} + y_z\hat{k} = \begin{vmatrix} \hat{i} & \hat{j} & \hat{k} \\ \cos xX & \cos xY & \cos xZ \\ \cos zX & \cos zY & \cos zZ \end{vmatrix} \tag{f}$$

Therefore,

$$\cos yX = \frac{y_x}{|Y|} \qquad \cos yY = \frac{y_y}{|Y|} \qquad \cos yZ \frac{y_z}{|Y|} \tag{g}$$

where

$$y_x = \cos xY \cos zZ - \cos xZ \cos zY$$
$$y_y = \cos xZ \cos zX - \cos xX \cos zZ \qquad \text{(h)}$$
$$y_z = \cos xX \cos zY - \cos xY \cos zX$$

and

$$|Y| = \sqrt{y_x^2 + y_y^2 + y_z^2} \qquad \text{(i)}$$

We have, therefore, shown that knowledge of points at the two ends of an element of a point P on the local plane x-y suffices to calculate the direction cosines of the transformation matrix [T_1] in eq.(5.6). These direction cosines are given by eq.(5.7), eq.(g) and eq.(c), respectively, for rows 1, 2 and 3 of matrix $[T_1]$. The choice of point P is generally governed by the geometry of the structure and the orientation of the principal directions of the cross-section of the member. Quite often point P is selected as a known point in the structure which is placed on the local axis y, although as it has been shown, the point P could be any point in the plane formed by the local x-y axes.

5.6 Practice Problems

Problem 5.4

For the space frame in Fig. P5.4 determine:

(a) Joint Displacements
(b) End-force members
(c) Support reactions

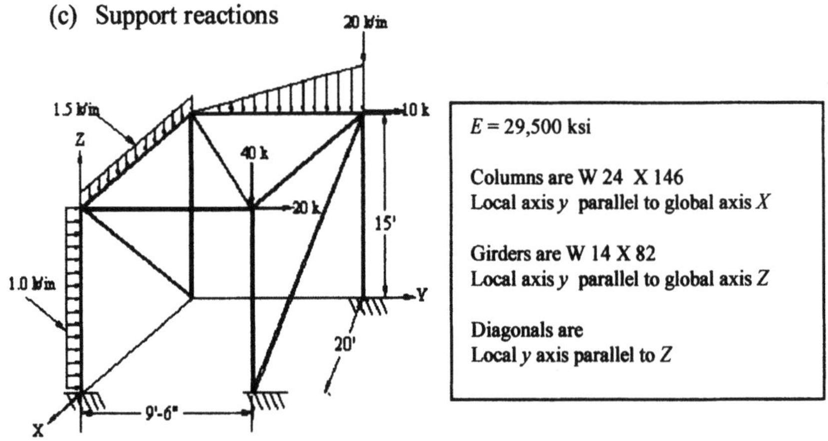

Fig. P5.4

Problem 5.5

For the space frame in Fig. P5.5 determine:

(a) Joint displacements
(b) End-force members
(c) Support reactions

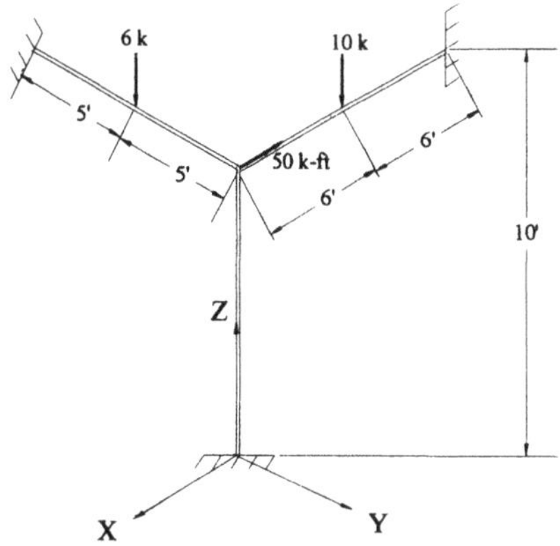

All Members:
$E = 29{,}500$ ksi
W 14 X 82

Girders have local axis y parallel to Z
Column has local axis y parallel to Y

Fig. P5.5

Problem 5.6

For the space frame in Fig. P5.6 determine:

(a) Joint displacements
(b) End-force members
(c) Support Reactions

Note: Girders have local axes y parallel to the global axis Z.
Columns have local axes y parallel to the global axis X.

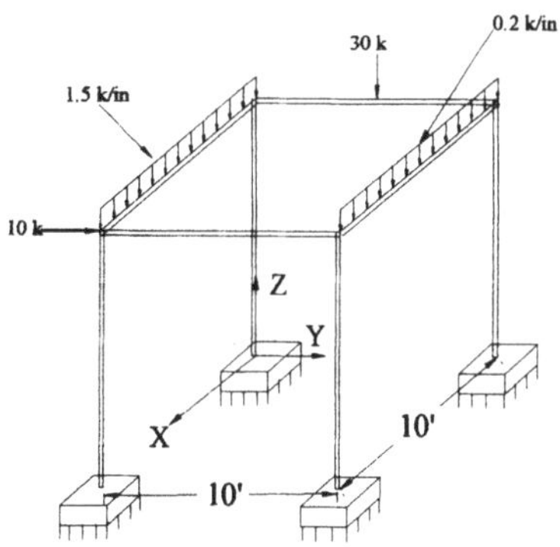

All Members:
$E = 29{,}500$ ksi
$G = 11650$ ksi
$A = 42.5$ in^2
$I_y = 324$ in^4
$I_z = 3850$ in^4
$J = 22.5$ in^4

Fig. P5.6

6 Plane Trusses

6.1 Introduction

Trusses are defined as structures assembled with longitudinal members connected at their ends by frictionless pins. Furthermore, it is assumed that loads are applied only at connecting joints between members of the truss; thus the self-weight, when considered in the analysis, is simply allocated to the joints at the end of the member. Under these assumptions, the elements of the truss are two-force members and the problem is reduced to the determination of the axial forces (tension or compression) in the members of the truss.

6.2 Element Stiffness Matrix in Local Coordinates

A member of a plane truss has a total of four nodal coordinates with two nodal coordinates at each joint as shown in Fig. 6.1. For small deflections, it may be assumed that the force-displacement relationship for the nodal coordinates along the axis of the member (coordinates 1 and 3 in Fig. 6.1) is independent of the transverse displacements along nodal coordinates 2 and 4. This assumption is equivalent to stating that a displacement along nodal coordinates 1 or 3 does not produce forces along nodal coordinates 2 or 4 and visa versa.

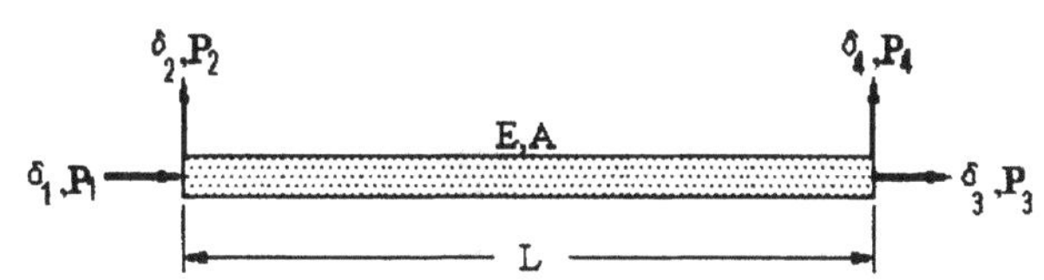

Fig. 6.1 Member of a plane truss showing nodal displacements and forces

The stiffness coefficients corresponding to axial nodal coordinates may be obtained using eq.(3.7) of Chapter 3. The application of this equation to a uniform element results in the following coefficients, using the notation in Fig. 6.1:

$$k_{11} = k_{33} = \frac{AE}{L} \qquad k_{13} = k_{31} = -\frac{AE}{L} \tag{6.1}$$

in which E is the modulus of elasticity, A the cross-sectional area and L the length of the element.

The stiffness coefficients for pin-ended elements, corresponding to the nodal coordinates 2 and 4, are all equal to zero, since a force is not required to produce displacements at these coordinates. Therefore, arranging the coefficients given by eq.(6.1), we obtain the stiffness equation for a uniform member of a truss as

$$\begin{Bmatrix} P_1 \\ P_2 \\ P_3 \\ P_4 \end{Bmatrix} = \frac{AE}{L}\begin{bmatrix} 1 & 0 & -1 & 0 \\ 0 & 0 & 0 & 0 \\ -1 & 0 & 1 & 0 \\ 0 & 0 & 0 & 0 \end{bmatrix}\begin{Bmatrix} \delta_1 \\ \delta_2 \\ \delta_3 \\ \delta_4 \end{Bmatrix} \tag{6.2}$$

or in condensed notation

$$\{P\} = [k]\{\delta\} \tag{6.3}$$

in which $[k]$ is the stiffness matrix for an element of a plane truss, $\{P\}$ and $\{\delta\}$ are, respectively, the force and the displacement vectors at the nodal coordinates.

6.3 Transformation of Coordinates

The stiffness matrix in eq.(6.2) was obtained in reference to the nodal coordinates associated with the local or element system of coordinates. As discussed in the preceding chapters on framed structures, it is necessary to transform these element matrices to a common system of reference, the global coordinate system. The transformation of displacements and forces at the nodal coordinates is accomplished, as was demonstrated in Chapter 3, by performing a rotation of coordinates. Deleting the angular coordinates in eq.(3.12) and re-labeling the remaining coordinates result in the following transformation for the nodal forces:

$$\begin{Bmatrix} P_1 \\ P_2 \\ P_3 \\ P_4 \end{Bmatrix} = \begin{bmatrix} \cos\theta & \sin\theta & 0 & 0 \\ -\sin\theta & \cos\theta & 0 & 0 \\ 0 & 0 & \cos\theta & \sin\theta \\ 0 & 0 & -\sin\theta & \cos\theta \end{bmatrix}\begin{Bmatrix} \overline{P}_1 \\ \overline{P}_2 \\ \overline{P}_3 \\ \overline{P}_4 \end{Bmatrix} \tag{6.4}$$

where θ is the angle between the global axis X and the local axis x as shown in Fig. 6.2. Equation 6.4 may be written in condensed notation as

$$\{P\} = [T]\{\bar{P}\} \tag{6.5}$$

in which $\{P\}$ and $\{\bar{P}\}$ are the nodal forces in reference to local and global coordinates, respectively, and $[T]$ the transformation matrix defined in eq.(6.4).

The same transformation matrix $[T]$ also serves to transform the nodal displacement vector $\{\bar{\delta}\}$ in the global coordinate system to the nodal displacement vector $\{\delta\}$ in local coordinates:

$$\{\delta\} = [T]\{\bar{\delta}\} \tag{6.6}$$

The substitution of eqs.(6.5) and (6.6) into the stiffness equation (6.3) gives

$$[T][\bar{P}] = [K][T]\{\bar{\delta}\}$$

Since $[T]$ is an orthogonal matrix ($[T]^{-1}=[T]^{T}$), it follows that

$$\{\bar{P}\} = [T]^T[K][T]\{\bar{\delta}\}$$

or

$$\{\bar{P}\} = [\bar{K}]\{\bar{\delta}\} \tag{6.7}$$

in which

$$[\bar{K}] = [\bar{T}]^T[K][T] \tag{6.8}$$

is the element stiffness matrix in the global coordinate system.

6.4 Element Stiffness Matrix in Global Coordinates

The substitution of the stiffness matrix $[k]$ from eq.(6.2) and of the transformation matrix $[T]$ and its transpose from eq.(6.4) into eq.(6.8), results in the element stiffness matrix in reference to the global system of coordinates:

$$[\bar{k}] = \frac{EA}{L}\begin{bmatrix} c^2 & cs & -c^2 & -cs \\ cs & s^2 & -cs & -s^2 \\ -c^2 & -cs & c^2 & cs \\ -cs & -s^2 & cs & s^2 \end{bmatrix} \tag{6.9}$$

In eq.(6.9), c and s designate $\cos\theta$ and $\sin\theta$, respectively.

6.5 Assemblage of System Stiffness Matrix

The system stiffness matrix for a plane truss is assembled by appropriately transferring the coefficients of the element stiffness matrices using exactly the same procedure described in the preceding chapters for beams or for frames. The following example (Illustrative Example 6.1) illustrates the application of the stiffness matrix for the analysis of a plane truss.

6.6 End-forces for an Element of a Truss

The end forces for an element or member of a truss may be determined as for case frame elements presented in the previous chapters. These end forces in global coordinates will then be calculated as

$$\{\overline{P}\} = [\overline{k}]\{\overline{\delta}\} \tag{6.10}$$

in which $\{\overline{P}\}$ and $\{\overline{\delta}\}$ are, respectively, the element nodal force vector and element nodal displacement vector and $[\overline{k}]$ is the element stiffness matrix. The vectors $\{\overline{P}\}$ and $\{\overline{\delta}\}$ as well as the matrix $[\overline{k}]$ are in reference to the global system of coordinates.

The end forces $\{P\}$ in reference to the local system of coordinates are then calculated by eq.(6.5). However, it is somewhat more convenient to calculate the element forces (tension or compression) by determining first the axial deformation and then the element axial force. Consider in Fig. 6.2 an element of a truss showing the nodal displacements $\overline{\delta}_1$ through $\overline{\delta}_4$ calculated in global coordinates.

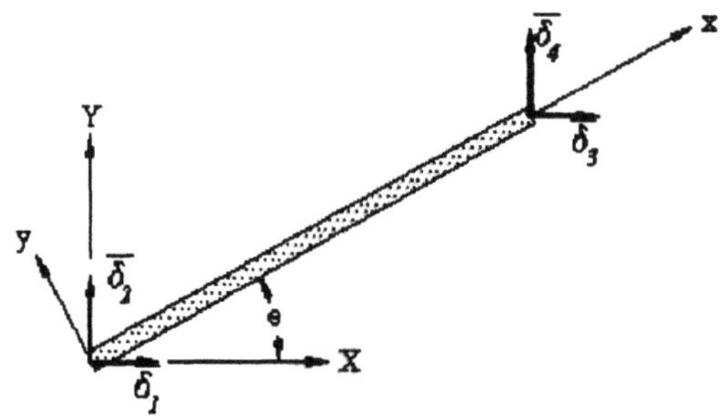

Fig. 6.2 Plane truss element showing the nodal displacements $\overline{\delta}_1$ through $\overline{\delta}_4$ in global coordinates

The elongation Δ of this element along its longitudinal axis x is given by

$$\Delta = \left(\bar{\delta}_3 - \bar{\delta}_1\right)\cos\theta + \left(\bar{\delta}_4 - \bar{\delta}_2\right)\sin\theta \tag{6.11}$$

Its strain is then ε=Δ/L the stress σ=E ε=EΔ/L and the axial force P=Aσ=EAΔ/L or using eq.(6.11):

$$P = \frac{EA}{L}\left[\left(\bar{\delta}_3 - \bar{\delta}_1\right)\cos\theta + \left(\bar{\delta}_4 - \bar{\delta}_2\right)\sin\theta\right] \tag{6.12}$$

in which

E = modulus of elasticity
A = cross-sectional area
L = length of the element
θ = angle between global axis X and the local axis x.

Illustrative Example 6.1

The plane truss shown in Fig. 6.3 (which has only three members) is used to illustrate the application of the stiffness method for trusses. Determine:

(a) Displacements at the joints,
(b) Axial forces in the members, and
(c) Reactions at the supports.

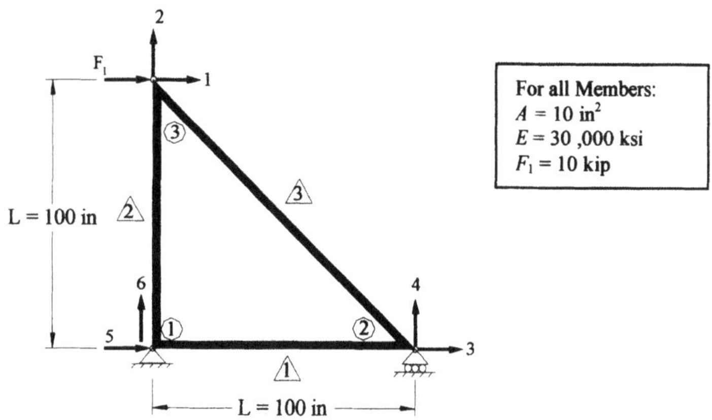

Fig. 6.3 Plane truss for Illustrative Example 6.1

Solution:

1. Mathematical Model.
   Figure 6.3 shows the model needed for this truss with 3 elements, 3 nodes and 6 nodal coordinates of which the first three are free coordinates and the last three are fixed.

2. Element stiffness matrices – Global coordinates.
ELEMENT 1: Use eq.(6.9) with $\theta = 0$

$$[\bar{k}]_1 = \frac{30\text{E}3\text{x}10}{100}\begin{bmatrix} 1 & 0 & -1 & 0 \\ 0 & 0 & 0 & 0 \\ -1 & 0 & 1 & 0 \\ 0 & 0 & 0 & 0 \end{bmatrix} = 3000 \begin{array}{c} \begin{array}{cccc} 5 & 6 & 3 & 4 \end{array} \\ \begin{bmatrix} 1 & 0 & -1 & 0 \\ 0 & 0 & 0 & 0 \\ -1 & 0 & 1 & 0 \\ 0 & 0 & 0 & 0 \end{bmatrix} \end{array} \begin{array}{c} \\ 5 \\ 6 \\ 3 \\ 4 \end{array} \tag{a}$$

ELEMENT 2: $\theta = 90^o$

$$[\bar{k}]_2 = 3000 \begin{array}{c} \begin{array}{cccc} 5 & 6 & 1 & 2 \end{array} \\ \begin{bmatrix} 0 & 0 & 0 & 0 \\ 0 & 1 & 0 & -1 \\ 0 & 0 & 0 & 0 \\ 0 & -1 & 0 & 1 \end{bmatrix} \end{array} \begin{array}{c} \\ 5 \\ 6 \\ 1 \\ 2 \end{array} \tag{b}$$

ELEMENT 3: $\theta = 135^o$

$$[\bar{k}]_3 = 1060 \begin{array}{c} \begin{array}{cccc} 3 & 4 & 1 & 2 \end{array} \\ \begin{bmatrix} 1 & -1 & -1 & 1 \\ -1 & 1 & 1 & -1 \\ -1 & 1 & 1 & -1 \\ 1 & -1 & -1 & 1 \end{bmatrix} \end{array} \begin{array}{c} \\ 3 \\ 4 \\ 1 \\ 2 \end{array} \tag{c}$$

3. Assemble reduced system stiffness matrix.
The coefficients in eq.(a), (b) and (c) corresponding to the first three nodal coordinates indicated at the top of the element stiffness matrices in these equations are transferred to the reduced system stiffness matrix:

$$[K]_R = \begin{bmatrix} 1060 & -1060 & -1060 \\ -1060 & 3000+1060 & 1060 \\ -1060 & 1060 & 3000+1060 \end{bmatrix} = \begin{bmatrix} 1060 & -1060 & -1060 \\ -1060 & 4060 & 1060 \\ -1060 & 1060 & 4060 \end{bmatrix} \tag{d}$$

4. Reduced system force vector.
The system force vector contains only the applied force 10 kips at the first nodal coordinate u_1:

$$\{F\}_R = \begin{Bmatrix} 10 \\ 0 \\ 0 \end{Bmatrix} \tag{e}$$

5. System stiffness matrix.

$$\begin{Bmatrix} 10 \\ 0 \\ 0 \end{Bmatrix} = \begin{bmatrix} 1060 & -1060 & -1060 \\ -1060 & 4060 & 1060 \\ -1060 & 1060 & 4060 \end{bmatrix} \begin{Bmatrix} u_1 \\ u_2 \\ u_3 \end{Bmatrix} \tag{f}$$

6. Nodal displacements.
Solution of eq.(f) results in

$u_1 = 0.01611$ (in)
$u_2 = 0.00333$ (in)
$u_3 = 0.00333$ (in)

7. Element end forces. Axial forces $P^{(i)}$ on each element of the truss are calculated by eq.(6.12).

ELEMENT 1:

$$P^{(1)} = \frac{30\text{E}3\text{x}10}{100}\left[(0.00333-0)\cos 0^\circ + (0-0)\sin 0^\circ\right]$$
$$= 10 \text{ kip}$$

ELEMENT 2

$$P^{(2)} = \frac{30\text{E}3\text{x}10}{100}\left[(0.01611-0)\ \cos 90^\circ + (0.00333-0)\sin 90^\circ\right]$$
$$= 10 \text{ kip}$$

ELEMENT 3

$$P^{(3)} = \frac{30\text{E}3\text{x}10}{100\sqrt{2}}\left[(0.01611-0.00333)\ \cos 135^\circ + (0.00333-0)\sin 135^\circ\right]$$
$$= -14.2 \text{ kip}$$

8. Support reactions:
Support reactions are calculated from the element forces as follows:

At joint ①:
$R_{1x} = -P^{(1)} = -10$ kips
$R_{1y} = -P^{(2)} = -10$ kips

At joint ②:
$R_{2x} = 0$
$R_{2y} = P_3 \cos 45^\circ = (14.2)(0.707) = 10$ kips

Illustrative Example 6.2

Use SAP2000 to perform the structural analysis of the plane frame shown in Fig. 6.4. For all the members of the truss, the modulus of elasticity is $E = 30{,}000$ ksi and the cross-sectional area $A = 2$ in^2.

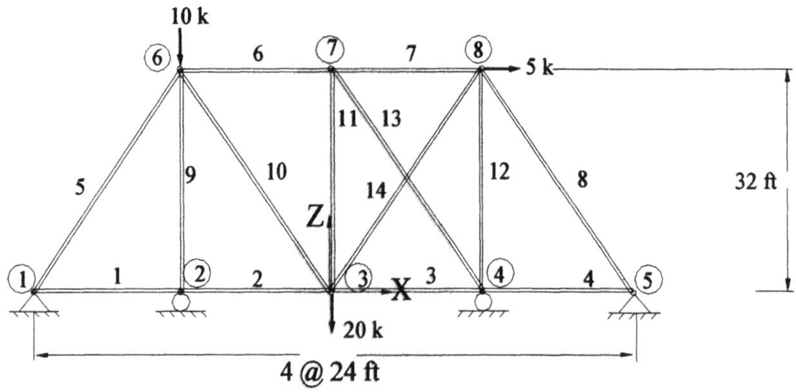

Fig. 6.4 Plane frame for Illustrative Example 6.2

Solution:

Begin: Open SAP2000.

Enter: OK to disable the "Tip of the Day".

Hint: Maximize both screens for a full view of all windows.

Select: In the lower right-hand corner of the screen use the drop-down menu to select "kip-in".

Select: From the Main Menu :
FILE>NEW MODEL
In the Coordinate System Definitions, enter:

Number of Grid Spaces:
X direction = 4
Y direction = 0
Z direction = 1

Grid Spacing:
X direction = 288
Y direction = 384
Y direction = 384. Then OK.

Edit: Minimize the 3-D screen and maximize the 2-D screen. Click on icon XZ to have a view of this plane. Use the PAN icon to center the coordinate axes on the screen.

Draw: From the Main Menu enter:
DRAW>DRAW FRAME ELEMENT
Click on lower left intersection of grid lines and drag the cursor horizontally to next grid line, click twice to create an element. Then click again at this last location and proceed horizontally to the next grid line and click twice to create a second element.

Proceed to create additional horizontal elements. Similarly, create all the other elements of the truss shown in Fig.6.4.

Label: From the main Menu enter:
VIEW>SET ELEMENTS
Check Labels on Joints and on Frames. Then OK.

Assign: Select joints ① and ⑤ of the truss by clicking at the joints. Then enter:
ASSIGN>JOINT>RESTRAINTS
Restrain the translations 1, 2 and 3. Then OK

Select joints ② and ④ at the two rollers then enter:
ASSIGN>JOINT>RESTRAINTS
Restrain only the translation 3. Then OK.

Select joints ③, ⑥, ⑦ and ⑧ and enter:
ASSIGN>JOINT>RESTRAINTS
Select no restraints by a click on the dot icon shown in the screen.
Then OK

Select: From the Main Menu enter:
DEFINE>MATERIALS
Select STEEL and Modify/Show Material.
Enter Modulus of Elasticity = 30000. Then OK, OK.

Define: From the Main Menu enter:
DEFINE>FRAME SECTIONS
Click on Add/Wide Flange and on Add General Section
Change the cross-section area to 2. Then OK.
Change all other section values to 0 (zero)
Change the label of the section from FSEC2 to TRUSS62. Then OK, OK.

Define: From the Main Menu enter:
DEFINE>STATIC LOAD CASES
Change DEAD to LIVE and set to 0 (zero) Self-Weight Multiplier.
Select "Change Load" . Then OK.

Loads: Select joint ③ at the center of the lower chord of the truss and enter:
ASSIGN>JOINT STATIC LOADS>FORCES
Enter Force Global $Z = -20$. Then OK.
Select joint ⑥ at the left end of the top chord of the truss and enter:
ASSIGN>JOINT STATIC LOADS>FORCES
Enter Force Global $Z = -10$. Then OK.
Note: Be sure to change other previous loads to zero

Select Joint ⑧ at the right end of the top chord of the truss and enter:
ASSIGN>JOINT STATIC LOADS>FORCES
Enter Force Global $X = 5$. Then OK.
Note: Be sure to change other previous loads to zero.

Set: From the Main Menu enter:
ANALYZE>SET OPTIONS
Select available Degrees of Freedom: *UX* and *UZ*. Then OK.

Analysis: From the Main Menu enter:
ANALYZE>RUN
Respond to the request for filename: "Example 6.2".
Then SAVE and OK.
When the analysis is concluded enter OK.

Print Input Tables: From the Main Menu enter:
FILE>PRINT INPUT TABLES
Select Coordinates, Frames and Joints. Then OK.
(Table 6.1 contains the edited input data for this Illustrative Example 6.2)

Table 6.1 Edited Input tables for Illustrative Example 6.2 (units: kips, inches)

JOINT DATA

JOINT	GLOBAL-X	GLOBAL-Y	GLOBAL-Z	RESTR	ANG-A	ANG-B	ANG-C
1	-576.00000	0.00000	0.00000	1 1 1 0 0 0	0.000	0.000	0.000
2	-288.00000	0.00000	0.00000	0 1 1 0 0 0	0.000	0.000	0.000
3	0.00000	0.00000	0.00000	0 0 0 0 0 0	0.000	0.000	0.000
4	288.00000	0.00000	0.00000	0 1 1 0 0 0	0.000	0.000	0.000
5	576.00000	0.00000	0.00000	1 1 1 0 0 0	0.000	0.000	0.000
6	-288.00000	0.00000	384.00000	0 0 0 0 0 0	0.000	0.000	0.000
7	0.00000	0.00000	384.00000	0 0 0 0 0 0	0.000	0.000	0.000
8	288.00000	0.00000	384.00000	0 0 0 0 0 0	0.000	0.000	0.000

Table 6.1 Continued

FRAME ELEMENT DATA

FRAME	JNT-1	JNT-2	SCTN	ANG	RLS	SGMNT	R1	R2	FCTR	LENGTH
1	1	2	TRUSS62	0.000	000000	4	0.000	0.000	1.000	288.000
2	2	3	TRUSS62	0.000	000000	4	0.000	0.000	1.000	288.000
3	3	4	TRUSS62	0.000	000000	4	0.000	0.000	1.000	288.000
4	4	5	TRUSS62	0.000	000000	4	0.000	0.000	1.000	288.000
5	1	6	TRUSS62	0.000	000000	2	0.000	0.000	1.000	480.000
6	6	7	TRUSS62	0.000	000000	4	0.000	0.000	1.000	288.000
7	7	8	TRUSS62	0.000	000000	4	0.000	0.000	1.000	288.000
8	8	5	TRUSS62	0.000	000000	2	0.000	0.000	1.000	480.000
9	2	6	TRUSS62	0.000	000000	2	0.000	0.000	1.000	384.000
10	3	6	TRUSS62	0.000	000000	2	0.000	0.000	1.000	480.000
11	3	7	TRUSS62	0.000	000000	2	0.000	0.000	1.000	384.000
12	4	8	TRUSS62	0.000	000000	2	0.000	0.000	1.000	384.000
13	4	7	TRUSS62	0.000	000000	2	0.000	0.000	1.000	480.000
14	3	8	TRUSS62	0.000	000000	2	0.000	0.000	1.000	480.000

JOINT FORCES Load Case LOAD1

JOINT	GLOBAL-X	GLOBAL-Y	GLOBAL-Z	GLOBAL-XX	GLOBAL-YY	GLOBAL-ZZ
3	0.000	0.000	-20.000	0.000	0.000	0.000
6	0.000	0.000	-10.000	0.000	0.000	0.000
8	5.000	0.000	0.000	0.000	0.000	0.000

Print Output Tables: From the Main Menu enter:

FILE>PRINT OUTPUT TABLES

(Table 6.2 contains the edited output tables for Illustrative Example 6.2)

Table 6.2 Output tables for Illustrative Example 6.2 (Units: kips, inches)

JOINT DISPLACEMENTS

JOINT	LOAD	UX	UY	UZ	RX	RY	RZ
1	LOAD1	0.0000	0.0000	0.0000	0.0000	-4.111E-05	0.0000
2	LOAD1	9.534E-03	0.0000	0.0000	0.0000	3.029E-04	0.0000
3	LOAD1	0.0191	0.0000	-0.1522	0.0000	-2.473E-05	0.0000
4	LOAD1	0.0221	0.0000	0.0000	0.0000	-2.958E-04	0.0000
5	LOAD1	0.0000	0.0000	0.0000	0.0000	9.215E-05	0.0000
6	LOAD1	0.0304	0.0000	-0.0737	0.0000	1.431E-04	0.0000
7	LOAD1	-4.448E-03	0.0000	-0.1074	0.0000	-9.122E-05	0.0000
8	LOAD1	-0.0141	0.0000	-0.0349	0.0000	-1.999E-04	0.0000

JOINT REACTIONS

JOINT	LOAD	F1	F2	F3	M1	M2	M3
1	LOAD1	1.0710	0.0000	4.0761	0.0000	0.0000	0.0000
2	LOAD1	0.0000	0.0000	11.5253	0.0000	0.0000	0.0000
4	LOAD1	0.0000	0.0000	12.4528	0.0000	0.0000	0.0000
5	LOAD1	-6.0710	0.0000	1.9457	0.0000	0.0000	0.0000

Table 6.2 Continued

FRAME ELEMENT FORCES

Note: The computer output shows extremely small values for the shear force V2 and the bending moment M3 (in the range of numerical accuracy). However, in editing this table these values have been set equal to zero.

FRAME	LOAD	LOC	P	V2	V3	T	M2	M3
1	LOAD1							
		0.00	1.99	0.00	0.00	0.00	0.00	0.00
		72.00	1.99	0.00	0.00	0.00	0.00	0.00
		144.00	1.99	0.00	0.00	0.00	0.00	0.00
		216.00	1.99	0.00	0.00	0.00	0.00	0.00
		288.00	1.99	0.00	0.00	0.00	0.00	0.00
2	LOAD1							
		0.00	1.99	0.00	0.00	0.00	0.00	0.00
		72.00	1.99	0.00	0.00	0.00	0.00	0.00
		144.00	1.99	0.00	0.00	0.00	0.00	0.00
		216.00	1.99	0.00	0.00	0.00	0.00	0.00
		288.00	1.99	0.00	0.00	0.00	0.00	0.00
3	LOAD1							
		0.00	-0.638	0.00	0.00	0.00	0.00	0.00
		72.00	-0.638	0.00	0.00	0.00	0.00	0.00
		144.00	-0.638	0.00	0.00	0.00	0.00	0.00
		216.00	-0.638	0.00	0.00	0.00	0.00	0.00
		288.00	-0.638	0.00	0.00	0.00	0.00	0.00
4	LOAD1							
		0.00	-4.61	0.00	0.00	0.00	0.00	0.00
		72.00	-4.61	0.00	0.00	0.00	0.00	0.00
		144.00	-4.61	0.00	0.00	0.00	0.00	0.00
		216.00	-4.61	0.00	0.00	0.00	0.00	0.00
		288.00	-4.61	0.00	0.00	0.00	0.00	0.00
5	LOAD1							
		0.00	-5.10	0.00	0.00	0.00	0.00	0.00
		240.00	-5.10	0.00	0.00	0.00	0.00	0.00
		480.00	-5.10	0.00	0.00	0.00	0.00	0.00
6	LOAD1							
		0.00	-7.26	0.00	0.00	0.00	0.00	0.00
		72.00	-7.26	0.00	0.00	0.00	0.00	0.00
		144.00	-7.26	0.00	0.00	0.00	0.00	0.00
		216.00	-7.26	0.00	0.00	0.00	0.00	0.00
		288.00	-7.26	0.00	0.00	0.00	0.00	0.00

Table 6.2 Continued

FRAME	LOAD	LOC	P	V2	V3	T	M2	M3
7	LOAD1							
		0.00	-2.01	0.00	0.00	0.00	0.00	0.00
		72.00	-2.01	0.00	0.00	0.00	0.00	0.00
		144.00	-2.01	0.00	0.00	0.00	0.00	0.00
		216.00	-2.01	0.00	0.00	0.00	0.00	0.00
		288.00	-2.01	0.00	0.00	0.00	0.00	0.00
8	LOAD1							
		0.00	-2.43	0.00	0.00	0.00	0.00	0.00
		240.00	-2.43	0.00	0.00	0.00	0.00	0.00
		480.00	-2.43	0.00	0.00	0.00	0.00	0.00
9	LOAD1							
		0.00	-11.52	0.00	0.00	0.00	0.00	0.00
		192.00	-11.52	0.00	0.00	0.00	0.00	0.00
		384.00	-11.52	0.00	0.00	0.00	0.00	0.00
10	LOAD1							
		0.00	7.00	0.00	0.00	0.00	0.00	0.00
		240.00	7.00	0.00	0.00	0.00	0.00	0.00
		480.00	7.00	0.00	0.00	0.00	0.00	0.00
11	LOAD1							
		0.00	7.00	0.00	0.00	0.00	0.00	0.00
		192.00	7.00	0.00	0.00	0.00	0.00	0.00
		384.00	7.00	0.00	0.00	0.00	0.00	0.00
12	LOAD1							
		0.00	-5.45	0.00	0.00	0.00	0.00	0.00
		192.00	-5.45	0.00	0.00	0.00	0.00	0.00
		384.00	-5.45	0.00	0.00	0.00	0.00	0.00
13	LOAD1							
		0.00	-8.75	0.00	0.00	0.00	0.00	0.00
		240.00	-8.75	0.00	0.00	0.00	0.00	0.00
		480.00	-8.75	0.00	0.00	0.00	0.00	0.00
14	LOAD1							
		0.00	9.25	0.00	0.00	0.00	0.00	0.00
		240.00	9.25	0.00	0.00	0.00	0.00	0.00
		480.00	9.25	0.00	0.00	0.00	0.00	0.00

Plot Displacements: From the Main Menu enter:
DISPLAY>SHOW DEFORMED SHAPE, then enter:
FILE>PRINT GRAPHICS
(The deformed shape shown in the screen is reproduced in Fig. 6.5)

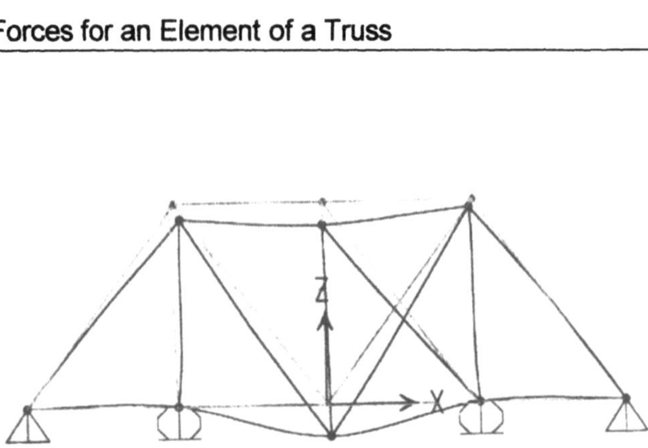

Fig. 6.5 Deformed shape for the plane truss of Illustrative Example 6.2

Plot Axial Forces: From the Main menu enter:
DISPLAY>SHOW ELEMENT FORCES/STRESSES>FRAMES
Select: Axial Force, then OK and enter:
FILE>PRINT GRAPHICS
(the Axial Force diagram is reproduced in Fig. 6.6)

Fig. 6.6 Axial Force diagram for Illustrative Example 6.2 (dark members are in compression and light members in tension)

Note: To view plots and values for any member of the truss, right click on the element and a pop-up window will depict the axial force of the selected element.

6.7 Problems

Problem 6.1

For the plane truss shown in Fig. P6.1, determine:

(a) Joint displacements
(b) Member axial forces
(c) Support reactions

All Members:
$E = 200$ GPa
$A = 48$ cm^2

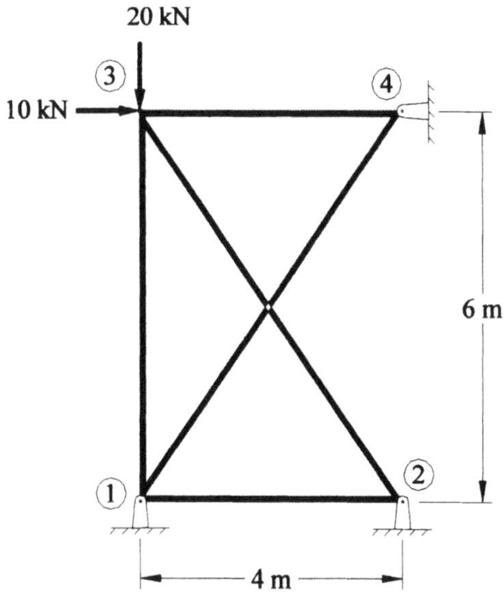

Fig. P6.1

Problem 6.2

Solve Problem 6.1 for the combined effect of the load shown in Fig P6.1 and a settlement of support ② of 10 mm down and 6 mm horizontally to the right.

Problem 6.3

For the plane truss shown in Fig. P6.3 determine:

(a) Joint displacements
(b) Member end-forces
(c) Support reactions

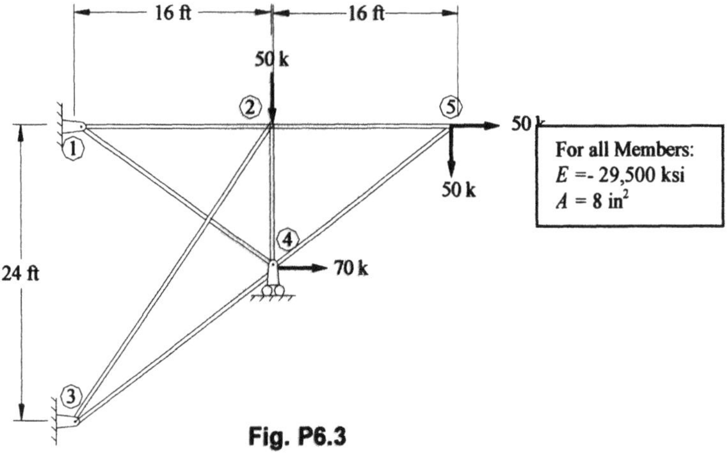

Fig. P6.3

Problem 6.4

Solve Problem 6.3 for the combined effect of the loads shown in Fig. P6.3 and a vertical settlement of 1.5 in at the support in joint ④.

Problem 6.5

For the plane frame shown in Fig. P6.5 determine:

(a) Joint displacements
(b) Axial member forces
(c) Support reactions

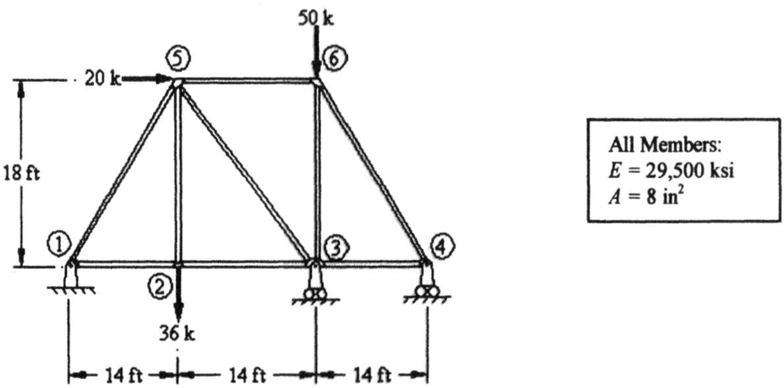

Fig. P6.5

Problem 6.6

Solve Problem 6.5 for the combined effect of the loads shown in Fig. P6.5 and a settlement of 1.0 in a downward direction and 2 in. to the right at the support in joint ①.

Problem 6.7

For the truss shown in Fig. P6.7 determine:

(a) Joint displacements
(b) Member axial forces
(c) Support reactions

All Members:
$E = 200$ GP$_a$
$A = 60$ cm^2

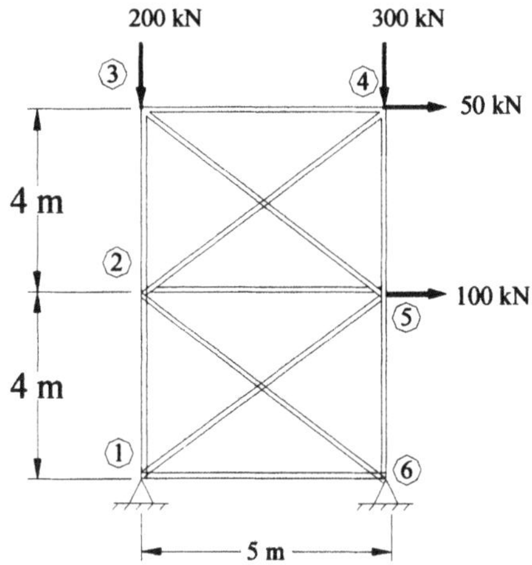

Fig P6.7

Problem 6.8

Solve Problem 6.7 for the combined effect of the loads shown in Fig. P6.7 and settlement of support at joint of 12 mm horizontally to the right and 18 mm down vertically.

Problem 6.9

For the truss shown in Fig. P6.9 determine:

(a) Joint displacements
(b) Axial member forces
(c) Support reactions

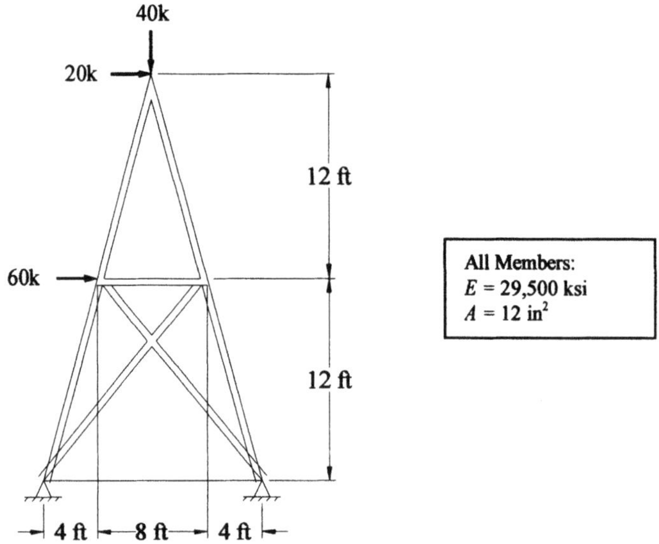

Fig. P6.9

7 Space Trusses

7.1 Introduction

Space trusses are three-dimensional structures with longitudinal members connected at their ends by hinges assumed to be frictionless. The loads on space trusses are applied only at the nodes or joints, thus the self-weight is allocated for each element at its two ends joining other elements of the truss. The conditions imposed on space trusses are certainly the same as those on plane trusses. Essentially, the only difference in the analysis of space trusses compared with plane trusses is that an element of a space truss has three nodal coordinates at each node while an element of a plane truss has only two.

7.2 Element Stiffness Matrix of a Space Truss – Local Coordinates

The stiffness matrix for an element of a space truss can be obtained as an extension of the corresponding matrix for the plane truss. Figure 7.1 shows the nodal coordinates in the local system (unbarred) and in the global system (barred) for a member of a space truss. The local axis x is directed along the longitudinal axes of the member while the y and z axes are set to agree with the principal directions of the cross-section of the member. The following matrix may then be written for a uniform member of a space truss as an extension of the stiffness equation for a member of a plane truss:

$$[k]=\begin{bmatrix} 1 & 0 & 0 & -1 & 0 & 0 \\ 0 & 0 & 0 & 0 & 0 & 0 \\ 0 & 0 & 0 & 0 & 0 & 0 \\ -1 & 0 & 0 & 1 & 0 & 0 \\ 0 & 0 & 0 & 0 & 0 & 0 \\ 0 & 0 & 0 & 0 & 0 & 0 \end{bmatrix} \tag{7.1}$$

Therefore, the element stiffness equation for an element of a space truss in reference to the local system of coordinate axes is given by

$$\{P\} = [k]\{\delta\} \tag{7.2}$$

in which $\{\delta\}$ is the element nodal displacement vector in reference to the local coordinates as shown in Fig.(7.1(a) and $\{P\}$ is the corresponding nodal force vector. These vectors are, respectively,

$$\{\delta\} = \begin{Bmatrix} \delta_1 \\ \delta_2 \\ \delta_3 \\ \delta_4 \\ \delta_5 \\ \delta_6 \end{Bmatrix} \quad \text{and} \quad \{P\} = \begin{Bmatrix} P_1 \\ P_2 \\ P_3 \\ P_4 \\ P_5 \\ P_6 \end{Bmatrix} \tag{7.3}$$

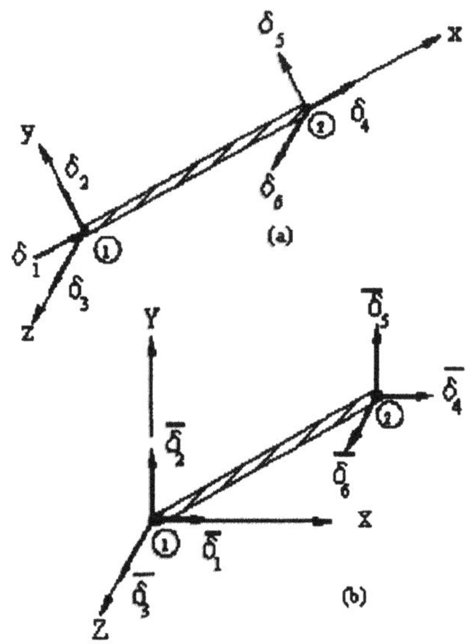

Fig. 7.1 Member of a space truss showing nodal displacement coordinates. (a) In the local system (unbarred). (b) In the global system (barred).

7.3 Transformation of the Element Stiffness Matrix

The transformation matrix $[T_l]$ corresponding to the three nodal coordinates at a node of a beam as developed in Chapter 5, is given by eq. (5.4) and it is repeated here for convenience.

$$[T_1] = \begin{bmatrix} \cos xX & \cos xY & \cos xZ \\ \cos yX & \cos yY & \cos yZ \\ \cos zX & \cos zY & \cos zZ \end{bmatrix} \tag{7.4}$$

where cos xY is the cosine of the angle between the local axis x and the global axis Y and analogous definitions for the other cosine functions in eq.(7.4). The direction cosines of local axis x in the first row of eq.(7.4), $c_1 = \cos xX$, $c_2 = \cos xY$, and $c_3 = \cos xZ$, are calculated from the coordinates of the two points $P_1(X_1, Y_1, Z_1)$ and $P_2(X_2, Y_2, Z_2)$ at the two ends of the truss element, that is

$$c_1 = \frac{X_2 - X_1}{L}, \quad c_2 = \frac{Y_2 - Y_1}{L} \quad \text{and} \quad c_3 = \frac{Z_2 - Z_1}{L} \tag{7.5}$$

with the length L of the element given by

$$L = \sqrt{(X_2 - X_1)^2 + (Y_2 - Y_1)^2 + (Z_2 - Z_1)^2} \tag{7.6}$$

The transformation matrix for the nodal coordinates for the two ends of a truss member is then given by

$$[T] = \begin{bmatrix} [T_1] & [0] \\ [0] & [T_1] \end{bmatrix} \tag{7.7}$$

in which $[T_1]$ is given by eq.(7.4). It follows that the relationship between the global displacement vector $\{\bar{\delta}\}$ and the local displacement vector $\{\delta\}$ at the two nodes of a truss element is

$$\{\delta\} = [T]\{\bar{\delta}\} \tag{7.8}$$

Analogously, the transformation of local force vector $\{P\}$ and the global force vector $\{\bar{P}\}$ is

$$\{P\} = [T]\{\bar{P}\} \tag{7.9}$$

The substitution of $\{\delta\}$ and $\{P\}$, respectively, from eqs.(7.8) and (7.9) into the element stiffness eq.(7.2) results in

$$[T]\{\bar{P}\} = [k][T]\{\bar{\delta}\} \tag{7.10}$$

Since the transformation matrix $[T]$ is orthogonal, ($[T]^{-1} = [T]^T$), eq.(7.10) may be written as

$$\{\bar{P}\} = [\bar{k}]\{\bar{\delta}\} \tag{7.11}$$

in which the element stiffness matrix $[\bar{k}]$ in reference to global coordinates is given by

$$[\bar{k}] = [T]^T[k][T] \tag{7.12}$$

In the evaluation of the element stiffness matrix $[\bar{k}]$ in global coordinates of a space truss, it is only necessary to calculate the direction cosines of the centroidal axis x of the element, which are given by eq.(7.5). The other direction cosines in eq.(7.4) corresponding to the axes y and z do not appear in the final expression for the element stiffness matrix $[\bar{k}]$ as may be verified by substituting eqs.(7.4) and (7.7) into eq.(7.12) and proceeding to multiply the matrices indicated in this last equation. The final result of this operation may be written as follows:

$$[\bar{K}] = \frac{AE}{L}\begin{bmatrix} c_1^2 & c_1c_2 & c_1c_3 & -c_1^2 & -c_1c_2 & -c_1c_3 \\ c_2c_1 & c_2^2 & c_2c_3 & -c_2c_1 & -c_2^2 & -c_2c_3 \\ c_3c_1 & c_3c_2 & c_3^2 & -c_3c_1 & -c_3c_2 & -c_3^2 \\ -c_1^2 & -c_1c_2 & -c_1c_3 & c_1^2 & c_1c_2 & c_1c_3 \\ -c_2c_1 & -c_2^2 & -c_2c_3 & c_2c_1 & c_2^2 & c_2c_3 \\ -c_3c_1 & -c_3c_2 & -c_3^2 & c_3c_1 & c_3c_2 & c_3^2 \end{bmatrix} \tag{7.13}$$

Consequently, the determination of the stiffness matrix for an element of a space truss, in reference to the global system of coordinates [eq.(7.13)] requires only the evaluation by eq.(7.5) of the direction cosines c_1, c_2 and c_3 of the local axis x along the element.

7.4 Element Axial Force

The expression to determine the axial force (tension or compression) in an element of a space truss may be obtained as an extension of eq.(6.12) for calculating the axial force in an element of a plane truss. Thus, extending eq.(6.12) to consider the three nodal displacements at the two ends of a space truss element, we obtain the expression for the axial force P as

$$P = \frac{EA}{L}\left[\left(\bar{\delta}_4 - \bar{\delta}_1\right)c_1 + \left(\bar{\delta}_5 - \bar{\delta}_2\right)c_2 + \left(\bar{\delta}_6 - \bar{\delta}_3\right)c_3\right] \tag{7.14}$$

where $\bar{\delta}_1$ through $\bar{\delta}_6$ are the nodal displacements in reference to the global system and $c_1 = \cos xX$, $c_2 = \cos xY$ and $c_3 = \cos xZ$ are the direction cosines of the axial axis x of the element. These direction cosines are calculated by eqs. (7.5) from the coordinates at the two points at the ends of the truss element.

7.5 Assemblage of the System Stiffness Matrix

The reduced system stiffness matrix is assembled, as explained in the previous chapters for frame type structures, by transferring to the appropriate location of the system stiffness matrix the coefficients in the element stiffness matrices. The following example illustrates the analysis of a space truss using matrix stiffness method of analysis.

Illustrative Example 7.1

Determine the reduced system stiffness matrix, the reduced system force vector and the response for the space truss shown in Fig. 7.2. The modulus of elasticity is $E = 30{,}000$ ksi and the cross-sectional area $A = 1.0$ in^2 for all members of the truss. The truss is supporting a force applied at joint ⑤ having equal components of magnitude 10 kips in the directions of the axes X, Y and Z as shown in Fig 7.2.

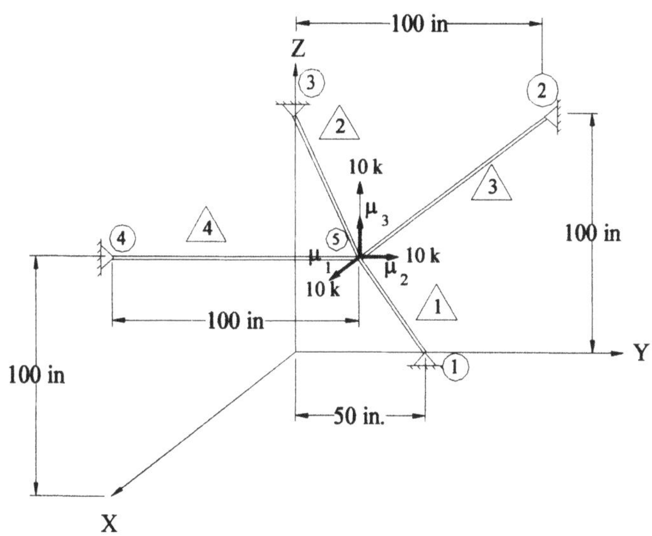

Fig. 7.2 Space truss for Illustrative Example 7.1

Solution:

1. Mathematical Model.

This truss is modeled as having 4 elements and 5 joints as shown in Fig. 7.3. Since only joint ⑤ is free to displace, the truss has only three free nodal coordinates u_1, u_2 and u_3 as indicated in Fig. 7.2. The coordinates X, Y and Z for the joints of the truss are conveniently arranged in the following table:

Joint Coordinates (inches)

Joint #	X	Y	Z
1	0	50	0
2	0	100	100
3	0	0	100
4	100	0	100
5	100	100	100

2. Element Stiffness Matrices – Global Coordinates.

The stiffness matrix for an element of a space truss is given by eq.(7.13) in which the direction cosines c_1, c_2 and c_3 are calculated by eq.(7.5).

ELEMENT 1 (Joint ⑤ to joint ①)

Direction Cosines: $c_1 = -\,0.667$ $c_2 = -\,0.333$ $c_3 = -\,0.667$

$$[\bar{k}]_1 = \begin{array}{c} \begin{array}{cccccc} \quad 1 & \qquad 2 & \qquad 3 & & & \end{array} \\ \left[\begin{array}{cccccc} 88.888 & 44.444 & 88.888 & -88.888 & -44.444 & -88.888 \\ 44.444 & 22.222 & 44.444 & -44.444 & -22.222 & -44.444 \\ 88.888 & 44.444 & 88.888 & -88.888 & -44.444 & -88.888 \\ -88.888 & -44.444 & -88.888 & 88.888 & 44.444 & 22.222 \\ -44.444 & -22.222 & -44.444 & 44.444 & 22.222 & 44.444 \\ -88.888 & -44.444 & -88.888 & 88.888 & 44.444 & 88.888 \end{array}\right] \end{array} \begin{array}{c} \\ 1 \\ 2 \\ 3 \\ \\ \\ \\ \end{array} \tag{a}$$

ELEMENT 2 (Joint ⑤ to joint ③)

Direction Cosines: $c_1 = -0.707 \quad c_2 = 0.707 \quad c_3 = 0$

$$[\bar{k}]_2 = \begin{array}{c@{}l} \begin{array}{rrrrrr} 1 & 2 & 3 & & & \end{array} & \\ \left[\begin{array}{rrrrrr} 106.066 & 106.066 & 0.000 & -106.066 & -106.066 & 0.000 \\ 106.066 & 106.066 & 0.000 & -106.066 & -106.066 & 0.000 \\ 0.000 & 0.000 & 0.000 & 0.000 & 0.000 & 0.000 \\ -106.066 & -106.066 & 0.000 & 106.066 & 106.066 & 0.000 \\ -106.066 & -106.066 & 0.000 & 106.066 & 106.066 & 0.000 \\ 0.000 & 0.000 & 0.000 & 0.000 & 0.000 & 0.000 \end{array}\right] & \begin{array}{l} 1 \\ 2 \\ 3 \\ \\ \\ \\ \end{array} \end{array} \tag{b}$$

ELEMENT 3 (Joint ⑤ to joint ②)

Direction Cosines: $c_1 = -1 \quad c_2 = 0 \quad c_3 = 0$

$$[\bar{k}] = \begin{array}{c@{}l} \begin{array}{rrrrrr} 1 & 2 & 3 & & & \end{array} & \\ \left[\begin{array}{rrrrrr} 300.000 & 0.000 & 0.000 & -300.000 & 0.000 & 0.000 \\ 0.000 & 0.000 & 0.000 & 0.000 & 0.000 & 0.000 \\ 0.000 & 0.000 & 0.000 & 0.000 & 0.000 & 0.000 \\ -300.000 & 0.000 & 0.000 & 300.000 & 0.000 & 0.000 \\ 0.000 & 0.000 & 0.000 & 0.000 & 0.000 & 0.000 \\ 0.000 & 0.000 & 0.000 & 0.000 & 0.000 & 0.000 \end{array}\right] & \begin{array}{l} 1 \\ 2 \\ 3 \\ \\ \\ \\ \end{array} \end{array} \tag{c}$$

ELEMENT 4 (Joint ⑤ to joint ④)

Direction cosines: $c_1 = 0 \quad c_2 = -1 \quad c_3 = 0$

$$[\bar{k}]_4 = \begin{array}{c@{}l} \begin{array}{rrrrrr} 1 & 2 & 3 & & & \end{array} & \\ \left[\begin{array}{rrrrrr} 0.000 & 0.000 & 0.000 & 0.000 & 0.000 & 0.000 \\ 0.000 & 300.000 & 0.000 & 0.000 & -300.000 & 0.000 \\ 0.000 & 0.000 & 0.000 & 0.000 & 0.000 & 0.000 \\ 0.000 & 0.000 & 0.000 & 0.000 & 0.000 & 0.000 \\ 0.000 & -300.000 & 0.000 & 0.000 & 300.000 & 0.000 \\ 0.000 & 0.000 & 0.000 & 0.000 & 0.000 & 0.000 \end{array}\right] & \begin{array}{l} 1 \\ 2 \\ 3 \\ \\ \\ \\ \end{array} \end{array} \tag{d}$$

3. Reduced System Stiffness Matrix.

The reduced system stiffness matrix $[K]_R$ is assembled by transferring the coefficients in the stiffness matrices of the elements according to the nodal coordinates indicated for the free coordinates 1, 2 and 3 at the top and on the right side of these matrices, (a), (b) and (d). The transfer of these coefficients to the appropriate locations in the reduced system stiffness matrix results in

$$
\begin{array}{c}
\begin{matrix} \quad\quad 1 & \quad\quad\quad\quad 2 & \quad\quad\quad 3 \end{matrix} \\
[K]_R = \begin{bmatrix} 88.89+106.07+300 & 44.44+106.07 & 88.89 \\ 44.44+106.07 & 22.22+106.07+300 & 44.44 \\ 88.89 & 44.44 & 88.89 \end{bmatrix} \begin{matrix} 1 \\ 2 \\ 3 \end{matrix}
\end{array}
$$

or

$$
[K]_R = \begin{bmatrix} 494.96 & 150.51 & 88.89 \\ 150.51 & 428.92 & 44.44 \\ 88.89 & 44.44 & 88.89 \end{bmatrix} \tag{e}
$$

4. Reduced system force vector.

For this Illustrative Example the reduced system vector $\{F\}_R$ contains the three forces of 10 kips applied to joint ⑤ of the truss. That is

$$
\{F\}_R = \begin{Bmatrix} 10 \\ 10 \\ 10 \end{Bmatrix} \tag{f}
$$

5. Reduced System Stiffness Matrix Equation.

The reduced system stiffness matrix equation obtained from eqs. (e) and (f) is

$$
\begin{Bmatrix} 10 \\ 10 \\ 10 \end{Bmatrix} = \begin{bmatrix} 494.96 & 150.51 & 88.89 \\ 150.51 & 428.92 & 44.44 \\ 88.89 & 44.44 & 88.89 \end{bmatrix} \begin{Bmatrix} u_1 \\ u_2 \\ u_3 \end{Bmatrix} \tag{g}
$$

6. Displacements at the Free Coordinates.

The solution of eq.(g) yields the displacements at the free nodal coordinates as

$$u_1 = -0.0032 \text{ (in)}$$

$$u_2 = 0.0133 \text{ (in)}$$

$$u_3 = 0.1080 \text{ (in)}$$

7. Element Axial Forces.

The element axial forces are calculated using eq.(7.14) after identifying the components of the element nodal displacements. For example, for element 1, the nodal displacement vector is identified as

$$\{\bar{\delta}\}_1 = \begin{Bmatrix} u_1 \\ u_2 \\ u_3 \\ 0 \\ 0 \\ 0 \end{Bmatrix} = \begin{Bmatrix} -0.0032 \\ 0.0133 \\ 0.1080 \\ 0 \\ 0 \\ 0 \end{Bmatrix}$$

The substitution into eqs.(7.5) and (7.6) of the coordinates for the two joints of element 1 ($X_1 = 100$, $Y_1 = 100$, $Z_1 = 100$) and ($X_2 = 0$, $Y_2 = 50$, $Z_2 = 0$) results in

$$L = \sqrt{(0-100)^2 + (50-100)^2 + (0-100)^2} = 150 \quad (in)$$

and

$$c_1 = \frac{0-100}{150} = -0.667 \qquad c_2 = \frac{0-50}{150} = -0.333 \qquad c_3 = \frac{0-100}{150} = -0.667$$

Then substituting $E = 30{,}000$ ksi, $A = 1.0$ in^2 and the above calculated results into eq.(7.14) gives the axial force $P^{(1)}$of element 1 as

$$P^{(1)} = \frac{30000 \times 1.0}{150}\left[(0.0032)(-0.667) + (-0.0133)(-0.333) + (-0.1080)(-0.667)\right]$$

$$P^{(1)} = 14.86 \;\; \text{kips (Tension)}$$

Analogously, the axial forces $P^{(2)}$, $P^{(3)}$ and $P^{(4)}$, calculated for the other elements of this truss are:

$P^{(2)} = 1.51$ kip (Tension)
$P^{(3)} = -0.96$ kip (Compression)
$P^{(4)} = 3.98$ kip (Tension)

Illustrative Example 7.2

Use SAP2000 to analyze the space truss of Illustrative Example 7.1. The model of this space truss is reproduced, for convenience, in Fig. 7.3.

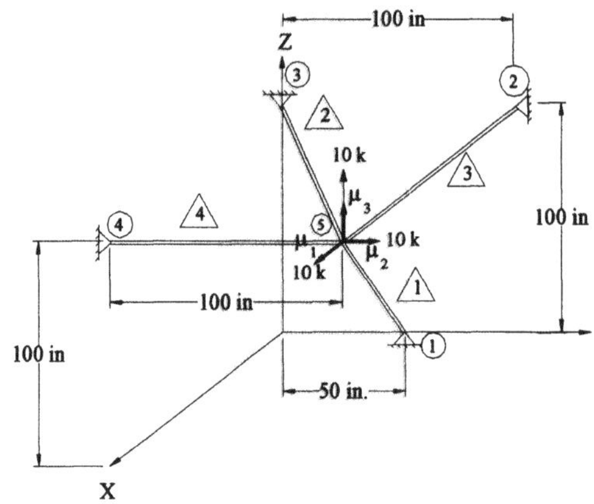

Fig. 7.3 Space Truss for Illustrative Example 7.2

Begin: Open SAP2000.

Enter: "OK" to disable the "Tip of the Day"

Hint: In the lower right-hand corner of the screen select "kip-in".

Select: From the Main Menu enter:
FILE>NEW MODEL
Number of Grid Spaces
X-Direction = 4
Y-Direction = 4
Z-Direction = 4
Grid Spacing:
X-Direction = 50
Y-Direction = 50
Z-Direction = 50. Then OK.

Set: Translate the system of coordinates on each window of the split screen: Click on the **3-D** view bar to select this window and enter from the Main Menu:

VIEW>SET 3-D VIEW
Change view direction angle to: Plan = 37°, Elevation = 23° and Aperture = 30°. Then OK.

VIEW>SET 2-D VIEW
Set *X-Y* plane at $Z = 0$
Set *X-Z* plane at $Y = 0$
Set *Y-Z* plane at $X = 0$. Then OK.

Select the *X-Y* plane window and from the Main Menu enter:
VIEW>SET 2-D VIEW
Set *X-Y* plane at $Z = 0$
Set *X-Z* plane at $Y = 0$
Set *Y-Z* plane at $X = 0$. Then OK.

Use the PAN icon to center the coordinate axes in the split screen.

Draw: Draw points in the *Y-Z* plan
Click on icon ***YZ*** to select this plane.
From the Main Menu enter:
DRAW>ADD SPECIAL JOINT
Click on the screen at point with coordinates $Y = 50$, $Z = 0$.
Click at Point $Y = 100$, $Z = 100$;
Click at Point $Y = 0$, $Z = 100$

Draw points in the *X-Z* plane
Click on icon *XZ*, then click on point $X = 100$, $Z = 100$
Finally, default to 3-D window and select or pick any point in the neighborhood of $X = 100$, $Y = 100$, $Z = 100$
Then right click on this selected point and change coordinates to:
$X = 100$, $Y = 100$, $Z = 100$

Label: From the Main Menu enter:
VIEW>SET ELEMENT
Click on Joint Labels. Then OK.

Draw: Draw beam elements:
From the Main Menu enter: DRAW>DRAW FRAME ELEMENT.

Click on point 5 and drag to point 1, then release and press Enter
Click on point 5 and drag to point, 2 then release and press Enter
Click on Point 5 and drag to point, 3 then release and press Enter
Click on Point 5 and drag to point, 4 then release and press Enter

Set: For a better view of the 3-D window minimize the *X-Z* plane window and maximize the 3-D view.
Use the PAN icon to center the view of the structure.

Label: From the Main Menu enter
VIEW>SET ELEMENTS
Click on Frame Labels. Then OK.

Material: From the Main Menu enter:
DEFINE >MATERIALS
Select STEEL and Modify/Show Material
Set E = 30,000. Then OK, OK.

Sections: From the Main Menu enter:
DEFINE>FRAME SECTIONS
Click on Add/Wide Flange and select "Add General" in the Property Data Screen set cross-section, set area = 1.00. Then OK.

For convenience change the name of the selected section from FSEC2 to TRUSS7-2. Then OK.

Assign: Mark or select all the elements and from the Main Menu enter:
ASSIGN>FRAME>SECTIONS
Select TRUSS7-2. Then OK.

Define: From the Main Menu enter
DEFINE>STATIC LOAD CASES
Change DEAD to LIVE
And set self-weight Multiplier to 0.
Then click on change load. Then OK.

Boundary: Mark joints ①, ②, ③ and ④, then from the Main Menu enter:
ASSIGN>JOINT>RESTRAINTS
Select restraints in translation 1, 2, and 3. OK
Mark joint ⑤ and from the Main Menu enter:
ASSIGN>JOINT>RESTRAINTS
Select no restraints. Then OK.

Loads: Mark joint ⑤ and from the Main Menu enter:
ASSIGN>JOINT STATIC LOADS>FORCES
In the Load Menu enter:
FORCE GLOBAL X = 10
FORCE GLOBAL Y = 10
FORCE GLOBAL Z = 10. Then OK.

Plot: From the Main Menu enter:
DISPLAY>UNDEFORMED SHAPE
DISPLAY>SHOW LOADS>JOINT
Click on show load values
FILE>PRINT GRAPHICS

(Fig. 7.3 reproduces the plot of the undeformed shape shown on the screen for this space truss)

Analyze: From the Main Menu enter:
ANALYZE>SET OPTIONS
Select available degrees of freedom translation in *X*, *Y*, and *Z*. Then OK.

From the Main Menu enter:
ANALYZE>RUN
Respond to filename request by entering "Example 7.2". Then SAVE.

Note: When the analysis is completed, click OK.

Print Input Tables: From the Main Menu enter:
FILE>PRINT INTPUT TABLES
(Edited input tables are reproduced as Table 7.1)

Print Output Tables: From the Main Menu enter:
FILE>PRINT OUTPUT TABLES
(Edited output tables are reproduced as Table 7.2)

Table 7.1 Edited Input Data for Illustrative Example 7.2 (Units: Kips, inches)

JOINT DATA

JOINT	GLOBAL-X	GLOBAL-Y	GLOBAL-Z	RESTRAINTS	ANG-A	ANG-B	ANG-C
1	0.00000	50.00000	0.00000	1 1 1 0 0 0	0.000	0.000	0.000
2	0.00000	100.00000	100.00000	1 1 1 0 0 0	0.000	0.000	0.000
3	0.00000	0.00000	100.00000	1 1 1 0 0 0	0.000	0.000	0.000
4	100.00000	0.00000	100.00000	1 1 1 0 0 0	0.000	0.000	0.000
5	100.00000	100.00000	100.00000	0 0 0 0 0 0	0.000	0.000	0.000

FRAME ELEMENT DATA

FRM	JNT-1	JNT-2	SCTN	ANG	RLS	SGMNTS	R1	R2	FACTOR	LENGTH
1	5	1	TRUSS7-2	0.000	000000	2	0.000	0.000	1.000	150.000
2	5	3	TRUSS7-2	0.000	000000	4	0.000	0.000	1.000	141.421
3	5	2	TRUSS7-2	0.000	000000	4	0.000	0.000	1.000	100.000
4	5	4	TRUSS7-2	0.000	000000	4	0.000	0.000	1.000	100.000

JOINT FORCES Load Case LOAD1

JOINT	GLOBAL-X	GLOBAL-Y	GLOBAL-Z	GLOBAL-XX	GLOBAL-YY	GLOBAL-ZZ
5	10.000	10.000	10.000	0.000	0.000	0.000

Table 7.2 Edited Output results for Illustrative Example 7.2 (Units: Kips. Inches)

JOINT DISPLACEMENTS

JOINT	LOAD	UX	UY	UZ	RX	RY	RZ
1	LOAD1	0.0000	0.0000	0.0000	0.0000	0.0000	0.0000
2	LOAD1	0.0000	0.0000	0.0000	0.0000	0.0000	0.0000
3	LOAD1	0.0000	0.0000	0.0000	0.0000	0.0000	0.0000
4	LOAD1	0.0000	0.0000	0.0000	0.0000	0.0000	0.0000
5	LOAD1	-3.204E-03	0.0133	0.1080	0.0000	0.0000	0.0000

JOINT REACTIONS

JOINT	LOAD	F1	F2	F3	M1	M2	M3
1	LOAD1	-9.8969	-4.9500	-9.9088	0.0000	0.0000	0.0000
2	LOAD1	0.9612	-4.758E-03	-0.0388	0.0000	0.0000	0.0000
3	LOAD1	-1.0654	-1.0675	-0.0137	0.0000	0.0000	0.0000
4	LOAD1	1.150E-03	-3.9777	-0.0388	0.0000	0.0000	0.0000

FRAME ELEMENT FORCES

FRAME	LOAD	LOC	P	V2	V3	T	M2	M3
1	LOAD1							
		0.00	14.85	0.00	0.00	0.00	0.00	0.00
		75.00	14.85	0.00	0.00	0.00	0.00	0.00
		150.00	14.85	0.00	0.00	0.00	0.00	0.00
2	LOAD1							
		0.00	1.51	0.00	0.00	0.00	0.00	0.00
		35.36	1.51	0.00	0.00	0.00	0.00	0.00
		70.71	1.51	0.00	0.00	0.00	0.00	0.00
		106.07	1.51	0.00	0.00	0.00	0.00	0.00
		141.42	1.51	0.00	0.00	0.00	0.00	0.00
3	LOAD1							
		0.00	-0.96	0.00	0.00	0.00	0.00	0.00
		25.00	-0.96	0.00	0.00	0.00	0.00	0.00
		50.00	-0.96	0.00	0.00	0.00	0.00	0.00
		75.00	-0.96	0.00	0.00	0.00	0.00	0.00
		100.00	-0.96	0.00	0.00	0.00	0.00	0.00
4	LOAD1							
		0.00	3.98	0.00	0.00	0.00	0.00	0.00
		25.00	3.98	0.00	0.00	0.00	0.00	0.00
		50.00	3.98	0.00	0.00	0.00	0.00	0.00
		75.00	3.98	0.00	0.00	0.00	0.00	0.00
		100.00	3.98	0.00	0.00	0.00	0.00	0.00

Plot Displacements: From the Main Menu enter:

DISPLAY>SHOW DEFORMED SHAPE, then enter:

FILE>PRINT GRAPICS (Fig. 7.4 shows the deformed shape of the truss)

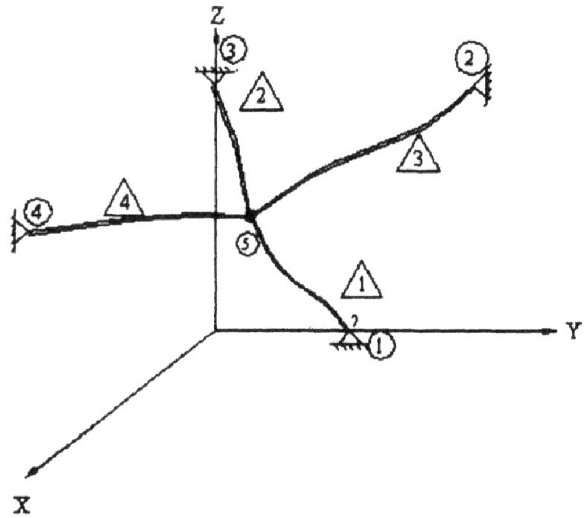

Fig. 7.4 Deformed shape for the space truss of Illustrative Example 7.2

Plot Axial Forces: From the Main Menu enter:
DISPLAY>SHOW ELEMENT FORCES/STRESSES>FRAMES
In the member force diagram for frames, select Axial Force and Fill Diagram. Then OK.
(Fig. 7.5 shows the axial force diagram for the members of the truss)

Note: The computer output shows extremely small values for the shear forces V2, V3 and the bending moments M2, M3 (in the range of numerical accuracy). However, in editing this table these values have been set equal to zero.

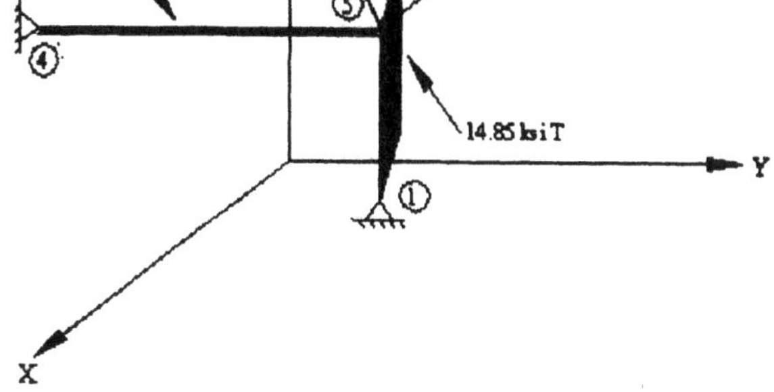

Fig. 7.5 Axial force diagram for the members of the space truss of Illustrative Example 7.2

7.6 Problems

Problem 7.1

For the space truss shown in Fig. P7.1 determine:

(a) Joint displacements
(b) Axial member forces
(c) Support reactions

All Members:
$E = 90$ GPa
$A = 25\ cm^2$

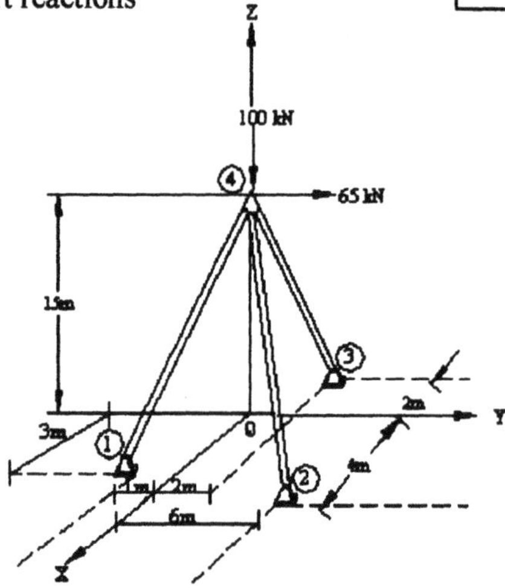

Fig. P7.1

Problem 7.2

For the space truss shown in Fig. P7.2 determine:

(a) Joint displacements
(b) Axial member forces
(c) Support reactions

All Members:
$E = 29500$ ksi
$A = 7.5\ in^2$

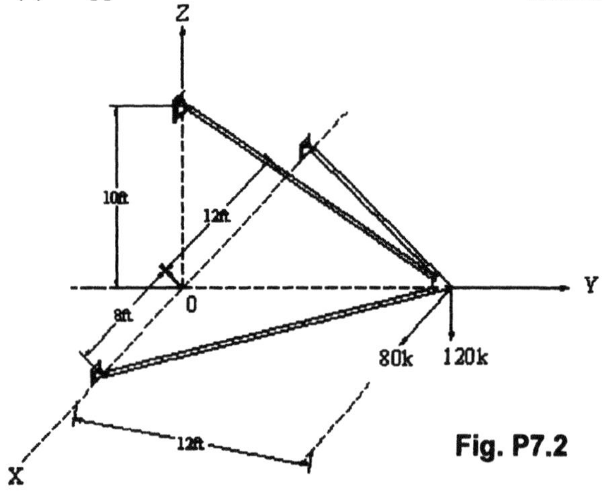

Fig. P7.2

Problem 7.3

For the space truss shown in Fig. P7.3 determine:

(a) Joint displacements
(b) Member axial forces
(c) Support reactions

All Members:
E = 29500 ksi
A = 8 in^2

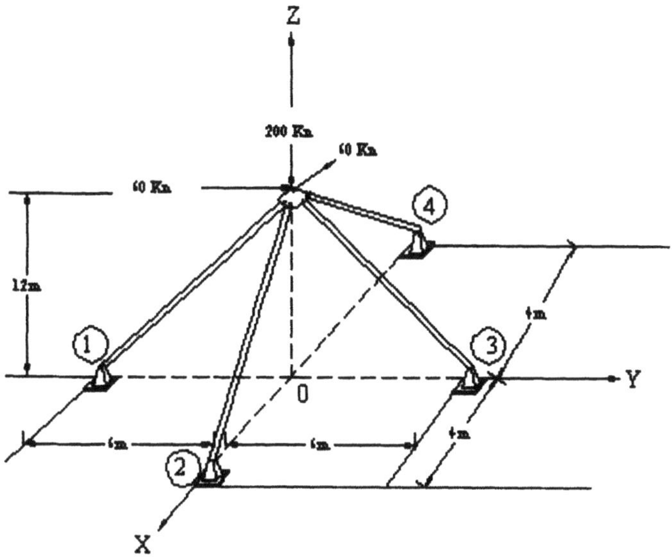

Fig. P7.3

Problem 7.4

Solve truss in Problem 7.3 for the combined action of vertical settlement of 2 in. at the support on joint ③.

Problem 7.5

For the space truss shown in Fig. P7.5 determine:

(a) Joint displacements
(b) Axial member forces
(c) Support reactions

All Members:
E = 200 GPa
A = 42 cm^2

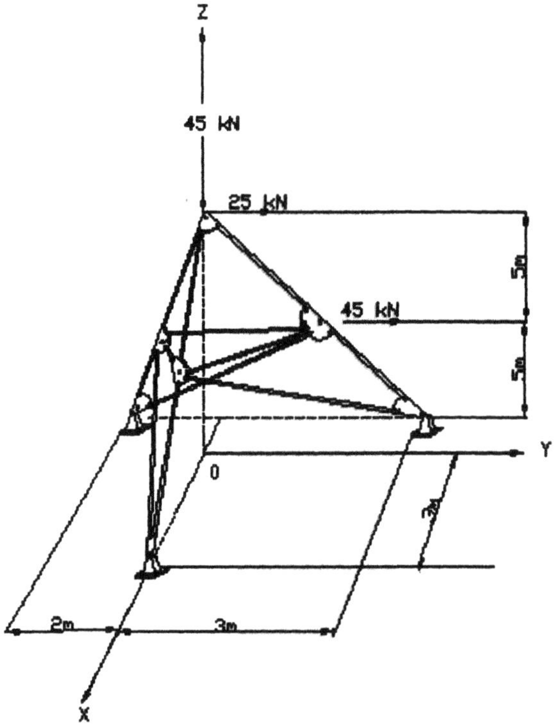

Fig. P7.5

8 Static Condensation and Substructuring

8.1 Introduction

In the previous chapters we established the relationship (through the system stiffness matrix) between forces applied at the nodal coordinates of the structure and the corresponding nodal displacements. There are instances, such as the use of substructuring for the analysis of large structures, in which it might be advantageous to reduce the number of nodal coordinates. Substructuring requires the reduction of nodal coordinates to allow the independent analysis of portions of the structure (substructuring). The process of reducing the number of free displacements or degrees of freedom is known as static condensation. The same process is also applied to dynamic problems although, in that case, it is only approximate and in general may result in large errors. The static condensation method has recently been modified for applications to dynamic problems. This method is known as the dynamic condensation method (Paz, M. 1997); its application to dynamic problems gives solutions that are virtually exact.

8.2 Static Condensation

A practical method of accomplishing the reduction of the number of degrees of freedom and hence the reduction of the stiffness matrix, is to identify those degrees of freedom to be condensed as secondary degrees of freedom, and to express them in terms of the remaining primary degrees of freedom. The relationship between secondary and primary degrees of freedom is found by establishing the static relation between them, hence the name Static Condensation Method. This relationship provides the means to reduce the number of unknowns in the system stiffness matrix equation. In order to describe the Static Condensation Method, assume that those secondary degrees of freedom to be reduced or condensed are arranged in the first s nodal coordinates and the remaining primary degrees of freedom are the last p nodal coordinates. With such an arrangement the stiffness equation for a structure may be written using partitioned matrices as

$$\left\{\begin{array}{c}\{F\}_s \\ \hline \{F\}_p\end{array}\right\} = \left[\begin{array}{c|c}[K]_{ss} & [K]_{sp} \\ \hline [K]_{ps} & [K]_{pp}\end{array}\right] \left\{\begin{array}{c}\{u\}_s \\ \hline \{u\}_p\end{array}\right\} \tag{8.1}$$

where $\{u\}_s$ is the displacement vector corresponding to the s secondary degrees of freedom to be reduced and $\{u\}_p$ is the vector containing the remaining p primary degrees of freedom. A simple multiplication of the partitioned system in eq.(8.1) yields the following two matrix equations:

$$[K]_{ss}\{u\}_s + [K]_{sp}\{u\}_p = \{F\}_s \tag{8.2}$$

$$[K]_{ps}\{u\}_s + [K]_{pp}\{u\}_p = \{F\}_p \tag{8.3}$$

Solving eq.(8.2) for the vector $\{u\}_s$ and subsequently substituting it in eq.(8.3) results in

$$\{u\}_s = [K]_{ss}^{-1}\left(\{F\}_s - [K]_{sp}\{u\}_p\right) \tag{8.4}$$

$$\left(\{F\}_p - [K]_{ps}[K]_{ss}^{-1}\{F\}_s\right) = \left([K]_{pp} - [K]_{ps}[K]_{ss}^{-1}[K]_{sp}\right)\{u\}_p \tag{8.5}$$

Equation (8.5) may conveniently be written as

$$\{\overline{F}\}_p = [\overline{K}]\{u\}_p \tag{8.6}$$

in which

$$\{\overline{F}\} = \{F\}_p - [K]_{ps}[K]_{ss}^{-1}\{F\}_s \tag{8.7}$$

and

$$[\overline{K}] = [K]_{pp} - [K]_{ps}[K]_{ps}[K]_{ss}^{-1}[K]_{sp} \tag{8.8}$$

Thus, eq.(8.6) is the condensed stiffness equation relating through the condensed stiffness matrix $[\overline{K}]$, the condensed force vector $\{\overline{F}\}$ and the primary nodal coordinates $\{u\}_p$.

The solution of the condensed stiffness equation (8.6) gives the displacement vector $\{u\}_p$ at the primary nodal coordinates. The displacements $\{u\}_s$

at the secondary nodal coordinates are then calculated by eq.(8.4) after substituting into this equation the displacement vector $\{u\}_p$ for the primary nodal degrees of freedom.

The development yielding eqs.(8.4) and (8.5) may appear to indicate that the use of the condensation method requires the inconvenient calculation of the inverse matrix $[K]_{ss}^{-1}$. However, the practical application of the Static Condensation Method does not actually require a matrix inversion. Instead, the standard Gauss-Jordan elimination process is applied systematically to the system stiffness equation up to the reduction of the secondary coordinates $\{u\}_s$. At this stage of the elimination process the stiffness equation (8.1) has been reduced to

$$\left\{\begin{array}{c}\{\bar{F}\}_s\\ \hline \{\bar{F}\}_p\end{array}\right\} = \left[\begin{array}{c|c}[I] & -[\bar{T}]\\ \hline [0] & [\bar{K}]\end{array}\right]\left\{\begin{array}{c}\{u\}_s\\ \hline \{u\}_p\end{array}\right\} \tag{8.9}$$

By expanding eq.(8.9), it may be seen that

$$\{\bar{F}\}_s = \{u\}_s - [\bar{T}]\{u\}_p$$

or

$$\{u\}_s = \{\bar{F}\}_s + [\bar{T}]\{u\}_p \tag{8.10}$$

and

$$\{\bar{F}\}_p = [\bar{K}]\{u\}_p \tag{8.11}$$

are equivalent to eqs.(8.4) and (8.5). Therefore, the partial use of Gauss-Jordan elimination process yields both the relationship [eq.(8.10)] between secondary nodal coordinate $\{u\}_s$ and the primary nodal coordinates $\{u\}_p$ as well as the stiffness equation for the condensed system [eq.(8.11)]. The following example illustrates the process of static condensation applied to a simple structural system.

Illustrative Example 8.1

Consider in Fig. 8.1(a) a step beam supporting a force P_0 = 8 kips and a moment M_0=240 kip-in.

Determine the displacement and rotation of the central section of this beam:

(a) no condensation, and

(b) condensation of the rotational degree of freedom at the central section.

E = 10,000 ksi
I_1 = 100 in^4
I_2 = 200 in^4

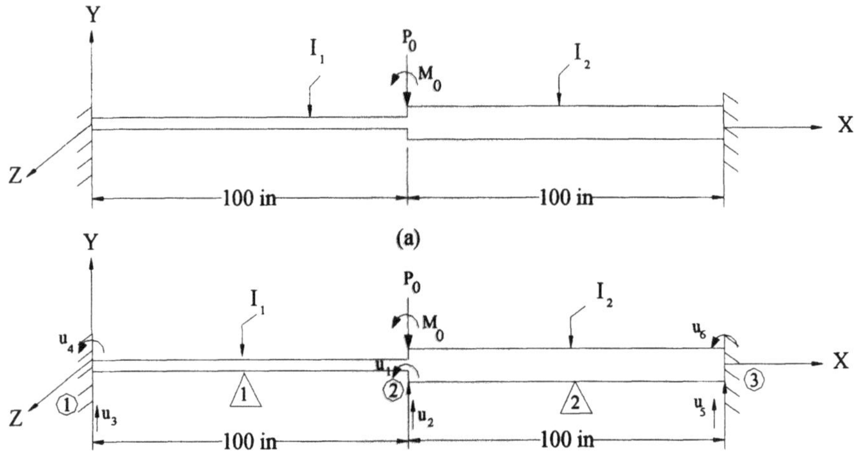

Fig. 8.1 (a) Step beam for Illustrative Example 8.1
(b) Analytical model

Solution a) – No condensation:

1. Analytical Model:

   Figure 8.1(b) shows the analytical model consisting of two beam elements with three joints and a total of six nodal coordinates of which the first two u_1 and u_2 are free nodal coordinates.

2. Element stiffness matrices:
   The stiffness matrix for a beam element is given by eq.(1.11).

   Substitution of numerical values for element 1 into eq.(1.11) results in

$$[k]_1 = \begin{matrix} 3 & 4 & 2 & 1 \end{matrix} \\ \begin{bmatrix} 12 & 600 & -12 & 600 \\ 600 & 40000 & -600 & 20000 \\ -12 & -600 & 12 & -600 \\ 600 & 20000 & -600 & 40000 \end{bmatrix} \begin{matrix} 3 \\ 4 \\ 2 \\ 1 \end{matrix} \qquad \text{(a)}$$

   and for element 2 :

$$[k]_2 = \begin{bmatrix} 24 & 1200 & -24 & 1200 \\ 1200 & 80000 & -1200 & 40000 \\ -24 & -1200 & 24 & -1200 \\ 1200 & 40000 & -1200 & 80000 \end{bmatrix} \begin{matrix} 2 \\ 1 \\ 5 \\ 6 \end{matrix} \quad \text{(columns: 2, 1, 5, 6)} \tag{b}$$

3. Reduced system stiffness matrix:

The reduced system stiffness matrix is assembled by transferring to appropriate locations the coefficients of the matrices in eqs.(a) and (b) corresponding to the free nodal coordinates 1 and 2 indicated at the top and on the right side of these matrices, namely

$$[K]_R = \begin{bmatrix} 40000+80000 & -600+1200 \\ -600+1200 & 12+24 \end{bmatrix}$$

or

$$[K]_R = \begin{bmatrix} 120000 & 600 \\ 600 & 36 \end{bmatrix} \tag{c}$$

4. Reduced System force vector:

The reduced system force vector $\{F\}_R$ contains the moment $M_0 = 240$ (kip-in) and the force $P_0 = -8$ kip applied, respectively, at the nodal coordinates 1 and 2 as shown in Fig. 8.1. Namely,

$$\{F\}_R = \begin{Bmatrix} 240 \\ -8 \end{Bmatrix} \tag{d}$$

5. Reduce system stiffness equation:
The reduced system stiffness matrix is

$$\{F\}_R = [K]_R(u)$$

Then substituting $[K]_R$ and $\{F\}_R$ respectively from eqs.(c) and (d),

$$\begin{Bmatrix} 240 \\ -8 \end{Bmatrix} = \begin{bmatrix} 120000 & 600 \\ 600 & 36 \end{bmatrix} \begin{Bmatrix} u_1 \\ u_2 \end{Bmatrix} \tag{e}$$

6. Nodal displacements at free coordinates:

The solution of eq.(e) yields

$$u_1 = 0.00339 \text{ (rad)}$$
$$u_2 = -0.279 \text{ (in)} \tag{f}$$

Solution b)– With Condensation:

1. The Gauss-Jordan process is applied to eq.(e) to condense the secondary degree of freedom u_1. Thus dividing the first row of eq.(e) by 120,000 gives

$$\begin{Bmatrix} 0.002 \\ -8 \end{Bmatrix} = \begin{bmatrix} 1 & 0.005 \\ 600 & 36 \end{bmatrix} \begin{Bmatrix} u_1 \\ u_2 \end{Bmatrix} \tag{g}$$

2. Then multiplying the first row of eq.(g) by 600 and subtracting this product from the second row yields:

$$\begin{Bmatrix} 0.002 \\ \hline -9.2 \end{Bmatrix} = \left[\begin{array}{c|c} 1 & 0.005 \\ \hline 0 & 33 \end{array}\right] \begin{Bmatrix} u_1 \\ \hline u_2 \end{Bmatrix} \tag{h}$$

3. Expanding the second row of eq.(h) provides the condensed stiffness equation as

$$-9.2 = 33u_2$$

or

$$u_2 = -0.279 \text{ (in)} \tag{i}$$

and from the first row of eq.(h), the relationship to calculate the secondary nodal coordinate u_1 as

$$0.002 = u_1 + 0.005u_2 \tag{j}$$

4. The substitution of $u_2 = -0.279$ from eq.(i) into eq.(j) results in

$$u_1 = 0.00339 \text{ (rad)} \tag{k}$$

5. As expected, the solution for the nodal displacements given by eq.(i) and (k) after condensation is equal to the solution obtained in eq.(f) with no condensation.

8.3 Substructuring

Substructuring or the separation of the structure into parts or substructures, is a convenient way to design and analyze structures that have been modeled with a large number of elements, hence resulting in a large number of equations. Some examples of this type of structure are aircraft, naval vessels, large high-rise buildings and domes.

In substructuring, use is made of the static condensation method to establish the analytical relationship between the various parts (substructures) of the structure.

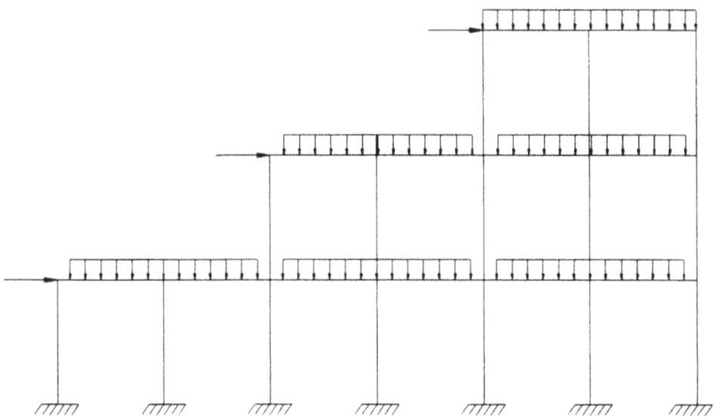

Fig. 8.2 Plane Frame

To illustrate the use of substructures consider the plane frame in Fig. 8.2, which has been divided into three substructures as shown in Fig. 8.3 where the connecting joints are indicated with dark dots.

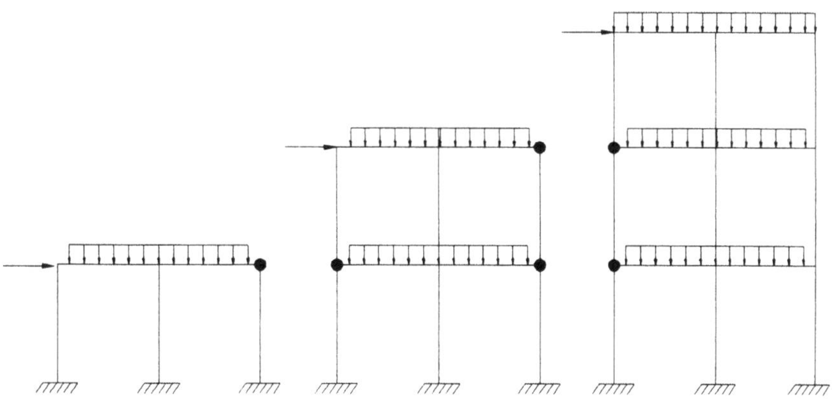

Fig. 8.3 Substructures for the plane frame in Fig. 8.2.

In analyzing the plane frame shown in Fig. 8.2 we may select as primary degrees of freedom, the displacements of the interface nodes between the substructures in Fig. 8.3. We then proceed to condense in each of the substructures the degrees of freedom not identified as primary selected degrees of freedom at the interface nodes.

Each condensed substructure will provide a condensed stiffness matrix and a condensed force vector, which then can be assembled to obtain the condensed stiffness equation of the structure. For the plane frame in Fig. 8.3 the Substructure 1 will be condensed, from 9 nodal coordinates to 3. Similarly, Substructures 2 and 3 will be reduced from 18 nodal coordinates to 9 and from 27 to 6, respectively. Then the subsequent combination of these condensed substructures considered to be super-elements will result in the condensed system stiffness equation with only 9 nodal coordinates reduced from a total of 45 free nodal coordinates in the plane frame of Fig. 8.2.

Illustrative Example 8.2

Consider the beam shown in Fig. 8.4(a) and perform its analysis using two substructures as indicated in Fig. 8.4(b).

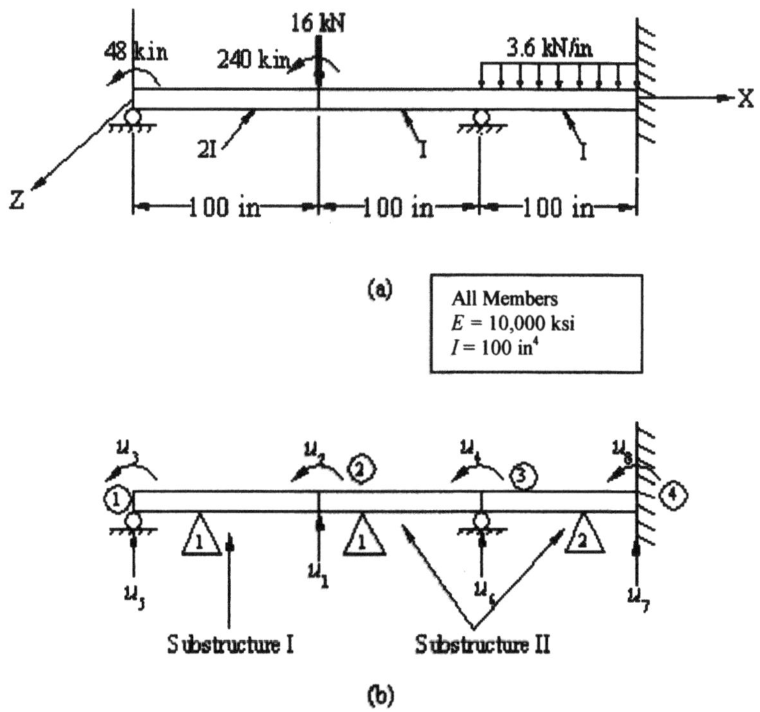

Fig. 8.4 (a) Beam for Illustrative Example 8.2 showing selected substructures; (b) Analytical model.

Solution:

The analytical model for this structure is shown in Fig.8.4(b) in which the primary coordinates located in the interface between substructures I and II are labeled as nodal coordinates 1 and 2 with displacements u_1 and u_2.

SUBSTRUCTURE I -- ELEMENT 1:

δ_1 secondary coordinate

δ_2 and δ_3 primary coordinates

$\delta_4 = 0$ fixed coordinate

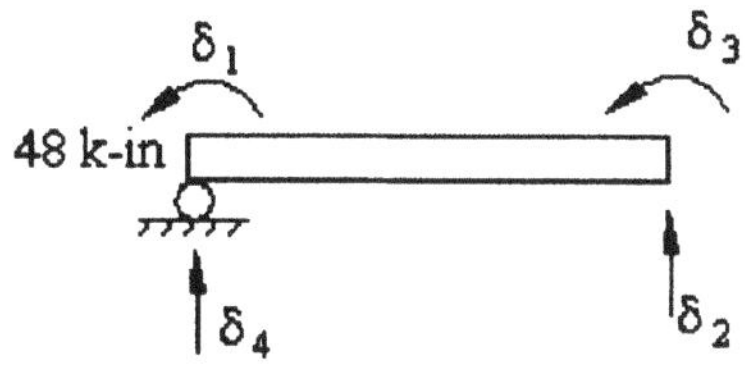

1. Element stiffness matrix $[k\}_1^{\mathrm{I}}$ (eq.1.11) and the element force vector $\{P\}_1^{\mathrm{I}}$ for element 1 of substructure I.

$$[k]_1^{\mathrm{I}} = \begin{array}{c} \begin{array}{cccc} 4 & 1 & 2 & 3 \end{array} \\ \left[\begin{array}{cccc} 24 & 1200 & -24 & 1200 \\ 1200 & 80000 & -1200 & 40000 \\ -24 & -1200 & 24 & -1200 \\ 1200 & 40000 & -1200 & 80000 \end{array}\right] \end{array} \begin{array}{c} 4 \\ 1 \\ 2 \\ 3 \end{array} \qquad \{P\}_1^{\mathrm{I}} = \begin{Bmatrix} 0 \\ 48 \\ 0 \\ 0 \end{Bmatrix} \begin{array}{c} 4 \\ 1 \\ 2 \\ 3 \end{array} \tag{a}$$

2. Assemble for substructure I the reduced system stiffness matrix $[K]_R^{\mathrm{I}}$ and the reduced system force vector $\{F\}_R^{\mathrm{J}}$ for the free nodal coordinates $\delta_1, \delta_2, \text{and}\, \delta_3$.

$$[K]_R^{\mathrm{I}} = \begin{bmatrix} 80000 & -1200 & 40000 \\ -1200 & 24 & -1200 \\ 40000 & -1200 & 80000 \end{bmatrix} \qquad \{F\}_R^{\mathrm{I}} = \begin{Bmatrix} 48 \\ 0 \\ 0 \end{Bmatrix}$$

3. Reduced system stiffness equation for substructure I.

$$\begin{Bmatrix} 48 \\ 0 \\ 0 \end{Bmatrix} = \begin{bmatrix} 80000 & -1200 & 40000 \\ -1200 & 24 & -1200 \\ 40000 & -1200 & 80000 \end{bmatrix} \begin{Bmatrix} \delta_1 \\ \delta_2 \\ \delta_3 \end{Bmatrix}^{\mathrm{I}} \tag{b}$$

4. Condensation of degree of freedom 1.

Divide by 80000 the first line in eq.(b):

$$\begin{Bmatrix} 0.0006 \\ 0 \\ 0 \end{Bmatrix} = \begin{bmatrix} 1 & -1.015 & 0.500 \\ -1200 & 24 & -1200 \\ 40000 & -1200 & 80000 \end{bmatrix} \begin{Bmatrix} \delta_1 \\ \delta_2 \\ \delta_3 \end{Bmatrix}^{\mathrm{I}}$$

Then add to the second row the first row multiplied by 1200 and add to the 3rd row the first row multiplied by (-40000):

$$\begin{Bmatrix} 0.0006 \\ \hline -0.72 \\ -24 \end{Bmatrix} = \left[\begin{array}{c|cc} 1 & -0.015 & 0.500 \\ \hline 0 & 6 & -600 \\ 0 & -600 & 60000 \end{array}\right] \begin{Bmatrix} \delta_1 \\ \hline \delta_2 \\ \delta_3 \end{Bmatrix} \qquad \text{(c)}$$

5. Expand the partitioned matrix (c) to obtain:

$$0.0006 = \delta_1 - 0.015\delta_2 + 0.500\delta_3 \qquad \text{(d)}$$

and

$$\begin{matrix} \\ 1 \\ 2 \end{matrix} \begin{Bmatrix} \\ +0.72 \\ -24 \end{Bmatrix} = \begin{Bmatrix} 1 & 2 \\ 6 & -600 \\ -600 & 60000 \end{Bmatrix} \begin{Bmatrix} \\ \delta_2 \\ \delta_3 \end{Bmatrix} \qquad \text{(e)}$$

SUBSTRUCTURE II – ELEMENTS 1 AND 2:

δ_1 secondary coordinate

δ_2 and δ_3 primary coordinates

δ_4, δ_5 and δ_6 fixed coordinates

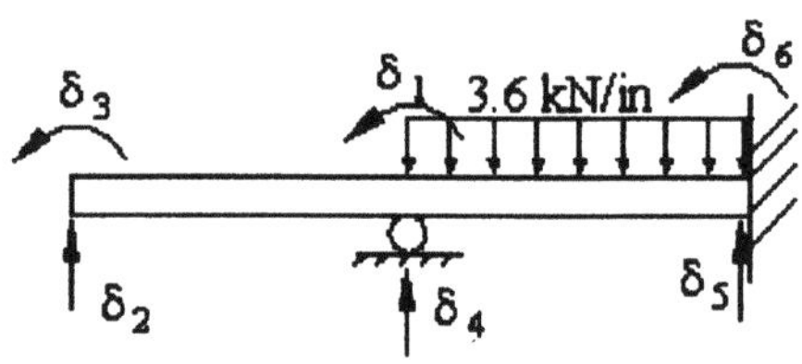

6. Element stiffness matrices for elements 1 and 2 of substructure II.

$$[k]_1^{\mathrm{II}} = [k]_2^{\mathrm{II}} = \begin{matrix} \begin{matrix} 4 & 1 & 5 & 6 \\ 2 & 3 & 4 & 1 \end{matrix} \\ \begin{bmatrix} 12 & 600 & -12 & 600 \\ 600 & 40000 & -600 & 20000 \\ -12 & -600 & 12 & -600 \\ 600 & 20000 & -600 & 40000 \end{bmatrix} \end{matrix} \begin{matrix} 2 & 4 \\ 3 & 1 \\ 4 & 5 \\ 1 & 6 \end{matrix} \qquad \text{(f)}$$

7. Element force vectors for substructure II.

$$\{P\}_1^{II} = \begin{Bmatrix} 0 \\ 0 \\ 0 \\ 0 \end{Bmatrix} \begin{matrix} 2 \\ 3 \\ 4 \\ 1 \end{matrix} \qquad \{P\}_2^{II} = \begin{Bmatrix} -180 \\ -3000 \\ -180 \\ +3000 \end{Bmatrix} \begin{matrix} 4 \\ 1 \\ 5 \\ 6 \end{matrix} \tag{g}$$

Assemble the reduced system stiffness equation for Substructure II from the element stiffness matrices in eq.(f) and the reduced system force vector from the element force vectors in eq.(g).

$$\begin{Bmatrix} -3000 \\ 0 \\ 0 \end{Bmatrix} = \begin{bmatrix} 40000+40000 & 600 & 20000 \\ 600 & 12 & 600 \\ 20000 & 600 & 40000 \end{bmatrix} \begin{Bmatrix} \delta_1 \\ \delta_2 \\ \delta_3 \end{Bmatrix}^{II}$$

8. Condense degree of freedom 1 (divide first row by 80,000).

$$\begin{Bmatrix} -0.0375 \\ 0 \\ 0 \end{Bmatrix} = \begin{bmatrix} 1 & 0.0075 & 0.250 \\ 600 & 12 & 600 \\ 20000 & 600 & 40000 \end{bmatrix} \begin{Bmatrix} \delta_1 \\ \delta_2 \\ \delta_3 \end{Bmatrix}^{II}$$

and (subtract from the second row, the first row multiplied by 600 and subtract from the third row the first row multiplied by 20,000):

$$\begin{Bmatrix} -0.0375 \\ \hline 22.5 \\ 750 \end{Bmatrix} = \left\{\begin{array}{c|cc} 1 & 0.0075 & 0.25 \\ \hline 0 & 7.5 & 450 \\ 0 & 450 & 35000 \end{array}\right\} \begin{Bmatrix} \delta_1 \\ \hline \delta_2 \\ \delta_3 \end{Bmatrix}^{II} \tag{h}$$

9. Expand the partitioned matrix eq.(h) to obtain:

$$-0.0375 = \delta_1 + 0.0075\delta_2 + 0.25\delta_3 \tag{i}$$

and

$$\begin{matrix} 1 \\ 2 \end{matrix} \begin{Bmatrix} 22.5 \\ 150 \end{Bmatrix} = \overset{\begin{matrix} 1 & \quad 2 \end{matrix}}{\begin{bmatrix} 7.5 & 450 \\ 450 & 35000 \end{bmatrix}} \begin{Bmatrix} \delta_2 \\ \delta_3 \end{Bmatrix} \tag{j}$$

10. Assemble the system stiffness equation and the system force vector.

The coefficients in the stiffness matrices in eqs.(e) and (j) are transferred to locations 1,2 of the system stiffness matrix as indicated on the top and on the right side of these matrices. Also, the forces in eqs.(e) and (j) are transferred to the system force vector locations 1 and 2 and added to the nodal force –16 kip and the moment 240 kip-in which are directly applied to nodal coordinates 1 and 2 as shown in Fig. 8.4.

$$\begin{Bmatrix} +0.72+22.5 \\ -24+750 \end{Bmatrix} + \begin{Bmatrix} -16 \\ 240 \end{Bmatrix} = \begin{bmatrix} 6+7.5 & -600+450 \\ -600+450 & 60000+35000 \end{bmatrix} \begin{Bmatrix} u_1 \\ u_2 \end{Bmatrix}$$

or

$$\begin{Bmatrix} 7.22 \\ 966 \end{Bmatrix} = \begin{bmatrix} 13.5 & -150 \\ -150 & 95000 \end{bmatrix} \begin{Bmatrix} u_1 \\ u_2 \end{Bmatrix} \tag{k}$$

11. Solve eq.(k):

$$u_1 = 0.659 \text{ (in)}$$
$$u_2 = 0.0112 \text{ (rad)}$$

12. Displacement at nodal coordinates of Substructure I.

From Fig 8.4(b) and the nodal coordinates assigned to substructure I, we identify $\delta_1 = u_3$; $\delta_2 = u_1 = 0.659$ and $\delta_3 = u_2 = 0.0112$. Then, the substitution of these nodal displacements into eq.(d) results in:

$$u_3 = \delta_1 = 0.0006 + (0.015)(0.659) - (0.500)(0.0112)$$
$$u_3 = 0.0049 \ (rad)$$

13. Displacement at the nodal coordinates of Substructure II.

We identify $\delta_1 = u_4$, $\delta_2 = u_1 = 0.659$ and $\delta_3 = u_2 = 0.0112$.

Then from eq.(i):

$$u_4 = \delta_1 = -0.0375 - (0.0075)(0.659) - (0.25)(0.0112)$$
$$u_4 = -0.0452 \text{ (rad)}$$

14. Element end forces.

Element end forces are given by eq.(1.20) as

$$\{P\} = [k]\{\delta\} - \{Q\} \tag{1.20) repeated (}$$

ELEMENT 1:

The displacement vector $\{\delta\}_1^{\mathrm{I}}$ and the equivalent force vector $\{Q\}_1^{\mathrm{I}}$ for element 1 of substructure I are identified from Fig. 8.4(b) as:

$$\{\delta\}_1^{\mathrm{I}} = \begin{Bmatrix} 0 \\ u_3 \\ u_1 \\ u_2 \end{Bmatrix}_1^{\mathrm{I}} = \begin{Bmatrix} 0 \\ 0.0049 \\ 0.659 \\ 0.0112 \end{Bmatrix} \quad \text{and} \quad \{Q\}_1^{\mathrm{I}} = \{0\} \text{ (no loads on this element)}$$

Then by eq.(1.20) and eq.(a):

$$\begin{Bmatrix} P_1 \\ P_2 \\ P_3 \\ P_4 \end{Bmatrix}_1^{\mathrm{I}} = \begin{bmatrix} 24 & 1200 & -24 & 1200 \\ 1200 & 80000 & -1200 & 40000 \\ -24 & -1200 & 24 & -1200 \\ 1200 & 40000 & -1200 & 80000 \end{bmatrix} \begin{Bmatrix} 0 \\ 0.0049 \\ 0.659 \\ 0.0112 \end{Bmatrix} = \begin{Bmatrix} 3.50 \\ 49.20 \\ -3.50 \\ 301.20 \end{Bmatrix}$$

The end forces for element 1 of substructure I are shown in Fig 8.5.

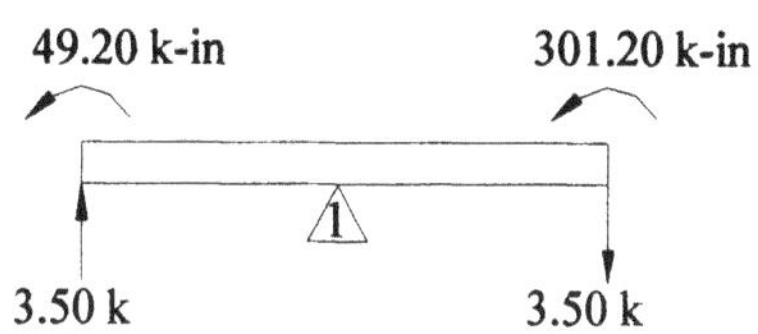

Fig. 8.5 End forces for element 1 of substructure I.

Element 1 of substructure II [the nodal displacements are identified from Fig. 8.4(b)]:

$$\{\delta\}_1^{\mathrm{II}} = \begin{Bmatrix} u_1 \\ u_2 \\ 0 \\ u_4 \end{Bmatrix} = \begin{Bmatrix} 0.659 \\ 0.0112 \\ 0 \\ -0.0452 \end{Bmatrix} \quad \text{and} \quad \{Q\}_1^{\mathrm{II}} = \{0\} \text{ (no loads)}$$

Then, by eq.(1.20):

$$\begin{Bmatrix} P_1 \\ P_2 \\ P_3 \\ P_4 \end{Bmatrix}_1^{\text{II}} = \begin{bmatrix} 12 & 600 & -12 & 600 \\ 600 & 40000 & -600 & 20000 \\ -12 & -600 & 12 & -600 \\ 600 & 20000 & -600 & 40000 \end{bmatrix} \begin{Bmatrix} 0.659 \\ 0.0112 \\ 0 \\ -0.0452 \end{Bmatrix} = \begin{Bmatrix} -12.49 \\ -60.60 \\ 12.49 \\ -1188.6 \end{Bmatrix}$$

ELEMENT 2 of substructure II [the nodal displacements are identified from Fig.8.4(b)]:

$$\{\delta\}_2^{\text{II}} = \begin{Bmatrix} 0 \\ u_4 \\ 0 \\ 0 \end{Bmatrix} = \begin{Bmatrix} 0 \\ -0.0452 \\ 0 \\ 0 \end{Bmatrix} \qquad \{Q\}_2^{\text{II}} = \begin{Bmatrix} -180 \\ -3000 \\ -180 \\ +3000 \end{Bmatrix}$$

Then, by eq.(1.20):

$$\begin{Bmatrix} P_1 \\ P_2 \\ P_3 \\ P_4 \end{Bmatrix}_2^{\text{II}} = \begin{bmatrix} 121 & 600 & -12 & 600 \\ 600 & 40000 & -600 & 20000 \\ -12 & -600 & 12 & -600 \\ 600 & 20000 & -600 & 40000 \end{bmatrix} \begin{Bmatrix} 0 \\ -0.0452 \\ 0 \\ 0 \end{Bmatrix} - \begin{Bmatrix} -180 \\ -3000 \\ -180 \\ +3000 \end{Bmatrix} = \begin{Bmatrix} 152.88 \\ 1192. \\ 207.1 \\ -3904. \end{Bmatrix}$$

End forces for elements 1 and 2 of substructure II are shown in Fig. 8.6.

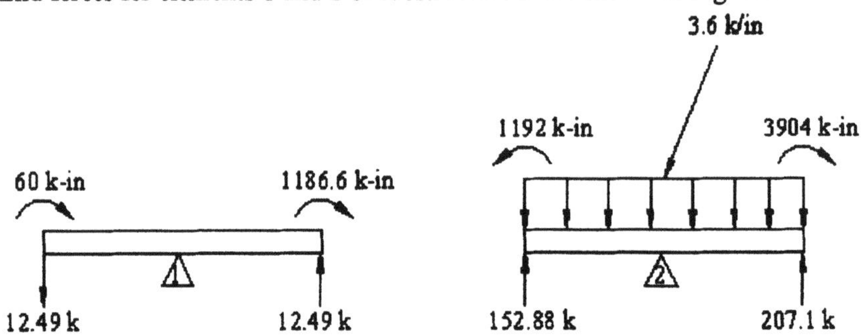

Fig. 8.6 Load and end forces for elements 1 and 2 of substructure II.

15. Reactions:

At node ①:

$$R_5 = P_{1,1} = 3.50 \text{ (kip)}$$

At node ③:

$$R_6 = P_{3,1} + P_{1,2} = 12.49 + 152.88 = 165.37 \text{ (kip)}$$

At node ④:

$$R_7 = P_{3,2} = 207.1 \text{ (kip)}$$
$$R_8 = P_{4,2} = -3904 \text{ (kip-in)}$$

Figure 8.7 shows the loads and the reactions for the beam in this example.

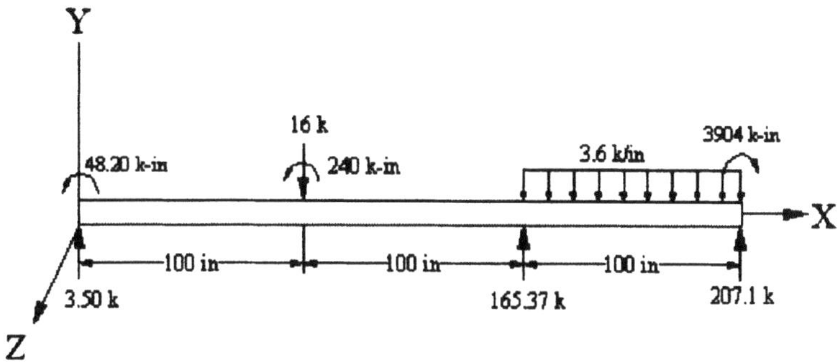

Fig. 8.7 Loads and reactions for the beam of Illustrative Example 8.2.

Illustrative Example 8.3

Use SAP2000 to analyze the beam of Illustrative Example 8.2 (no substructuring), which is shown in Fig. 8.4. Compare the results with those in Illustrative Example 8.2 obtained using substructuring. Edited input data for Illustrative Example 8.3 is given in Table 8.1.

Solution:

Begin: Open SAP2000.

Hint: Maximize both screens for a full views of all windows.

Units: Select kip-in in the drop-down menu located n the lower right hand corner of the screen.

Model: FILE>NEW MODEL FROM TEMPLATE
Select "Beam"
Change the number of spans to 3 and span length to 100. Then OK.

Edit: Minimize the 3-D screen.
Maximize the 2-D screen (X-Z screen). Then drag the figure to the center of the screen using the PAN icon on the toolbar.

Translate axes: Enter:
SELECT>SELECT ALL
EDIT>MOVE
Delta X = 150. Then OK
Use icon PAN to center figure.

Grid: Enter: VIEW
Remove check on "Show Grid"

Labels: Enter: VIEW>SET ELEMENTS
Check: Joint Labels
Check: Frame Labels. Then OK.

Restraints: Click (mark): Joint ④ and enter:
ASSIGN>JOINT>RESTRANTS
Check: Restrain all directions. Then OK.

Click: Joints ① and ③
Enter: ASSIGN>JOINT RESTRAINTS
Check restraints in the Z direction only. Then OK.

Click: Joint ②
ASSIGN>JOINT>RESTRAINTS
Set free in all directions. Then OK.

Material: Enter: DEFINE MATERIAL>STEEL
Click: Modify/Show Materials
Set: Modulus of Elasticity = 30,000. Then OK.

Sections: Enter: DEFINE>FRAME SECTIONS
Click: Add/Wide Flange
Select: Add General (Section)
Set: Moment of Inertia about axis 3 = 100
Shear area in 2 directions = 0. Then OK.

Enter: DEFINE >FRAME SECTIONS
Click: Add/Wide Flange
Select: Add General (Section)
Set: Moment of Inertia about axis 3 = 200
Shears are in 2 directions = 0. Then OK.

Click on: Frame 1
Enter: ASSIGN>FRAME>SECTIONS
Select: FSFC3. Then OK.

Click on: Frames 2 and 3
Enter: ASSIGN>FRAME >SECTIONS
Select: FSEC2. Then OK.

Loads: Enter: DEFINE>STATIC LOAD CASES
Change: DEAD to LIVE
Self weight multiplier = 0
Click: Change Load. Then OK.

Click on Frame 3 and enter:
ASSIGN>FRAME STATIC LOAD>POINT LOAD AND UNIFORM
Check: Forces Z direction
Uniform Load = -3.6

Click on Joint ①
Enter: ASSIGN>JOINT STATIC LOADS>FORCES
Warning: Be certain to zero out the values of previous entries.
Enter: Moment Global YY = -48. Then OK.

Click on Joint ②
Enter: ASSIGN>JOINT STATIC LOADS>FORCES
Warning: Be certain to zero out the values of previous entries.
Enter: Force Global Z = -16
Enter: Moment YY = -240. Then OK.

Options: Enter: ANALYZE>SET OPTIONS
Mark: Available Degrees of Freedom UX, UZ, RY. Then OK.

Analyze: Enter: ANALYZE>RUN
Filename "Example 8.3". Then SAVE.
At the conclusion of the solution process enter OK.

Print Input Tables: Enter: FILE>PRINT INPUT TABLES
Check: Joint Data: Coordinates
Element data: Frames
Static Loads: Joints, Frames. Then OK.

(Table 8.1 reproduces the edited Input Data Tables for Illustrative Example 8.3)

Table 8.1 Edited Input data for Illustrative Example 8.3 (Units: kip,inches)

JOINT DATA

JOINT	GLOBAL-X	GLOBAL-Y	GLOBAL-Z	RESTRAINTS
1	0.00000	0.00000	0.00000	0 0 1 0 0 0
2	100.00000	0.00000	0.00000	0 0 0 0 0 0
3	200.00000	0.00000	0.00000	0 0 1 0 0 0
4	300.00000	0.00000	0.00000	1 1 1 1 1 1

FRAME ELEMENT DATA

FRAME	JNT-1	JNT-2	SECTION	RELEASES	SEGMENTS	LENGTH
1	1	2	FSEC3	0.000	4	100.000
2	2	3	FSEC2	0.000	4	100.000
3	3	4	FSEC2	0.000	4	100.000

JOINT FORCES Load Case LOAD1

JOINT	GLOBAL-X	GLOBAL-Y	GLOBAL-Z	GLOBAL-XX	GLOBAL-YY	GLOBAL-ZZ
1	0.000	0.000	0.000	0.000	-48.000	0.000
2	0.000	0.000	-16.000	0.000	-240.000	0.000

FRAME SPAN DISTRIBUTED LOADS Load Case LOAD1

FRAME	TYPE	DIRECTION	DISTANCE-A	VALUE-A	DISTANCE-B	VALUE-B
3	FORCE	GLOBAL-Z	0.0000	-3.6000	1.0000	-3.6000

Print Output Tables: FILE>PRINT OUTPUT TABLE

Check: Displacements
Reactions
Frame Forces

Then OK.

(Table 8.2 reproduces the edited Output Tables for Illustrative Example 8.3)

Table 8.2 Edited Output tables for Illustrative Example 8.3* (Units: kip, inches)

JOINT DISPLACEMENTS

JOINT	LOAD	UX	UY	UZ	RX	RY	RZ
1	LOAD1	0.0000	0.0000	0.0000	0.0000	-4.886E-03	0.0000
2	LOAD1	0.0000	0.0000	0.6594	0.0000	-0.0112	0.0000
3	LOAD1	0.0000	0.0000	0.0000	0.0000	0.0452	0.0000
4	LOAD1	0.0000	0.0000	0.0000	0.0000	0.0000	0.0000

JOINT REACTIONS

JOINT	LOAD	F1	F2	F3	M1	M2	M3
1	LOAD1	0.0000	0.0000	3.4895	0.0000	0.0000	0.0000
3	LOAD1	0.0000	0.0000	165.3619	0.0000	0.0000	0.0000
4	LOAD1	0.0000	0.0000	207.1486	0.0000	3904.9524	0.0000

* This table has been edited to conform to the formatting requirements of the text.

Table 8.2 Continued

FRAME ELEMENT FORCES

FRAME	LOAD	LOC	P	V2	V3	T	M2	M3
1	LOAD1							
		0.00	0.00	-3.49	0.00	0.00	0.00	-48.00
		25.00	0.00	-3.49	0.00	0.00	0.00	39.24
		50.00	0.00	-3.49	0.00	0.00	0.00	126.48
		75.00	0.00	-3.49	0.00	0.00	0.00	213.71
		100.00	0.00	-3.49	0.00	0.00	0.00	300.95
2	LOAD1							
		0.00	0.00	12.51	0.00	0.00	0.00	60.95
		25.00	0.00	12.51	0.00	0.00	0.00	-251.81
		50.00	0.00	12.51	0.00	0.00	0.00	-564.57
		75.00	0.00	12.51	0.00	0.00	0.00	-877.33
		100.00	0.00	12.51	0.00	0.00	0.00	-1190.10
3	LOAD1							
		0.00	0.00	-152.85	0.00	0.00	0.00	-1190.10
		25.00	0.00	-62.85	0.00	0.00	0.00	1506.19
		50.00	0.00	27.15	0.00	0.00	0.00	1952.48
		75.00	0.00	117.15	0.00	0.00	0.00	148.76
		100.00	0.00	207.15	0.00	0.00	0.00	-3904.95

Deformed Shape: Enter: DISPLAY>SHOW DEFORMED SHAPE
Check: Wire Shadow. Then OK.
Enter: FILE>PRINT GRAPHICS
(The deformed shape of the beam shown on screen is reproduced in Fig. 8.8.)

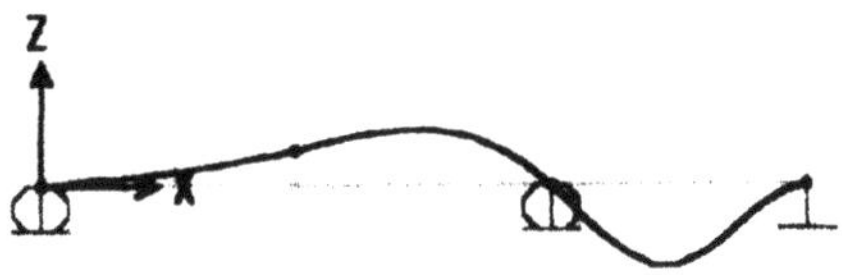

Fig. 8.8 Deformed shape of the beam of Illustrative Example 8.3.

Shear Force Diagram: Enter: DISPLAY>SHOW ELEMENT FORCES/STRESSES>FRAMES
Check: Shear 2-2. Then OK.

(The Shear Force Diagram shown on the screen is reproduced in Fig. 8.9).

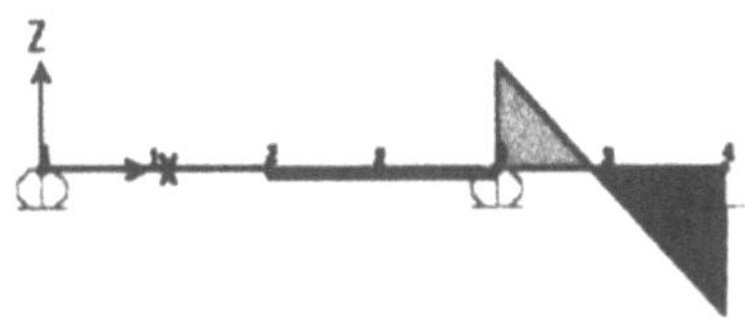

Fig. 8.9 Shear Force Diagram of the beam n Illustrative Example 8.3.

Bending Moment Diagram: Enter:
DISPLAY>SHOW ELEMENT FORCES/STRESSES>FRAMES
Check: Moment 3-3. Then OK.

(The Bending Moment diagram shown on screen is reproduced in Fig. 8.10.)

Fig. 8.10 Bending Moment diagram of the beam in Illustrative Example 8.3.

Comparison of results: Table 8.3 compares results obtained using substructuring in Example 8.2 with the solution provided for the same structure by SAP2000.

Table 8.3 Comparison of results obtained in Illustrative Examples 8.2 and 8.3.

Nodal Displacements			
Node	*Direction*	*Illustrative Example 8.2*	*Illustrative Example 8.3*
1	UZ	0	0
1	RY	0.0049 rad*	-0.00488 rad*
2	UZ	0.659 in	0.6594 in
2	RY	0.0112 rad*	-0.0112 rad*
3	UZ	0	0
3	RY	-0.0452 rad*	0.0452 rad*
4	UZ	0	0
4	RY	0	0

Reactions			
Joint	*Direction*	*Illustrative Example 8.2*	*Illustrative Example 8.3*
1	F3	3.5 kip	3.4895 kip
3	F3	165.37 kip	165.3619 kip
4	F3	207.1 kip	207.1488 kip
4	M2	3904 kip	3904.9524 kip

*Note: For the system of coordinates in Fig. 8.4 counter-clockwise rotations are positive while for the system of coordinates adopted for the solution using SAP2000, clockwise rotations are positive.

8.4 Problems

Problem 8.1

A three story building is modeled as a shear building in which the only degrees of freedom are the horizontal displacements at the three levels of the building as shown in Fig. P8.1. Obtain the system stiffness matrix and then condense degree of freedom u_1.

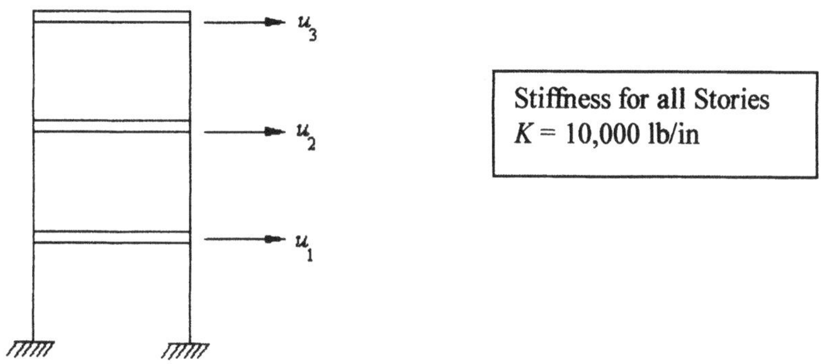

Fig. P8.1

Problem 8.2

Obtain the system stiffness matrix for the uniform shear building shown in Fig. P8.2. the condense degrees of freedom u_1 and u_3.

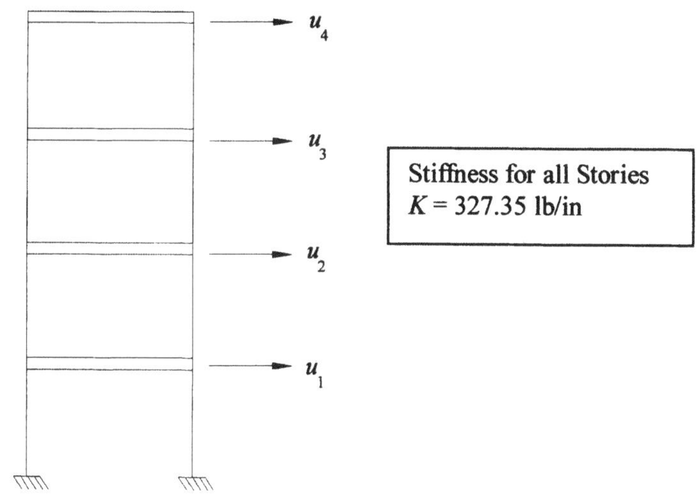

Fig. P.8.2

Problem 8.3
Assuming a horizontal force of magnitude 10,000 lb is applied at the top level of the shear building of Problem 8.2, determine the displacements at the various levels of the building:

a) With no condensation of degrees of freedom
b) After condensation of degrees of freedom u_1 and u_3.

Problem 8.4
Consider the step beam supporting a force P_0 = 10 kips, a force $2P_o$ = 20 kips and moment M_0 = 200 kip-in as shown in Fig. P8.4. Condense the rotational degrees of freedom and then determine the displacements and rotations at sections B and C.

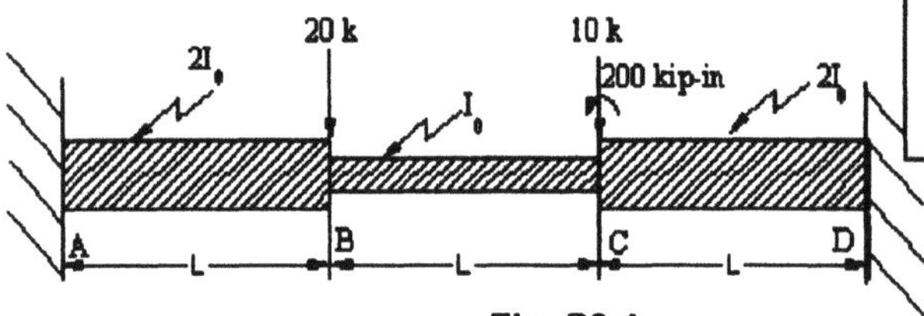

L = 100 in
I_0 = 1,000 in^4
E = 10,000 ksi

Fig. P8.4

Problem 8.5
Consider the plane truss shown in Fig. P8.5(a) separated into two substructures as shown in Fig. P8.5(b). Perform the analysis using the method of substructuring.

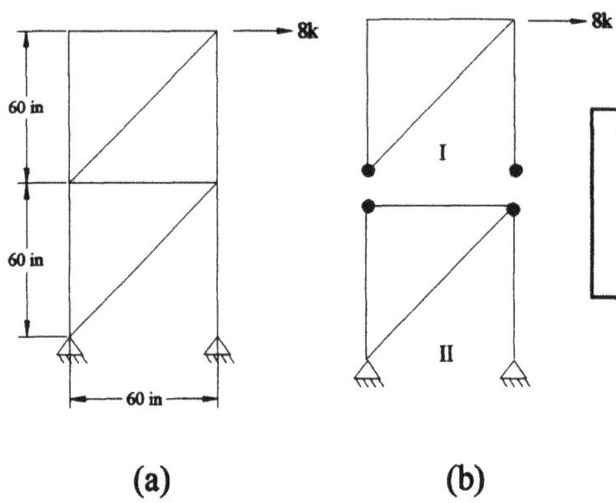

For all Members
E = 30,000 ksi
A = 2 in^2

Fig. P8.5

9 Introduction to Finite Element Method

9.1 Introduction

In the preceding chapters, we have considered the matrix analysis of structures modeled as beams, frames, or trusses. The elements of all these types of structures are described by a single coordinate along their longitudinal axis. These are structures with unidirectional elements, called skeletal structures. In general, they consist of individual members or elements connected at points designated as nodes or joints. For these types of structures, the behavior of each element is considered independently through the calculation of the element stiffness matrices. The element stiffness matrices are then assembled into the system stiffness matrix in such a way that the equilibrium of forces and the compatibility of displacements are satisfied at each nodal point. The analysis of such structures is commonly known as the Matrix Structural Analysis and could be applied equally to static and dynamic problems.

The structures presented in this chapter are continuous structures which are conveniently idealized as consisting of two-dimensional elements assumed to be connected only at the selected nodal points. For example, Fig. 9.1 shows a thin plate idealized with plane triangular elements. The method of analysis for such idealized structures is known as the Finite Element Method (FEM). This is a powerful method for the analysis of structures with complex geometrical configurations, material properties or loading conditions. The FEM is analogous to Matrix Structural Analysis for skeletal structures (beams, frames, and trusses) presented in the preceding chapters. The Finite Element Method differs from the Matrix Structural Method only in two respects: (1) the selection of elements and nodal points are not naturally or clearly established by the geometry as it is for skeletal structures, and (2) the displacements at interior points of an element are expressed in the FEM by interpolating functions and not by an exact analytical relationship as it is in the Matrix Structural Method. Furthermore, for skeletal structures, the displacement of an interior point of an element is governed by ordinary differential equations, while for continuous two-dimensional element it is governed by partial differential equations of much greater complexity.

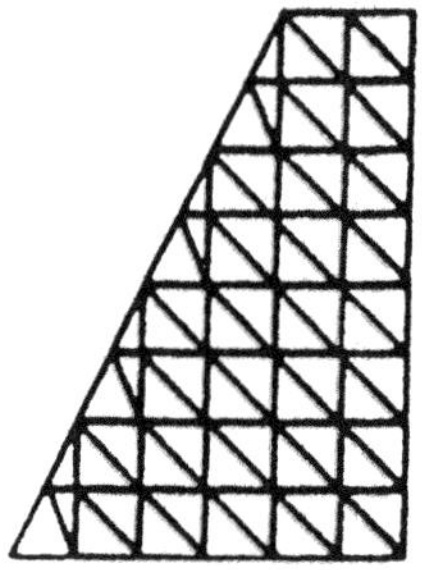

Fig. 9.1 Finite Element Method modeling of a thin plate with triangular elements.

9.2 Plane Elasticity Problems

Plane elasticity problems refer to plates that are loaded in their own planes. Out-of-plane displacements are induced when plates are loaded by normal forces that are perpendicular to the plane of the plate, such problems are generally referred to as plate bending. (Plate bending is considered in Section 9.3.)

There are two different types of plane elasticity problems: (1) plane stress and (2) plane strain. In the plane stress problems, the plate is thin relative to the other dimensions and the stresses normal to the plane of the plate are not considered. Figure 9.2 shows a perforated strip-plate in tension as an example of a plane stress problem.

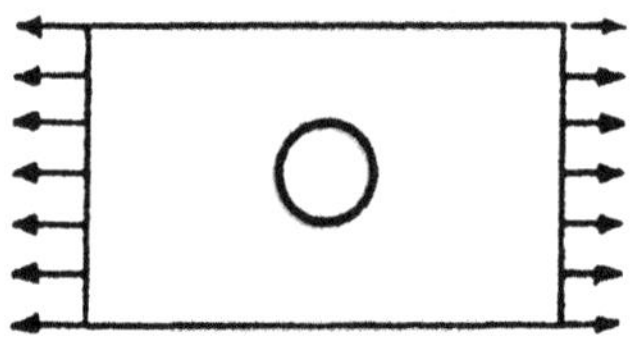

Fig. 9.2 Perforated plate tension element as an example of an structural member loaded in plane stress.

For plane strain problems, the strain normal to the plane of loading is suppressed and assumed to be zero. Figure 9.3 shows a transverse slice of a retaining wall as an example of a plane strain problem.

In the analysis of plane elasticity problems, the continuous plate is idealized as finite elements interconnected at their nodal points. The displacements at these nodal points are the basic unknowns as are the displacements at the joints in the analysis of beams, frames or trusses. Consequently, the first step in the application of the FEM is to model the continuous system into discrete elements. The most common geometric elements used for plane elasticity problems are triangular, rectangular or quadrilateral, although other geometrical shapes could be used as well.

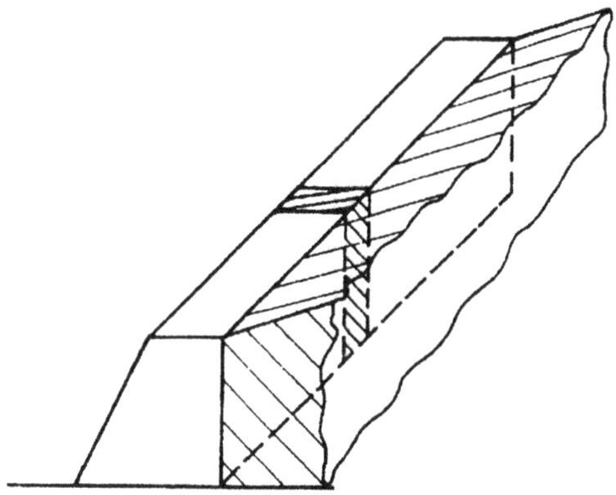

Fig. 9.3 Retaining wall showing a plate slice as an example of plane strain conditions.

The following steps are presented to describe the application of the FEM for the analysis of structural problems:

1. Modeling the structure.

   Figure 9.4 shows a thin cantilever plate modeled with triangular plane stress triangular elements. The plate is supporting an external force P at its upper right end in addition to a distributed force on the plane of the plate. This distributed force is generally known as <u>body force</u> and is expressed as a force per unit of volume. Its components along the x and y directions, b_x and b_y are conveniently arranged as a vector $\{b\}_e$.

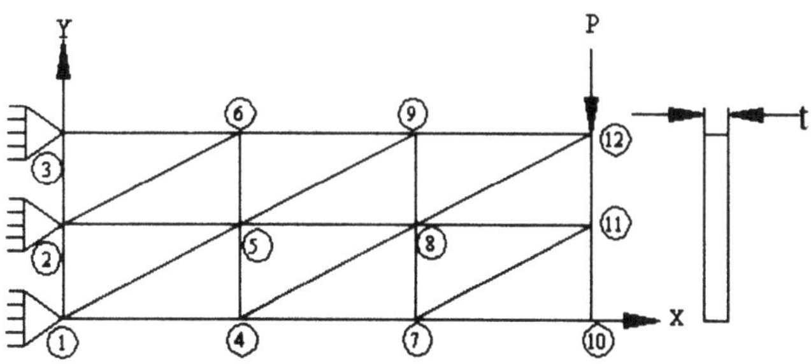

Fig. 9.4 Thin cantilevered plate modeled with plane stress triangular elements.

Consider in Fig. 9.5 a triangular element isolated from a plane stress plate. This figure shows the element nodal forces $\{P\}_e$ and corresponding nodal displacements $\{q\}_e$ with components in the x and y directions at the three joints of this triangular element. Figure 9.5 also shows the components $b_x dV$ and $b_y dV$ of the body force applied at an interior location of the plate of volume dV.

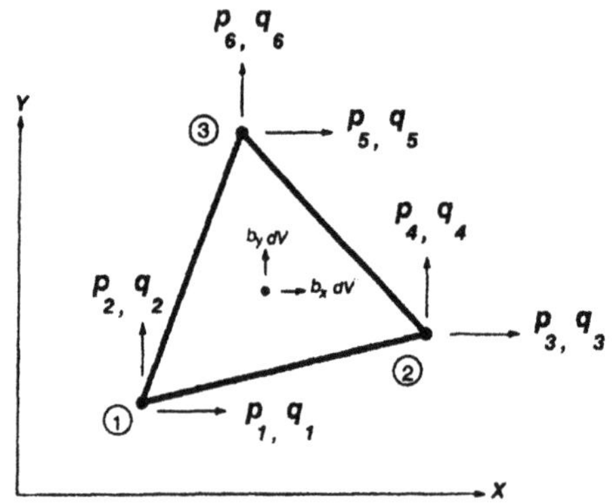

Fig. 9.5 Triangular plate element showing nodal forces P_i and corresponding nodal displacements q_i at the three joints.

The triangular element in plane elasticity problems with two nodal coordinates at each of the three joints, results in a total of six nodal coordinates. Therefore, for this element, the nodal displacement vector $\{q\}_e$ and corresponding nodal force vector $\{P\}_e$ have six components. These vectors may be written as:

$$\{q\{_e = \begin{Bmatrix} q_1 \\ q_2 \\ q_3 \\ q_4 \\ q_5 \\ q_6 \end{Bmatrix} \qquad \{P\}_e = \begin{Bmatrix} P_1 \\ P_2 \\ P_3 \\ P_4 \\ P_5 \\ P_6 \end{Bmatrix} \tag{9.1}$$

Consequently, for the plane triangular element with three nodes, the element stiffness matrix [k] relating the nodal forces and the nodal displacements is of dimension 6x6.

2. Selection of a suitable displacement function.

The displacements, $u = u(x,y)$ and $v = v(x,y)$, respectively, in the x and y directions at any interior point $P(x,y)$ of the triangular element, are expressed approximately by polynomial functions with a total of six coefficients equal in number to the possible nodal displacements. In this case, the simplest expressions for the displacement functions, $u(x,y)$ and $v(x,y)$, at an interior point of the triangular element are:

$$\begin{aligned} u(x,y) &= c_1 + c_2 x + c_3 y \\ v(x,y) &= c_4 + c_5 x + c_6 y \end{aligned} \tag{9.2}$$

or

$$\{q(x,y)\} = [g(x,y)](c) \tag{9.3}$$

where $\{c\}$ is a vector containing the six coefficients c_i in eq.(9.2); $[g(x,y)]$ a matrix function of the position of a point in the element having coordinates (x,y) and $\{q(x,y)\}$ a vector with the displacement components $u(x,y)$ and $v(x,y)$ at an interior point along x and y directions, respectively.

3. Displacements $\{q(x,y)\}$ at a point in the element are expressed in terms of the nodal displacements $\{q\}_e$.

The evaluation of eq.(9.3) for the displacements of the three nodes of the triangular element results in

$$\begin{Bmatrix} q_1 \\ q_2 \\ q_3 \\ q_4 \\ q_5 \\ q_6 \end{Bmatrix} = \begin{bmatrix} 1 & x_1 & y_1 & 0 & 0 & 0 \\ 0 & 0 & 0 & 1 & x_1 & y_1 \\ 1 & x_2 & y_2 & 0 & 0 & 0 \\ 0 & 0 & 0 & 1 & x_2 & y_2 \\ 1 & x_3 & y_3 & 0 & 0 & 0 \\ 0 & 0 & 0 & 1 & x_3 & y_3 \end{bmatrix} \begin{Bmatrix} c_1 \\ c_2 \\ c_3 \\ c_4 \\ c_5 \\ c_6 \end{Bmatrix} \tag{9.4}$$

We now solve eq.(9.4) for the coefficients c_1 through c_6 in terms of the nodal displacements q_1 through q_6. Then we substitute back these coefficients into eq. (9.2) to obtain in matrix notation:

$$\begin{Bmatrix} u(x,y) \\ \mathrm{v}(x,y) \end{Bmatrix} = \begin{bmatrix} N_1 & N_2 & N_3 & 0 & 0 & 0 \\ 0 & 0 & 0 & N_4 & N_5 & N_6 \end{bmatrix} \begin{Bmatrix} q_1 \\ q_2 \\ q_3 \\ q_4 \\ q_5 \\ q_6 \end{Bmatrix} \tag{9.5}$$

where the shape functions N_1, N_2 and N_3 are given by

$$N_1 = \frac{1}{D_0}\left[(x_2 y_3 - x_3 y_2) + (y_2 - y_3)x + (x_3 - x_2)y\right]$$

$$N_2 = \frac{1}{D_0}\left[(x_3 y_1 - x_1 y_3) + (y_3 - y_1)x + (x_1 - x_3)y\right] \tag{9.6}$$

$$N_3 = \frac{1}{D_0}\left[(x_1 y_2 - x_2 y_1) + (y_1 - y_2)x + (x_2 - x_1)y\right]$$

where

$$D_0 = x_2 y_3 - x_3 y_2 + x_1(y_2 - y_3) + y_1(x_3 - x_2) \tag{9.7}$$

The expression for D_0 is equal to twice the area of the triangular element. Equation (9.5) may be written in condensed notation as

$$\{q(x,y)\} = \begin{Bmatrix} u(x,y) \\ \mathrm{v}(x,y) \end{Bmatrix} = [f(x.y)]\{q\}_e \tag{9.8}$$

where $f(x,y)$ is a function of the coordinates at a point in the plate element.

4. Relationship between strain, $\varepsilon\{(x,y)\}$ at any point within the element to the displacements $\{q(x,y)\}$ and hence to the nodal displacements $\{q\}_e$.

It is shown in Theory of Elasticity (S. Timoshenko and J.N. Goodier, 1970), that the strain vector, $\{\varepsilon(x,y)\}$ with axial components ε_x, ε_y, and shearing strain γ_{xy} is given in terms of derivatives of the displacement function by

$$\{\varepsilon(\mathrm{x},\mathrm{y})\} = \begin{Bmatrix} \varepsilon_{\mathrm{x}} \\ \varepsilon_{\mathrm{y}} \\ \gamma_{xy} \end{Bmatrix} = \begin{Bmatrix} \dfrac{\partial u}{\partial x} \\ \dfrac{\partial \mathrm{v}}{\partial y} \\ \dfrac{\partial u}{\partial y} + \dfrac{\partial \mathrm{v}}{\partial x} \end{Bmatrix} \tag{9.9}$$

in which $u = u(x,y)$ and $\mathrm{v} = \mathrm{v}(x,y)$ refer to displacement functions in the x and y directions, respectively.

The strain components may be expressed in terms of the nodal displacements $\{q\}_e$ by substituting into eq.(9.9) the derivatives of $u(x,y)$ and $\mathrm{v}(x,y)$ given by eqs.(9.5) and (9.6) to obtain:

$$\begin{Bmatrix} \varepsilon_{\mathrm{x}} \\ \varepsilon_{\mathrm{y}} \\ \gamma_{xy} \end{Bmatrix} = \frac{1}{D_0} \begin{bmatrix} (y_2 - y_3) & 0 & (y_3 - y_1) & 0 & (y_1 - y_2) & 0 \\ 0 & (x_3 - x_2) & 0 & (x_1 - x_3) & 0 & (x_2 - x_1) \\ (x_3 - x_2) & (y_2 - y_3) & (x_1 - x_3) & (y_3 - y_1) & (x_2 - x_1) & (y_1 - y_2) \end{bmatrix} \begin{Bmatrix} q_1 \\ q_2 \\ q_3 \\ q_4 \\ q_5 \\ q_6 \end{Bmatrix} \tag{9.10}$$

or in condensed notation

$$\{\varepsilon(x,y)\} = [B]\{q\}_e \tag{9.11}$$

in which the matrix $[B]$ is defined in eq.(9.10).

5. Relationship between internal stresses $\{\sigma(x,y)\}$ to strains $\{\varepsilon(x,y)\}$ and hence to the nodal displacements $\{q\}_e$.

For plane elasticity problems, the relationship between the normal stresses σ_x, σ_y and shearing stress τ_{xy} and the corresponding strains ε_x, ε_y, and γ_{xy} may be expressed, in general, as follows:

$$\begin{Bmatrix} \sigma_x \\ \sigma_y \\ \tau_{xy} \end{Bmatrix} = \begin{bmatrix} d_{11} & d_{12} & 0 \\ d_{21} & d_{22} & 0 \\ 0 & 0 & d_{33} \end{bmatrix} \begin{Bmatrix} \varepsilon_x \\ \varepsilon_y \\ \gamma_{xy} \end{Bmatrix} \tag{9.12}$$

or in a short matrix notation as:

$$\{\sigma(x,y)\} = \{D\}\{\varepsilon(x,y)\} \tag{9.13}$$

The substitution into eq.(9.13) of $\{\varepsilon(x,y)\}$ from eq.(9.11) gives the desired relationship between stresses $\{\sigma(x,y)\}$ at a point in the element and the displacements $\{q\}_e$ at the nodes as:

$$\{\sigma\{x,y)\} = [D][B]\{q\}_e \tag{9.14}$$

The coefficients of the matrix $[D]$ in eq.(9.14) have different expressions for plane stress problems and for plane strain problems. These expressions are as follows:

a) Plane stress problems:

$$[D] = \frac{E}{1-\nu^2}\begin{bmatrix} 1 & \nu & 0 \\ \nu & 1 & 0 \\ 0 & 0 & \dfrac{1-\nu}{2} \end{bmatrix} \tag{9.15}$$

b) Plane strain problems:

$$[D] = \frac{E}{(1+\nu)(1-2\nu)}\begin{bmatrix} 1-\nu & \nu & 0 \\ \nu & 1-\nu & 0 \\ 0 & 0 & \dfrac{1-2\nu}{2} \end{bmatrix} \tag{9.16}$$

in which E is the modulus of elasticity and ν is the Poisson's ratio.

6. Element stiffness matrix

Use is made of the Principle of Virtual Work to establish the expressions for the element stiffness matrix $[k]_e$. This principle states that for structures in equilibrium subjected to small compatible virtual displacements, $\delta\{q\}$, the virtual work δW_E of the external forces is equal to the virtual work of internal stresses δW_I, that is

$$\delta W_I = \delta W_E \tag{9.17}$$

In applying this principle, we assume a virtual displacement vector $\delta\{q(x,y)\}$ of the displacement function $\{q(x,y)\}$. Hence, by eq.(9.8)

$$\delta\{q(x,y)\} = [f(x,y)]\delta\{q\}_e \tag{9.18}$$

The virtual internal work δW_I during this applied virtual displacement is then given by the product of the virtual strain and the stresses integrated over the volume of the element:

$$\delta W_1 = \int_V \delta\{\varepsilon\}^T \{\sigma(x,y)\} dV \tag{9.19}$$

The virtual external work δW_E includes the work of the body forces $\{b\}_e\, dV$ and that of the nodal forces $\{P\}_e$ shown in Fig. 9.5. The total external virtual work is then equal to the product of these forces times the corresponding virtual displacements, that is

$$\delta W_E = \int_V \delta\{q\}^T \{b\}_e dV + \delta\{q\}_e^T \{P\}_e \tag{9.20}$$

The substitution into eq.(9.17) of δW_I and δW_E, respectively, from eqs. (9.19) and (9.20) results in

$$\int_V \delta\{\varepsilon\}^T \{\sigma(x,y) dV = \int_V \delta\{q\}^T \{b\}_e dV + \delta\{q\}_e^T \{P\}_e \tag{9.21}$$

Finally, substituting into eq.(9.21) $\{\sigma(x,y)\}$, $\delta\{q\}^T$ and $\delta\{\varepsilon\}^T$, respectively, from eqs. (9.13), (9.18) and (9.11) we obtain, after cancellation of the common factor $\delta\{q\}_e^T$, the stiffness equation of the element:

$$[k]\{q\}_e = \{P\}_e + \{P_b\}_e \tag{9.22}$$

in which the element stiffness matrix $[k]$ is given in general by

$$[k] = \int_V [B]^T [D][B] dV \tag{9.23}$$

and the vector of equivalent nodal forces $\{P_b\}_e$ due to the body forces by

$$\{P_b\}_e = \int_V [f(x,y)]\{b\}_e dV \tag{9.24}$$

For plane elasticity problems for which the matrices $[B]$ and $[D]$ result in constant expressions. Consequently, by eq.(9.23), the element stiffness matrix is given by

$$[k]=[B]^T[D][B]tA \tag{9.25}$$

where t is the thickness of the element and A its area.

7. Assemblage of the system stiffness matrix $[K]$ and the system equivalent nodal force vector $\{P_b\}$due to the body forces.

   The system stiffness matrix $[K]$ is assembled by transferring the coefficients of the element stiffness matrices to appropriate locations in the system stiffness matrix by exactly the same process used in the previous chapters to assemble the system stiffness matrix for skeletal-type structures. The system force vector is assembled from the element equivalent nodal force vector in addition to external forces applied directly to the nodal coordinates. Hence, we may symbolically write

$$[K]=\Sigma[k] \tag{9.26}$$

   and

$$\{P_b\}=\Sigma\{P_b\}_e \tag{9.27}$$

8. Solution of the system stiffness equation.

   The system stiffness equation is then given by

$$[K]_R\{u\}=\{F\}_R \tag{9.28}$$

   in which $[K\}_R$ is the reduced system stiffness matrix $F\}_R$ is the reduced system force vector and $\{u\}$ is the vector of the unknown displacements at the free nodal coordinates. Thus, the solution of eq.(9.28) provides the displacement vector {u} at these free nodal coordinates.

9. Determination of nodal stresses.

   The final step is the calculation of stresses. These stresses can be calculated from the element nodal displacements $\{q\}_e$ identified from the system nodal displacements $\{u\}$ already determined. The element stresses are given by eq.(9.14) as

$$\{\sigma(x,y)\}=[D][B]\{q\}_e \tag{9.29}$$

For plane elasticity problems both matrices $[D]$ and $[B]$ are constants. Therefore, in these problems the stress through the element or at its nodal coordinates is constant.

Illustrative Example 9.1

Consider the deep cantilever beam of rectangular cross-section supporting a load of 3 kN as shown in Fig. 9.6(a). Model this structure with two plane stress triangular elements as shown in Fig. 9.6(b) and perform the finite element analysis.

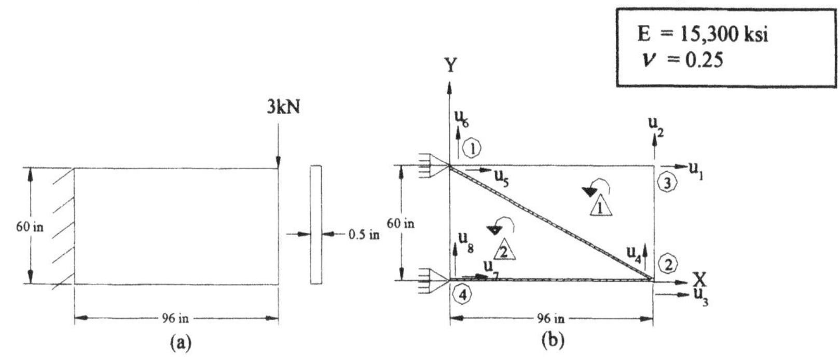

Fig. 9.6 Deep cantilever beam for Illustrative Example 9.1 modeled with two plane stress triangular elements.

Solution:

1. Element stiffness matrices.

ELEMENT 1

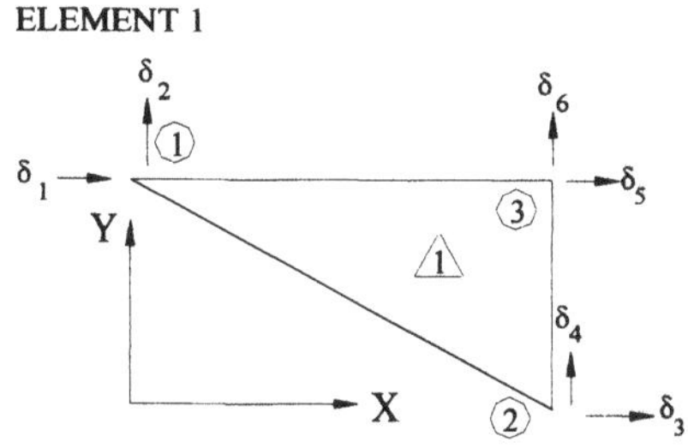

$$x_1 = 0 \qquad y_1 = 60 \text{ in}$$
$$x_2 = 96 \text{ in} \qquad y_2 = 0 \qquad \text{(a)}$$
$$x_3 = 96 \text{ in} \qquad y_3 = 60 \text{ in}$$

Fig. 9.7 Element 1 showing nodal coordinates

From eq.(9.7):

$$D_0 = (60)(96) = 5760 \text{ in}^2 = 2 \text{ x area} \qquad \text{(b)}$$

From eqs.(9.10) and (9.11):

$$[B]_1 = \frac{1}{5760}\begin{bmatrix} -60 & 0 & 0 & 0 & 60 & 0 \\ 0 & 0 & 0 & -96 & 0 & 96 \\ 0 & -60 & -96 & 0 & 96 & 60 \end{bmatrix} \tag{c}$$

From eq. (9.15):

$$[D]_1 = \frac{15300}{0.9375}\begin{bmatrix} 1 & 0.25 & 0 \\ 0.25 & 1 & 0 \\ 0 & 0 & 0.375 \end{bmatrix} \tag{d}$$

The element stiffness matrices are obtained by substituting into eq.(9.25) the matrices $[B]$, its transpose and $[D]$, respectively, from eqs.(c) and (d):

ELEMENT 1

$$[k]_1 = \frac{15300}{5760^2 \times 0.9375}\begin{bmatrix} -60 & 0 & 0 \\ 0 & 0 & -60 \\ 0 & 0 & -96 \\ 0 & -96 & 0 \\ 60 & 0 & 96 \\ 0 & 96 & 60 \end{bmatrix}\begin{bmatrix} 1 & 0.25 & 0 \\ 0.25 & 1 & 0 \\ 0 & 0 & 0.375 \end{bmatrix}$$

$$\begin{bmatrix} -60 & 0 & 0 & 0 & 60 & 0 \\ 0 & 0 & 0 & -96 & 0 & 96 \\ 0 & -60 & -96 & 0 & 96 & 60 \end{bmatrix}(2880)(0.5)$$

or

$$[k]_1 = \begin{array}{c} \begin{matrix} 5 & \quad 6 & \quad 3 & \quad 4 & \quad 1 & \quad 2 \end{matrix} \\ \begin{bmatrix} 2550 & 0 & 0 & 1020 & -2550 & -1020 \\ 0 & 956 & 1530 & 0 & -1530 & -956 \\ 0 & 1530 & 2448 & 0 & -2448 & -1530 \\ 1020 & 0 & 0 & 6528 & -1020 & -6528 \\ -2550 & -1530 & -2448 & -1020 & 4948 & 2550 \\ -1020 & -956 & -1530 & -6528 & 2550 & 7484 \end{bmatrix} \end{array} \begin{matrix} 5 \\ 6 \\ 3 \\ 4 \\ 1 \\ 2 \end{matrix} \tag{e}$$

ELEMENT 2

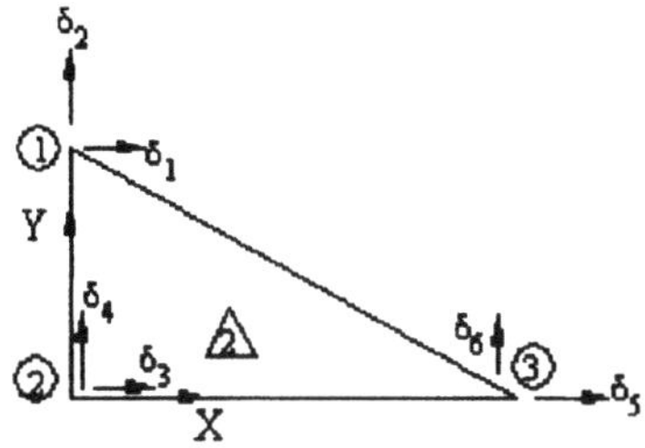

$$x_1 = 0 \qquad y_1 = 60\text{ in}$$
$$x_2 = 0 \qquad y_2 = 0 \qquad \text{(f)}$$
$$x_3 = 96\text{ in} \qquad y_3 = 0$$

Fig. 9.8 Element 2 showing nodal coordinates.

Analogously, we obtain:

$$[B]_2 = \frac{1}{5760}\begin{bmatrix} 0 & 0 & -60 & 0 & 60 & 0 \\ 0 & 96 & 0 & -96 & 0 & 0 \\ 96 & 0 & -96 & -60 & 0 & 60 \end{bmatrix}$$

$$[D]_2 = \frac{15300}{0.975}\begin{bmatrix} 1 & 0.25 & 0 \\ 0.25 & 1 & 0 \\ 0 & 0 & 0.375 \end{bmatrix}$$

and

$$[k]_2 = [B]_2^T [D]_2 [B]_2\, A\, t$$

which results in

$$[k]_2 = \begin{array}{c} \begin{matrix} 5 & 6 & 7 & 8 & 3 & 4 \end{matrix} \\ \begin{bmatrix} 2448 & 0 & -2448 & -1530 & 0 & 1530 \\ 0 & 6528 & -1020 & -6528 & 1020 & 0 \\ -2448 & -1020 & 4998 & 2550 & 2550 & -1530 \\ -1530 & -6528 & 2550 & 7484 & -1020 & -956 \\ 0 & 1020 & -2550 & -1020 & 2550 & 0 \\ 1530 & 0 & -1530 & -956 & 0 & 956 \end{bmatrix} \begin{matrix} 5 \\ 6 \\ 7 \\ 8 \\ 3 \\ 4 \end{matrix} \end{array} \qquad \text{(g)}$$

2. Reduced system stiffness matrix.

Transferring to the system stiffness matrix the coefficients from eqs.(e) and (g) corresponding to the free nodal coordinates (1, 2, 3 and 4) as indicated on the top and on the left of these matrices, yields

$$[K]_R = \begin{array}{c} \begin{array}{cccc} 1 & 2 & 3 & 4 \end{array} \\ \begin{bmatrix} 4948 & 2550 & -2448 & -1020 \\ 2550 & 7484 & -1530 & -6528 \\ -2448 & -1530 & 2448+2550 & 0 \\ -1020 & -6528 & 0 & 6528+956 \end{bmatrix} \begin{array}{c} 1 \\ 2 \\ 3 \\ 4 \end{array} \end{array} \qquad \text{(g)}$$

3. Reduced system force vector.

$$\{F\}_R = \begin{Bmatrix} 0 \\ -3 \\ 0 \\ 0 \end{Bmatrix} \qquad \text{(h)}$$

4. Reduced system stiffness equation.

$$\{F\}_R = [K]_R\{u\} \qquad \text{(i)}$$

Substituting into eq.(i) the reduced system stiffness matrix $[K]_R$ and the reduced system force vector $\{F\}_R$, respectively, from eqs.(g) and (h) gives:

$$\begin{Bmatrix} 0 \\ -3 \\ 0 \\ 0 \end{Bmatrix} = \begin{bmatrix} 4948 & 2550 & -2448 & -1020 \\ 2550 & 7484 & -1530 & -6528 \\ -2448 & -1530 & 4998 & 0 \\ -1020 & -6528 & 0 & 7484 \end{bmatrix} \begin{Bmatrix} u_1 \\ u_2 \\ u_3 \\ u_4 \end{Bmatrix} \qquad \text{(j)}$$

5. System nodal displacements.

Solution of eq.(j) yields

$$\begin{aligned} u_1 &= 6.934\ 10^{-4}\ in \\ u_2 &= -27.481\ 10^{-4}\ in \\ u_3 &= -5.016\ 10^{-4}\ in \\ u_4 &= -23.025\ 10^{-4}\ in \end{aligned} \qquad \text{(k)}$$

6. Element stresses.

Element stresses are given by eq.(9.29) as

$$\sigma(x, y) = [D][B]\{q\}_e \qquad \text{(9.29) repeated}$$

in which the element displacements $\{q\}_e$ are identified from Fig. 9.4 and the values of the system nodal displacements given by eq.(k)

ELEMENT 1

$$q_1 = 0 \quad q_2 = 0 \quad q_3 = -5.016\ 10^{-6} \quad q_4 = -23.025\ 10^{-4} \quad q_5 = 6.934\ 10^{-4} \quad q_6 = -27.025\ 10^{-4}$$

$$\begin{Bmatrix} \sigma_x \\ \sigma_y \\ \tau_{xy} \end{Bmatrix} = \frac{1530}{0.975x5760} \begin{bmatrix} 1 & 0.25 & 0 \\ 0.25 & 1 & 0 \\ 0 & 0 & 0.375 \end{bmatrix} \begin{bmatrix} -60 & 0 & 0 & 0 & 60 & 0 \\ 0 & 0 & 0 & -96 & 0 & 96 \\ 0 & -60 & -96 & 0 & 96 & 60 \end{bmatrix} \begin{Bmatrix} 0 \\ 0 \\ -5.016 \\ -23.025 \\ 6.934 \\ -27.025 \end{Bmatrix} 10^{-4}$$

or

$$\sigma_x = 0.00871\ ksi$$
$$\sigma_y = -0.00762$$
$$\tau_{xy} = -0.00484$$

ELEMENT 2

$$q_1 = 0 \quad q_2 = 0 \quad q_3 = 0 \quad q_4 = 0 \quad q_5 = -5.016\ 10^{-4} \quad q_6 = -23.025\ 10^{-4}$$

$$\begin{Bmatrix} \sigma_x \\ \sigma_y \\ \tau_{xy} \end{Bmatrix} = \frac{1530}{0.975x5760} \begin{bmatrix} 1 & 0.25 & 0 \\ 0.25 & 1 & 0 \\ 0 & 0 & 0.375 \end{bmatrix} \begin{bmatrix} 0 & 0 & -60 & 0 & 60 & 0 \\ 0 & 96 & 0 & -96 & 0 & 0 \\ 96 & 0 & -96 & -60 & 0 & 60 \end{bmatrix} \begin{Bmatrix} 0 \\ 0 \\ 0 \\ 0 \\ -5.016 \\ -23.025 \end{Bmatrix} 10^{-4}$$

or

$$\sigma_x = -0.00819\ \text{ksi}$$
$$\sigma_y = -0.00205$$
$$\tau_{xy} = -0.0141$$

Illustrative Example 9.2

Use SAP 2000 to solve Illustrative Example 9.1.

Solution:
The following commands are implemented in SAP2000:

Begin: Open SAP2000.

Hint: Maximize both screens for a full view of all windows.

Units: Select kip-in in the drop down menu located on the lower right corner of the screen.

Model: FILE>NEW MODEL
Number of grid spaces:*
X direction = 2
Y direction = 0
Z direction = 2

Grid Spacing:
X direction = 96
Y direction = 60
Z direction = 60

Edit: Minimize the 3-D screen
Maximize the 2-D screen (X-Z screen). Then drag the figure to the center of the screen using the PAN icon on the toolbar.

Draw: DRAW>DRAW SHELL ELEMENT
Click at grid intersection $x = 0$ and $z = 60$, drag to the grid intersection $x = 96$ and $z = 0$ and click. Then click again at this intersection and drag to grid intersection $x = 96$ and $z = 60$. Finally, click again at this last point and drag to grid intersection $x = 0$ and $z = 60$ to complete the first shell element.

Now draw the second triangular element by clicking and dragging successively starting at location of node ① in Fig. 9.6 continuing to node ④, then to node ②, and finally to close at node ① to complete the second shell element.

Label: VIEW>SET ELEMENTS
Click on joint labels and on shell labels. Then OK.

---

*Plane X-Z is chosen as the plane of the truss.

Material: DEFINE >MATERIALS>OTHER
Click on modify/Show materials
Select Material Name OTHER.
Set: Modulus of elasticity = 15,300
Poisson's Ratio = 0.25. Then OK, OK.

Shell Sections: DEFINE>SHELL SECTIONS
Click on Add New Sections
Set: Material = OTHER
Thickness = 0.5
Type = shell. Then OK, OK.

Assign Shells: Select shell 1 and 2 by clicking on the area of these shells.
Enter: ASSIGN>SHELL >SECTIONS
Click on SSEC2. Then OK.

Boundary: Select joints ① and ④.
Enter: ASSIGN>JOINT >RESTRAINTS
Click to select restraints in all directions. Then OK.

Load: DEFINE>STATIC LOAD CASES
Change Dead Load to LIVE.
Set self weight multiplier = 0.
Click on Change Load. Then OK.

Assign Load: Select joint ③ and enter:
ASSIGN>JOINT STATIC LOADS>FORCES
Set: Force Global Z = -3.0. then OK.

Analyze: ANALYZE>SET OPTIONS
Set available degrees of freedom UX, UZ and RY. Then OK.

Note: By providing the rotational degree of freedom RY, the element will improve performance.

ANALYZE>RUN
Assign: Filename "Example 9.2". Then Save.
When the calculation is complete check for errors. Then OK.

Note: The displacement value may be seen on the screen by right-clicking on a joint of the deformed shape diagram.

Plot: FILE>PRINT GRAPHICS

(Fig. 9.9 reproduces the deformed plot of the plate for Illustrative Example 9.2)

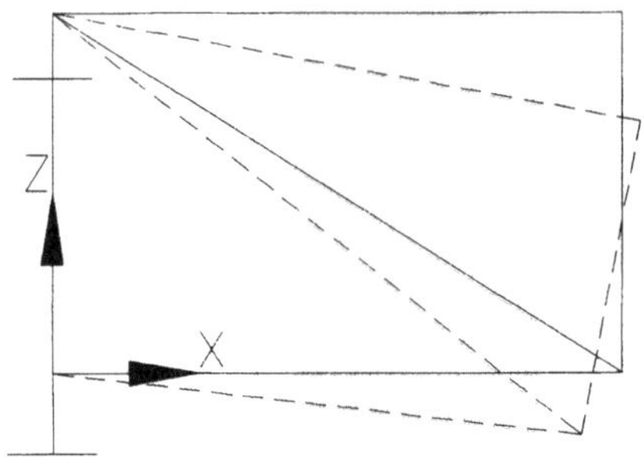

Fig. 9.9 Deformed plate of Illustrative Example 9.2 (Units: kips-inches)

Input Tables: FILE>PRINT INPUT TABLES
Click on Print to File. Then OK.
Use a word editor (such as Word or Notepad) to edit the input tables and then print them.

(Table 9.1 contains the edited input tables for the plate of Illustrative Example 9.2)

Table 9.1 Edited Input Tables for Illustrative Example 9.2 (Units: Kips-inches)

JOINT DATA

JOINT	GLOBAL-X	GLOBAL-Y	GLOBAL-Z	RESTRAINTS
1	0.00000	0.00000	60.00000	1 1 1 1 1 1
2	96.00000	0.00000	0.00000	0 0 0 0 0 0
3	96.00000	0.00000	60.00000	0 0 0 0 0 0
4	0.00000	0.00000	0.00000	1 1 1 1 1 1

SHELL ELEMENT DATA

SHELL	JNT-1	JNT-2	JNT-3	JNT-4	SECTION	ANGLE	AREA
1	2	3	1	1	SSEC2	0.000	2880.000
2	4	2	1	1	SSEC2	0.000	2880.000

JOINT FORCES Load Case LOAD1

JOINT	GLOBAL-X	GLOBAL-Y	GLOBAL-Z	GLOBAL-XX	GLOBAL-YY	GLOBAL-ZZ
3	0.000	0.000	-3.000	0.000	0.000	0.000

Output Tables: FILE>PRINT OUTPUT TABLES
Click on Print to File and on Append.

(Table 9.2 contains the edited output tables for the plate of Illustrative Example 9.2)

Table 9.2 Edited output tables for Illustrative Example 9.2 (Units: Kips-inches)

JOINT DISPLACEMENTS

JOINT	LOAD	UX	UY	UZ	RX	RY	RZ
1	LOAD1	0.0000	0.0000	0.0000	0.0000	0.0000	0.0000
2	LOAD1	-5.014E-04	0.0000	-2.281E-03	0.0000	3.802E-05	0.0000
3	LOAD1	6.774E-04	0.0000	-2.722E-03	0.0000	3.877E-05	0.0000
4	LOAD1	0.0000	0.0000	0.0000	0.0000	0.0000	0.0000

SHELL ELEMENT STRESSES

SHELL/LOAD/JNT	S11-BOT	S22-BOT	S12-BOT	S11-TOP	S22-TOP	S12-TOP
1 LOAD1						
2	-1.371E-02	-1.164E-01	-5.476E-02	-1.371E-02	-1.164E-01	-5.476E-02
3	1.853E-01	-6.662E-02	-5.247E-02	1.853E-01	-6.662E-02	-5.247E-02
1	1.877E-01	-5.687E-02	6.692E-02	1.877E-01	-5.687E-02	6.692E-02
2 LOAD1						
4	-8.578E-02	-2.145E-02	-8.783E-02	-8.578E-02	-2.145E-02	-8.783E-02
2	-8.578E-02	-2.145E-02	-2.049E-01	-8.578E-02	-2.145E-02	-2.049E-01
1	1.094E-01	2.735E-02	-8.783E-02	1.094E-01	2.735E-0	-8.783E-02

Note: The values given in this example by SAP2000 for Joint Displacements and for Shell Element Stresses are slightly different than those obtained by hand calculation in Illustrative Example 9.1. This difference is explained by the fact that in using SAP2000 we added an additional rotational degree of freedom at each joint of the plate which did not exist in the hand calculation. This additional degree of freedom helps somewhat in improving the solution.

Plot Stresses: DISPLAY>SHOW ELEMENT FORCES/STRESSES/SHELLS
Click on Stresses and on S11 (Stress parallel to direction X)
The screen shows a color contour distribution of stresses parallel to the direction *X*. Right clicking on an element will display on the screen an enlarged stress contour plot for that element showing numerical values as the cursor is moved along the area of the element. Analogous plots may be obtained by selecting desired stress components.

Illustrative Example 9.3

Use SAP2000 to analyze a cantilever rectangular steel plate of dimensions 6 in by 4 in supporting a concentrated force of 10 kips applied at its free end as shown in Fig 9.10. Model the plate with rectangular shell elements.

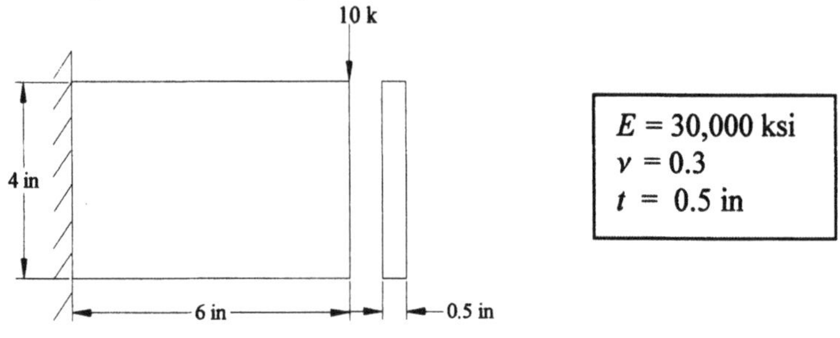

Fig. 9.10 Rectangular steel plate of Illustrative Example 9.3.

Solution:

1. Modeling the structure

   The plate is modeled using shell plate elements of dimensions 2 in by 2 in as shown in Fig. 9.11.

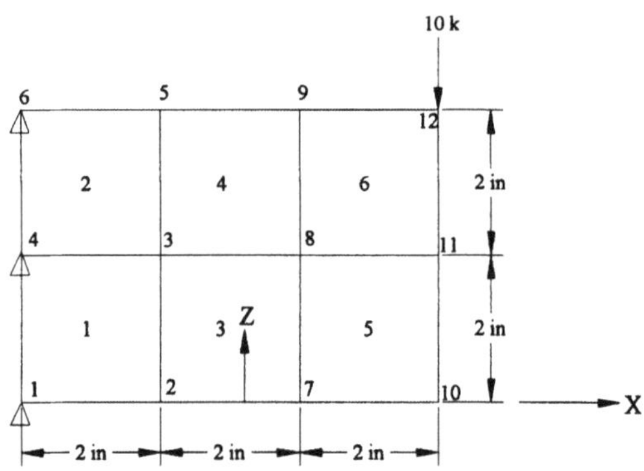

Fig. 9.11 Plate of Illustrative Example 9.3 modeled with plate shell elements of dimension 2 in by 2 in.

The following commands are implemented in SAP2000:

Begin: Open SAP2000.

Hint: Maximize both screens for a full view of all windows.

Units: Select kip-in in the drop down menu located on the lower right corner of the screen.

Model: FILE>NEW MODEL FROM TEMPLATE
Select the plate shown with a rectangular grid.
Set:
Number of spaces along X = 3
Number of spaces along Z = 2
Space width along X = 2
Space width along Z = 2. Then OK.

Editing: Maximize the 2-D screen (X-Z screen)
Then drag the figure to the center of the screen using the PAN icon on the toolbar.

Material: DEFINE>MATERIALS
Select STEEL, then click on Modify/Show Material
Set: Modulus of Elasticity = 30,000
Poisson's Ratio = 0.3. Then OK, OK.

Shell Sections: DEFINE>SHELL SECTIONS
Click: Add New Section
Select: STEEL
Thickness = 0.5. Then OK, OK.
Click to select the shell elements on the screen and enter.

Assign Shells: ASSIGN>SHELL>SECTIONS
Select: SSEC2. Then OK.

Label: VIEW>SET ELEMENTS
Click to select Joint Labels and on Shell Labels. Then OK.

Boundary: Select joints ③, ④ and ⑥ on the left end of the plate and enter:
ASSIGN>JOINT>RESTRAINTS
Click on restraints in all directions. Then OK.

Loads: DEFINE>STATIC LOAD CASES
Change DEAD load to LIVE
Self weight multiplier = 0
Click on change Load. Then OK.

Assign Load: Select joint ⑫ and enter
ASSIGN>JOINT STATIC LOADS>FORCES
Set: Force Global Z = -10.0. Then OK.

Analyze: ANALYZE>OPTIONS
Set available degrees of freedom UX, UZ and RY. Then OK.

ANALYZE>RUN
Assign: Filename "Example 9.3". Then SAVE.
When the calculation is completed check for errors. Then OK.

Note: The displacement values may be seen on the screen by right-clicking on a joint of the deformed plot of the plate shown on the screen.

Plot deformed shape: DISPLAY>SHOW DEFORMED SHAPE
Check: Wire shadow. Then OK.
Enter: FILE>PRINT GRAPHICS

(Fig. 9.12 reproduces the deformed plot of the plate for Illustrative Example 9.3)

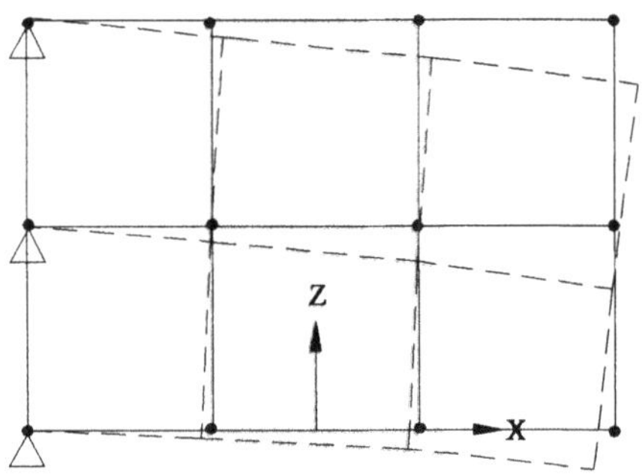

Fig. 9.12 Deformed Plate of Illustrative Example 9.3

Input Tables: FILE>PRINT INPUT TABLES
Click on Print to File. Then OK.
Use a word editor (such as Word or Notepad) to edit and print the Input Tables.

(Table 9.3 contains the edited input tables for Illustrative Example 9.3)

Table 9.3 Edited input tables for Illustrative Example 9.3 (Units: Kips-inches)

JOINT DATA

JOINT	GLOBAL-X	GLOBAL-Y	GLOBAL-Z	RESTRAINTS
1	-3.00000	0.00000	0.00000	1 1 1 1 1 1
2	-1.00000	0.00000	0.00000	0 0 0 0 0 0
3	-1.00000	0.00000	2.00000	0 0 0 0 0 0
4	-3.00000	0.00000	2.00000	1 1 1 1 1 1
5	-1.00000	0.00000	4.00000	0 0 0 0 0 0
6	-3.00000	0.00000	4.00000	1 1 1 1 1 1
7	1.00000	0.00000	0.00000	0 0 0 0 0 0
8	1.00000	0.00000	2.00000	0 0 0 0 0 0
9	1.00000	0.00000	4.00000	0 0 0 0 0 0
10	3.00000	0.00000	0.00000	0 0 0 0 0 0
11	3.00000	0.00000	2.00000	0 0 0 0 0 0
12	3.00000	0.00000	4.00000	0 0 0 0 0 0

SHELL ELEMENT DATA

SHELL	JNT-1	JNT-2	JNT-3	JNT-4	SECTION	ANGLE	AREA
1	1	2	3	4	SSEC2	0.000	4.000
2	4	3	5	6	SSEC2	0.000	4.000
3	2	7	8	3	SSEC2	0.000	4.000
4	3	8	9	5	SSEC2	0.000	4.000
5	7	10	11	8	SSEC2	0.000	4.000
6	8	11	12	9	SSEC2	0.000	4.000

JOINT FORCES Load Case LOAD1

JOINT	GLOBAL-X	GLOBAL-Y	GLOBAL-Z	GLOBAL-XX	GLOBAL-YY	GLOBAL-ZZ
12	0.000	0.000	-10.000	0.000	0.000	0.000

Output Tables: FILE>PRINT OUTPUT TABLES
Check on Print to File and on Append. Then OK.

(Table 9.4 contains the edited output tables for the plate of Illustrative Example 9.3)

Table 9.4 Edited output tables for Illustrative Example 9.3 (Units: Kips-inches)

JOINT DISPLACEMENTS

JOINT	LOAD	U1	U2	U3	R1	R2	R3
1	LOAD1	0.0000	0.0000	0.0000	0.0000	0.0000	0.0000
2	LOAD1	-2.440E-03	0.0000	-2.285E-03	0.0000	1.661E-03	0.0000
3	LOAD1	2.265E-05	0.0000	-1.801E-03	0.0000	1.265E-03	0.0000
4	LOAD1	0.0000	0.0000	0.0000	0.0000	0.0000	0.0000
5	LOAD1	2.381E-03	0.0000	-2.239E-03	0.0000	1.477E-03	0.0000
6	LOAD1	0.0000	0.0000	0.0000	0.0000	0.0000	0.0000
7	LOAD1	-3.801E-03	0.0000	-6.161E-03	0.0000	1.619E-03	0.0000
8	LOAD1	-1.485E-04	0.0000	-6.071E-03	0.0000	2.464E-03	0.0000
9	LOAD1	3.985E-03	0.0000	-5.877E-03	0.0000	2.012E-03	0.0000
10	LOAD1	-4.098E-03	0.0000	-0.0104	0.0000	2.358E-03	0.0000
11	LOAD1	-1.957E-04	0.0000	-0.0108	0.0000	1.869E-03	0.0000
12	LOAD1	4.879E-03	0.0000	-0.0124	0.0000	3.621E-03	0.0000

Table 9.4 Continued

JOINT REACTIONS

JOINT	LOAD	F1	F2	F3	M1	M2	M3
1	LOAD1	14.3843	0.0000	4.0266	0.0000	-0.6096	0.0000
4	LOAD1	-0.0503	0.0000	1.6783	0.0000	-1.2678	0.0000
6	LOAD1	-14.3339	0.0000	4.2951	0.0000	-0.6862	0.0000

SHELL ELEMENT STRESSES

SHELL	LOAD	JOINT	S11-BOT	S22-BOT	S12-BOT	S11-TOP	S22-TOP	S12-TOP
1	LOAD1							
		1	-40.23	-12.07	-3.60	-40.23	-12.07	-3.60
		2	-37.83	-4.08	-6.27	-37.83	-4.08	-6.27
		4	3.733E-01	1.120E-01	-3.09	3.733E-01	1.120E-01	-3.09
		3	2.77	8.10	-5.76	2.77	8.10	-5.76
2	LOAD1							
		4	3.733E-01	1.120E-01	-3.09	3.733E-01	1.120E-01	-3.09
		3	-1.79	-7.11	-5.31	-1.79	-7.11	-5.31
		6	39.25	11.77	-4.40	39.25	11.77	-4.40
		5	37.08	4.55	-6.61	37.08	4.55	-6.61
3	LOAD1							
		2	-20.04	1.25	-6.11	-20.04	1.25	-6.11
		7	-21.99	-5.24	-5.92	-21.99	-5.24	-5.92
		3	-4.262E-01	7.14	-5.79	-4.262E-01	7.14	-5.79
		8	-2.38	6.410E-01	-5.60	-2.38	6.410E-01	-5.60
4	LOAD1							
		3	-4.99	-8.07	-5.33	-4.99	-8.07	-5.33
		8	-1.86	2.36	-5.10	-1.86	2.36	-5.10
		5	24.27	7.110E-01	-3.07	24.27	7.110E-01	-3.07
		9	27.40	11.14	-2.83	27.40	11.14	-2.83
5	LOAD1							
		7	-4.45	1.917E-02	-4.15	-4.45	1.917E-02	-4.15
		10	-6.62	-7.23	-3.55	-6.62	-7.23	-3.55
		8	-3.315E-01	1.25	-4.63	-3.315E-01	1.25	-4.63
		11	-2.51	-6.00	-4.03	-2.51	-6.00	-4.03
6	LOAD1							
		8	1.834E-01	2.97	-4.13	1.834E-01	2.97	-4.13
		11	-8.74	-26.77	-4.55	-8.74	-26.77	-4.55
		9	15.69	7.62	-7.04	15.69	7.62	-7.04
		12	6.77	-22.11	-7.46	6.77	-22.11	-7.46

Note: In order to save space, two additional tables, "Shell Element Principal Stresses" and "Shell Element Resultants", which are given as output tables by SAP2000 are not reproduced as a part of Table 9.4.

Plot Stresses: DISPLAY>SHOW ELEMENTS-FORCES/STRESSES/SHELLS
Click on desired Force or Stress and then on the desired direction.
Then OK.

For example click on STRESS and on direction FF1
The screen will show a color contour plot of the stress distribution on the plate. By right clicking on a shell element the selected element with its contour stresses will be shown on the screen. Numerical values for the stresses at any location in the element will be shown on the screen where the cursor is located.

9.3 Plate Bending

The application of the finite element method is now considered for the analysis of plate bending, that is, plates loaded by forces that are perpendicular to the plane of the plate. The presentation that follows is based on two assumptions: 1) the thickness of the plate is assumed to be small compared to other dimensions of the plate, and 2) the deflections of the plate under the load are assumed to be small relative to its thickness. These assumptions are not particular to the application of the finite element method, they are also made in the classical theory of elasticity for bending of thin plates. These two assumptions are necessary because if the thickness of the plate is large, the plate has to be analyzed as a three-dimensional problem. Also, if the deflections are also large, then in-plane membrane forces are developed and should be accounted for in the analysis. The analysis of plates can be undertaken by the finite element method without these two assumptions. The program SAP2000 used in this book may be applied for the analysis of either thin plates undergoing small deflections or to thick plates in which these two assumptions are not required. However, the theoretical presentation in this section is limited to thin plates that undergo small deflections.

9.4 Rectangular Finite Element for Plate Bending

The derivation of the element stiffness matrix as well as the vector of equivalent nodal forces for body forces or any other forces distributed on the plane element is obtained by following the same steps used for derivation presented for a triangular element subjected to in-plane loads presented in section 9.2.

1. Modeling the structure.

   A suitable system of coordinates and node numbering for a rectangular plate is defined in Fig. 9.13(a) with the x,y axes along the continuous sides of the rectangular plate element, and the z axis normal to the plane of the plate, completing a right-hand system of Cartesian coordinates. This rectangular element has three nodal coordinates at each of its four nodes, a rotation about the x axis θ_x, a rotation about the y axis θ_y, and a displacement (w) along the z axis transverse to the plane of the plate. These nodal displacements labeled

q_i (i = 1,2...12) and the nodal forces P_i (i = 1, 2....12) are shown in their positive sense in Fig. 9.13(b).

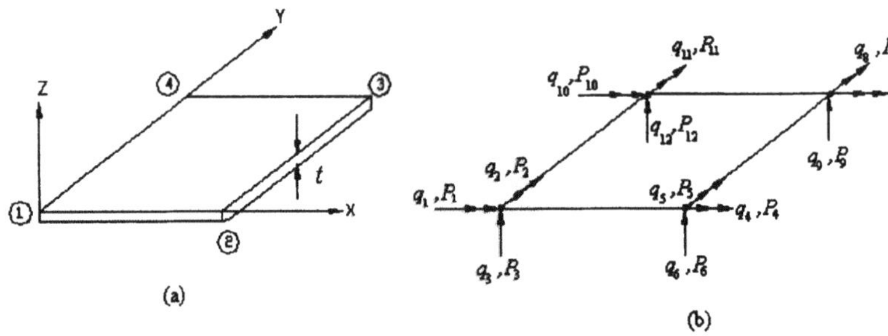

Fig. 9.13 Rectangular plate bending element. (a) Coordinate system and node numbering. (b) Nodal coordinates q_i and corresponding nodal forces P_i.

Corresponding to the three nodal displacements at each node, two moments and a force which are labeled P_j, are also shown in this figure. These 12 nodal displacements and 12 nodal forces are conveniently arranged in two vectors of 12 components, $\{q\}_e$ and $\{P\}_e$. Therefore, the stiffness matrix relating the nodal forces and the nodal displacements for this rectangular plate bending element with four nodes is of dimension 12 x 12.

The angular displacements θ_x and θ_y at any point $P(x,y)$ of the plate element are related to the normal displacement w by the following expressions:

$$\theta_x = -\frac{\partial w}{\partial y} \quad \text{and} \quad \theta_y = \frac{\partial w}{\partial x} \tag{9.30}$$

The positive directions of θ_x and θ_y are shown to agree with the angular nodal displacements q_1, q_2, q_4, q_5, etc. as indicated in Fig. 9.13. Therefore, after a function $w = w(x,y)$ is chosen for the lateral displacement w, the angular displacements are determined through the relations in eq.(9.30).

2. Selection of a suitable displacement function.

Since the rectangular element in plate bending has twelve degrees of freedom, the polynomial expression chosen for the normal displacements w must contain 12 constants. A suitable polynomial function is given by

$$w = c_1 + c_2 x + c_3 y + c_4 x^2 + c_5 xy + c_6 y^2 + c_7 x^3 + c_8 x^2 y + c_9 xy^2 + c_{10} y^3 + c_{11} x^3 y + c_{12} xy^3 \tag{9.31}$$

The displacement functions of the rotations θ_x and θ_y are then obtained from eqs.(9.30) and (9.31) as

$$\theta_x = -\frac{\partial w}{\partial y} = -\left(c_3 + c_5 x + 2c_6 y + c_8 x^2 + 2c_9 xy + 3c_{10} y^2 + c_{11} x^3 + 3c_{12} xy^2\right) \quad (9.32)$$

and

$$\theta_y = \frac{\partial w}{\partial x} = c_2 + 2c_4 x + c_5 y + 3c_7 x^2 + 2c_8 xy + c_9 y^2 + 3c_{11} x^2 y + c_{12} y^2 \quad (9.33)$$

By considering the displacements at the edge of one element, that is, in the boundary between adjacent elements, it may be demonstrated that there is continuity of normal lateral displacements and of the rotational displacement in the direction of the boundary line, but not in the direction transverse to this line as it is shown graphically in Fig. 9.14. The displacement function in eq.(9.31) is called a non-conforming function because it does not satisfy the condition of continuity at the boundaries between elements for all three displacements w, θ_x ,and θ_y .

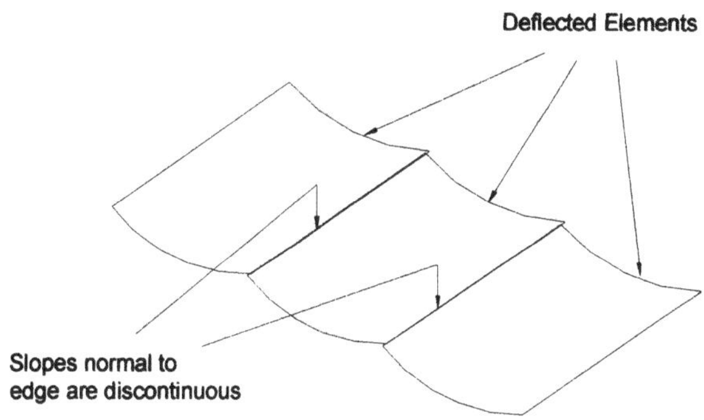

Fig. 9.14 Deflected continuous rectangular elements.

3. Displacements $\{q(x,y)\}$ at a point within the element are expressed in terms of the nodal displacements $\{q\}_e$.

Equations (9.31), (9.32), and (9.33) may be written in matrix notation as

$$\{q(x,y)\} = \begin{Bmatrix} \theta_x \\ \theta_y \\ w \end{Bmatrix} = [g(x,y)]\{c\} \quad (9.34)$$

in which $g(x,y)$ is a function of the coordinates x,y at a point in the element. Equation (9.34) is now used to evaluate the displacements at the nodal coordinates of the element $\{q\}_e$ to obtain:

$$\{q\}_e = [A]\{c\} \tag{9.35}$$

in which the matrix [**A**] is a function of the coordinates at the nodes of the element. The vector $\{c\}$ containing the constant coefficients is then given by

$$\{c\} = [A]^{-1}\{q\}_e \tag{9.36}$$

where $[A]^{-1}$ is the inverse of the matrix $[A]$ defined in eq.(9.35). Finally, the substitution of the vector of constants $\{c\}$ from eq.(9.36) into eq.(9.34) provides the required relationship for displacements $\{q(x,y)\}$ at any interior point in the rectangular element and the displacements $\{q\}_e$ at the nodes:

$$\{q(x,y)\} = \begin{Bmatrix} -\dfrac{\partial w}{\partial y} \\ \dfrac{\partial w}{\partial x} \\ w \end{Bmatrix} = [f(x,y)]\{q\}_e \tag{9.37}$$

in which $[f(x,y)]=[g(x,y)][A]^{-1}$ is solely a function of the coordinates x,y at a point within the element.

4. Relationship between strains $[\varepsilon(x,y)]$ at a point within the element to displacements $\{q(x,y)\}$ and hence to the nodal displacements $\{q\}_e$.

   For plate bending, the state of strain at any point of the element may be represented by three components: 1) the curvature in the x direction, 2) the curvature in the y direction, and 3) a component representing torsion in the plate. The curvature in the x direction is equal to the rate of change of the slope in that direction, that is, to the derivative of the slope,

$$-\frac{\partial}{\partial x}\left(\frac{\partial w}{\partial x}\right) = -\frac{\partial^2 w}{\partial x^2} \tag{9.38}$$

   Similarly, the curvature in the y direction is

$$-\frac{\partial}{\partial y}\left(\frac{\partial w}{\partial y}\right)=-\frac{\partial^2 w}{\partial y^2} \tag{9.39}$$

Finally, the torsional strain component is equal to the rate of change, with respect to y, of the slope in the x direction, that is

$$\frac{\partial}{\partial y}\left(\frac{\partial w}{\partial x}\right)=-\frac{\partial^2 w}{\partial x \partial y} \tag{9.40}$$

The bending moments M_x and M_y and the torsional moments M_{xy} and M_{yx} each act on two opposite sides of the element, but since M_{xy} is numerically equal to M_{yx} , one of these torsional moments, M_{xy} , can be considered to act in all four sides of the element, thus allowing for simply doubling the torsional strain component. Hence, the "strain" vector, $[\varepsilon(x,y)]$, for a plate bending element can be expressed by

$$\{\varepsilon(x,y)\}=\begin{Bmatrix} -\dfrac{\partial^2 w}{\partial x^2} \\ -\dfrac{\partial^2 w}{\partial y^2} \\ 2\dfrac{\partial^2 w}{\partial x \partial y} \end{Bmatrix} \tag{9.41}$$

The substitution of the second derivatives obtained by differentiation of eq.(9.37) into eq.(9.41) yields

$$\{\varepsilon(x,y)\}=[B]\{q\}_e \tag{9.42}$$

in which $[B]$ is a function of the coordinates (x,y) only.

5. Relationship between internal stresses $\{\sigma(x,y)\}$ to internal strains $[\varepsilon(x,y)]$ and hence to nodal displacements $\{q\}_e$.

In a plate bending, the internal "stresses" are expressed as bending and twisting moments, and the "strains" are the curvatures and the twist. Thus, for plate bending, the state of internal "stresses" can be represented by

$$\{\sigma(x,y)\} = \begin{Bmatrix} M_x \\ M_y \\ M_{xy} \end{Bmatrix} \tag{9.43}$$

where M_x and M_y are internal bending moments per unit of length and M_{xy} is the internal twisting moment per unit of length. For a small rectangular element of the plate bending, these internal moments are shown in Fig. 9.15.

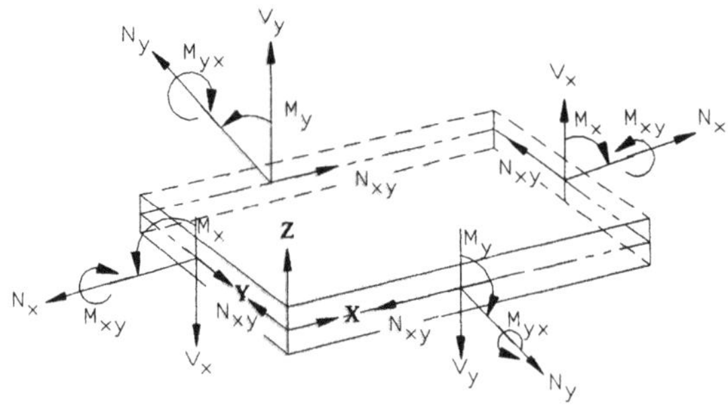

Fig. 9.15 Direction of force and moment components per unit of length as defined for thin plates by SAP2000.

The moment-curvature relationships obtained from plate bending theory (S. Timoshenko and J.N. Goodier, 1970) are:

$$\begin{aligned} M_x &= -\left(D_x \frac{\partial^2 w}{\partial x^2} + D_1 \frac{\partial^2 w}{\partial y^2}\right) \\ M_y &= -\left(D_y \frac{\partial^2 w}{\partial y^2} + D_1 \frac{\partial^2 w}{\partial x^2}\right) \\ M_{xy} &= 2D_{xy} \frac{\partial^2 w}{\partial x \partial y} \end{aligned} \tag{9.44}$$

These relations are written in general for an orthotropic plate, i.e., a plate which has different elastic properties in two perpendicular directions, in which D_x and D_y are flexural rigidities in the x and y directions, respectively, D_1 is a "coupling" rigidity coefficient representing the Poisson's ratio type of effect and D_{xy} is the torsional rigidity. For an isotropic plate which has the same properties in all directions, the flexural and twisting rigidities are given by

$$D_x = D_y = \frac{Et^3}{12(1-v^2)}$$
$$D_1 = v\,D_x \qquad (9.45)$$
$$D_{xy} = \frac{(1-v)}{2} D_x$$

in which E is the modulus of elasticity, v the Poission's Ratio coefficient and t the thickness of the plate. Equation (9.44) may be written in matrix form as

$$\{\sigma(x,y)\} = \begin{Bmatrix} M_x \\ M_y \\ M_{xy} \end{Bmatrix} = \begin{bmatrix} D_x & D_1 & 0 \\ D_1 & D_y & 0 \\ 0 & 0 & D_{xy} \end{bmatrix} \begin{Bmatrix} -\dfrac{\partial^2 w}{\partial x^2} \\ -\dfrac{\partial^2 w}{\partial y^2} \\ 2\dfrac{\partial^2 w}{\partial x \partial y} \end{Bmatrix} \qquad (9.46)$$

or symbolically as

$$\{\sigma(x,y)\} = [D]\{\varepsilon(x,y)\} \qquad (9.47)$$

where the matrix $[D]$ is defined in eq.(9.46). The substitution of $[\varepsilon(x,y)]$ from eq.(9.42) into eq.(9.47), gives the required relationship between the element stresses and nodal displacements as

$$\{\sigma(x,y)\} = [D][B]\{q\}_e \qquad (9.48)$$

6. Element stiffness matrix.

The stiffness matrix for an element of plate bending is obtained by applying the Principle of Virtual Work resulting in equations analogous to eqs.(9.22) and (9.23), and for the equivalent forces due to the applied body forces on the element in eq.(9.24). For plate bending the matrices $[f(x,y)\}$, $[B]$, and $[D]$ required in these equations are defined, respectively, in eqs.(9.37), (9.42), and (9.47). The calculation of these matrices and also of the integrals indicated in eqs.(9.22) and (9.24), are usually undertaken by numerical methods implemented in the coding of computer programs.

7. Assemblage of the system stiffness matrix $[K]$ and of the vector of the external forces $\{F\}$.

   The system stiffness matrix and the system nodal force vector are assembled from the corresponding element stiffness matrices and element force vectors given, respectively, by eqs.(9.25) and (9.27).

8. Solution of the differential equations of motion.

   The system stiffness equation is given by eq.(9.28) in which $[K]$ is the assembled stiffness matrices and $\{F\}$ is the system vector of the external forces. The solution of eq.(9.28) thus provides the system nodal displacement vector $\{u\}$.

9. Nodal stresses.

   The element nodal stresses, $\{\sigma(x_j, y_j)\}_e$ for node j of an element are given from eq.(9.48) as

$$\{\sigma(x_j, y_j)\} = [D]_j [B]_j \{q\}_e \qquad (9.49)$$

   in which the matrices $[D]_j$ and $[B]_j$ are evaluated for the coordinates of node j of the element.

Illustrative Example 9.4

A square steel plate 40 in. by 40 in. and thickness 0.10 in., assumed to be fixed at the supports on its four sides (Fig. 9.16), is loaded with a uniformly distributed force of magnitude 0.01 kip/in^2 applied downward normal to the plane of the plate. Use SAP2000 to analyze the plate and determine deflections and stresses.

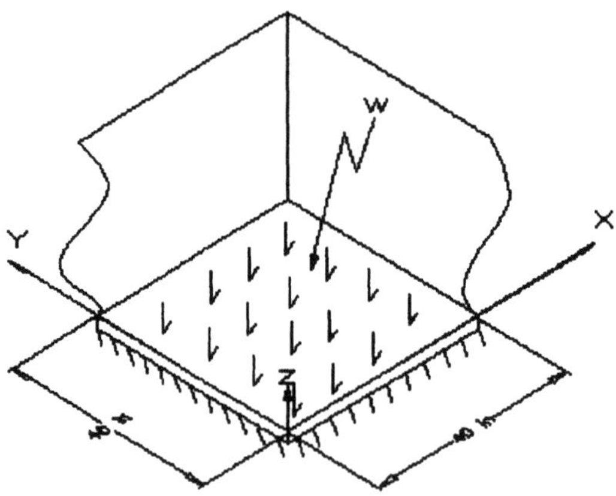

E = 29,500 ksi
ν = 0.3
t = 0.10 in.
w = 0.01 kip/in^2

Fig. 9.16 Steel plate supporting normal distributed load of Illustrative Example 9.4.

Solution:
The plate is modeled with rectangular plate elements of dimension 10 in. by 10 in. as shown in Fig. 9.17:

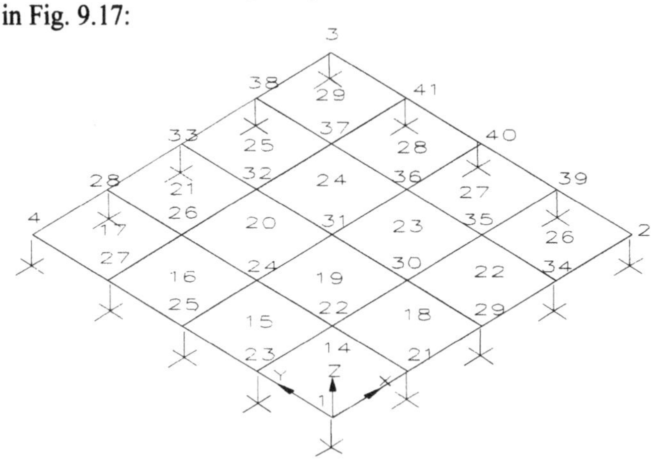

Fig. 9.17 Finite Element Modeling of the plate in Illustrative Example 9.4.

The following commands are implemented in SAP2000:

Begin: Open SAP2000.

Hint: Maximize both screens for a full view of all windows.

Units: Select kip-in in the drop down menu located on the lower right corner of the screen.

Model: FILE>NEW MODEL
Number of grid spaces:
X direction = 4
Y direction = 4
Z direction = 0

Grid Spacing:
X direction = 10
Y direction = 10
Z direction = 10

Select: Click on the top of the right window to default to the *X-Y* plane at *Z* = 0 view.

Draw: DRAW>DRAW SHELL ELEMENT
Click at grid intersection $X = -20$, $Y = -20$ and drag to grid intersection $X = 20$, $Y = 0$. Then click again at this intersection and drag to intersection $X = 20$, $Y = 20$. It should be observed that a square plate is drawn on the screen.

Mesh Shells: Select (click) on the shell shown on the screen. Then enter:
EDIT>MESH SHELLS
Change number of divisions in both entries to 4. Then OK.
The screen will now show a mesh of 16 square shells.

Materials: DEFINE>MATERIALS>STEEL
Modify/ Show Material
Modulus of Elasticity = 29,500
Poisson's Ratio = 0.3. Then OK, OK.

Sections: DEFINE>SHELL SECTIONS
Add new section
In the drop-down menu select STEEL
Thickness:
Membrane = 0.1
Bending = 0.1. Then OK, OK.

Click on all 16 shells and enter:
ASSIGN>SHELL>SECTIONS
Select SSEC2. Then OK.

Boundary: Select all the joints on the boundary of the plate and enter:
ASSIGN>JOINT>RESTRAINTS
Restrain in all directions. Then OK.

Load: DEFINE>STATIC LOAD CASES
Change type of Load to LIVE load
Self-weight multiplier to 0 (zero)
Click on Change Load. Then OK.

Select all the 16 elements of the plate and enter:
ASSIGN>SHELL STATIC LOADS>UNIFORM
Load = -0.01
Global Z. Then OK.

Analyze: ANALYZE>SET OPTIONS
Check available degrees of freedom: UZ, RX and RY.
Then OK and enter:
ANALYZE>RUN
Filename: "Example 9.4". Then SAVE.
At the conclusion of the calculations enter OK.

Note: Displacement values may be seen by right-clicking on a node of the deformed plate shown on the screen.

Plot: DISPLAY>SHOW DEFORMED SHAPE
Click on wire shadow
FILE>PRINT GRAPHICS

(Fig. 9.18 shows the deformed shape for the plate of Illustrative Example 9.4)

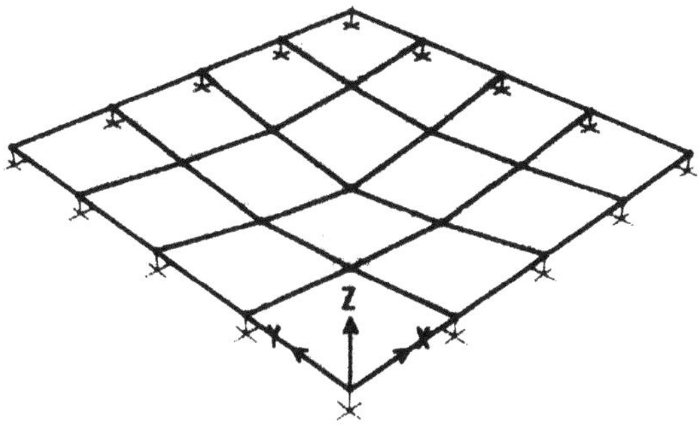

Fig. 9.18 Deformed shape for the plate of Illustrative Example 9.4.

Input Tables: FILE>PRINT INPUT TABLES
Click on Print to File. Then OK.
Use a word editor (such as Word or Notepad) to edit the input tables and then print them.

(Table 9.5 contains the edited input tables for the plate of Illustrative Example 9.4)

Table 9.5 Edited Input Tables for Illustrative Example 9.4 (Units: Kips-inches)

JOINT DATA

JOINT	GLOBAL-X	GLOBAL-Y	GLOBAL-Z	RESTRAINTS
1	-20.00000	-20.00000	0.00000	1 1 1 1 1 1
2	20.00000	-20.00000	0.00000	1 1 1 1 1 1
3	20.00000	20.00000	0.00000	1 1 1 1 1 1
4	-20.00000	20.00000	0.00000	1 1 1 1 1 1
21	-10.00000	-20.00000	0.00000	1 1 1 1 1 1
22	-10.00000	-10.00000	0.00000	0 0 0 0 0 0
23	-20.00000	-10.00000	0.00000	1 1 1 1 1 1
24	-10.00000	0.00000	0.00000	0 0 0 0 0 0
25	-20.00000	0.00000	0.00000	1 1 1 1 1 1
26	-10.00000	10.00000	0.00000	0 0 0 0 0 0
27	-20.00000	10.00000	0.00000	1 1 1 1 1 1
28	-10.00000	20.00000	0.00000	1 1 1 1 1 1
29	0.00000	-20.00000	0.00000	1 1 1 1 1 1
30	0.00000	-10.00000	0.00000	0 0 0 0 0 0
31	0.00000	0.00000	0.00000	0 0 0 0 0 0

Table 9.5 Continued

JOINT	GLOBAL-X	GLOBAL-Y	GLOBAL-Z	RESTRAINTS
32	0.00000	10.00000	0.00000	0 0 0 0 0 0
33	0.00000	20.00000	0.00000	1 1 1 1 1 1
34	10.00000	-20.00000	0.00000	1 1 1 1 1 1
35	10.00000	-10.00000	0.00000	0 0 0 0 0 0
36	10.00000	0.00000	0.00000	0 0 0 0 0 0
37	10.00000	10.00000	0.00000	0 0 0 0 0 0
38	10.00000	20.00000	0.00000	1 1 1 1 1 1
39	20.00000	-10.00000	0.00000	1 1 1 1 1 1
40	20.00000	0.00000	0.00000	1 1 1 1 1 1
41	20.00000	10.00000	0.00000	1 1 1 1 1 1

SHELL ELEMENT DATA

SHELL	JNT-1	JNT-2	JNT-3	JNT-4	SECTION	ANGLE	AREA
14	1	21	22	23	SSEC1	0.000	100.000
15	23	22	24	25	SSEC1	0.000	100.000
16	25	24	26	27	SSEC1	0.000	100.000
17	27	26	28	4	SSEC1	0.000	100.000
18	21	29	30	22	SSEC1	0.000	100.000
19	22	30	31	24	SSEC1	0.000	100.000
20	24	31	32	26	SSEC1	0.000	100.000
21	26	32	33	28	SSEC1	0.000	100.000
22	29	34	35	30	SSEC1	0.000	100.000
23	30	35	36	31	SSEC1	0.000	100.000
24	31	36	37	32	SSEC1	0.000	100.000
25	32	37	38	33	SSEC1	0.000	100.000
26	34	2	39	35	SSEC1	0.000	100.000
27	35	39	40	36	SSEC1	0.000	100.000
28	36	40	41	37	SSEC1	0.000	100.000
29	37	41	3	38	SSEC1	0.000	100.000

Output Tables: FILE>OUTPUT TABLES
Click on Print to File and Append.

(Table 9.6 contains the edited output tables for the plate of Illustrative Example9.4).

Table 9.6 Edited output tables for Illustrative Example 9.4 (Units: Kips-inches)

JOINT DISPLACEMENTS

JOINT	LOAD	UX	UY	UZ	RX	RY	RZ
1	LOAD1	0.0000	0.0000	0.0000	0.0000	0.0000	0.0000
2	LOAD1	0.0000	0.0000	0.0000	0.0000	0.0000	0.0000
3	LOAD1	0.0000	0.0000	0.0000	0.0000	0.0000	0.0000
4	LOAD1	0.0000	0.0000	0.0000	0.0000	0.0000	0.0000
21	LOAD1	0.0000	0.0000	0.0000	0.0000	0.0000	0.0000
22	LOAD1	0.0000	0.0000	-4.239E-03	-5.327E-04	5.327E-04	0.0000
23	LOAD1	0.0000	0.0000	0.0000	0.0000	0.0000	0.0000
24	LOAD1	0.0000	0.0000	-7.103E-03	0.0000	8.869E-04	0.0000
25	LOAD1	0.0000	0.0000	0.0000	0.0000	0.0000	0.0000
26	LOAD1	0.0000	0.0000	-4.239E-03	5.327E-04	5.327E-04	0.0000
27	LOAD1	0.0000	0.0000	0.0000	0.0000	0.0000	0.0000
28	LOAD1	0.0000	0.0000	0.0000	0.0000	0.0000	0.0000

Table 9.6 Continued

JOINT	LOAD	UX	UY	UZ	RX	RY	RZ
29	LOAD1	0.0000	0.0000	0.0000	0.0000	0.0000	0.0000
30	LOAD1	0.0000	0.0000	-7.103E-03	-8.869E-04	0.0000	0.0000
31	LOAD1	0.0000	0.0000	-0.0119	0.0000	0.0000	0.0000
32	LOAD1	0.0000	0.0000	-7.103E-03	8.869E-04	0.0000	0.0000
33	LOAD1	0.0000	0.0000	0.0000	0.0000	0.0000	0.0000
34	LOAD1	0.0000	0.0000	0.0000	0.0000	0.0000	0.0000
35	LOAD1	0.0000	0.0000	-4.239E-03	-5.327E-04	-5.327E-04	0.0000
36	LOAD1	0.0000	0.0000	-7.103E-03	0.0000	-8.869E-04	0.0000
37	LOAD1	0.0000	0.0000	-4.239E-03	5.327E-04	-5.327E-04	0.0000
38	LOAD1	0.0000	0.0000	0.0000	0.0000	0.0000	0.0000
39	LOAD1	0.0000	0.0000	0.0000	0.0000	0.0000	0.0000
40	LOAD1	0.0000	0.0000	0.0000	0.0000	0.0000	0.0000
41	LOAD1	0.0000	0.0000	0.0000	0.0000	0.0000	0.0000

JOINT REACTIONS

JOINT	LOAD	F1	F2	F3	M1	M2	M3
1	LOAD1	0.0000	0.0000	0.0182	0.0714	-0.0714	0.0000
2	LOAD1	0.0000	0.0000	0.0182	0.0714	0.0714	0.0000
3	LOAD1	0.0000	0.0000	0.0182	-0.0714	0.0714	0.0000
4	LOAD1	0.0000	0.0000	0.0182	-0.0714	-0.0714	0.0000
21	LOAD1	0.0000	0.0000	0.1061	0.4393	-0.0111	0.0000
23	LOAD1	0.0000	0.0000	0.1061	0.0111	-0.4393	0.0000
25	LOAD1	0.0000	0.0000	0.1697	0.0000	-0.6797	0.0000
27	LOAD1	0.0000	0.0000	0.1061	-0.0111	-0.4393	0.0000
28	LOAD1	0.0000	0.0000	0.1061	-0.4393	-0.0111	0.0000
29	LOAD1	0.0000	0.0000	0.1697	0.6797	0.0000	0.0000
33	LOAD1	0.0000	0.0000	0.1697	-0.6797	0.0000	0.0000
34	LOAD1	0.0000	0.0000	0.1061	0.4393	0.0111	0.0000
38	LOAD1	0.0000	0.0000	0.1061	-0.4393	0.0111	0.0000
39	LOAD1	0.0000	0.0000	0.1061	0.0111	0.4393	0.0000
40	LOAD1	0.0000	0.0000	0.1697	0.0000	0.6797	0.0000
41	LOAD1	0.0000	0.0000	0.1061	-0.0111	0.4393	0.0000

Note: The output of SAP2000 also includes the following tables:
Shell Element Resultant
Shell element Stresses
Shell Element Principal Stresses
These additional tables are not reproduced as part of Table 9.6.

Plot Stresses: DISPLAY>SHOW ELEMENT FORCES/STRESSES>SHELLS

Note: The screen shown allows for selection of forces or stresses for color contour plots of components, maximum or principal values. These plots are not reproduced for this Illustrative Example 9.4.

9.5 Problems

Problem 9.1

Determine the displacements and stresses for the triangular plate shown in Fig. P9.1. Assume a state of plane stress.

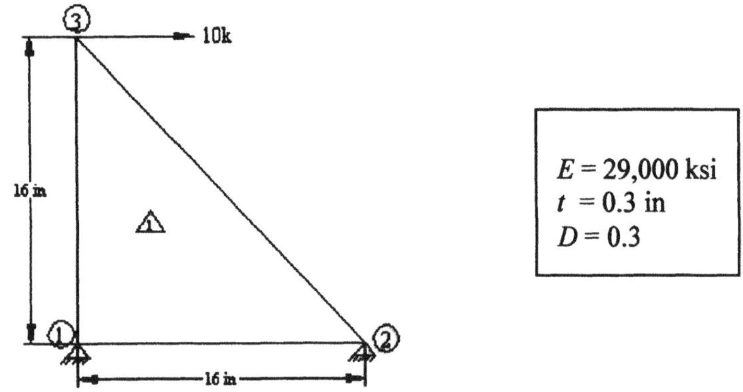

Fig. P9.1

Problem 9.2

The plate shown in Fig. P9.2 is modeled with two triangular elements. Determine the displacements and stresses.

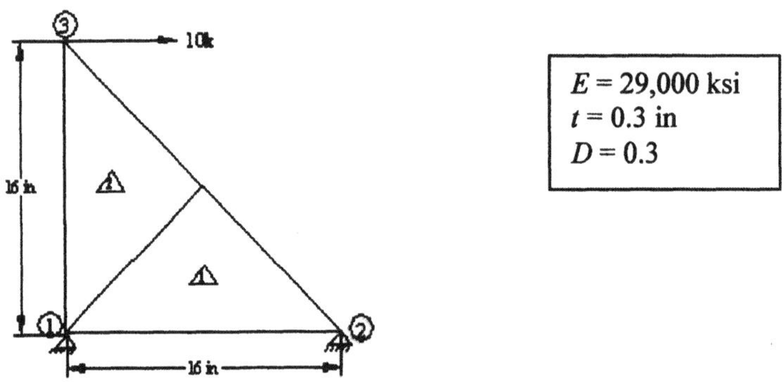

Fig. P9.2

Problem 9.3

The plate shown in Fig. P9.3 is modeled with two triangular plate elements. Determine displacements and stresses. Assume a state of plane stress.

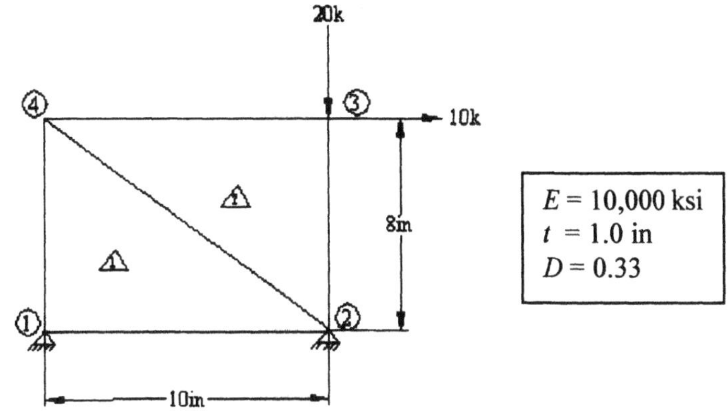

Fig. P9.3

Problem 9.4

Use SAP2000 to solve Illustrative Example 9.1 modeling the plate with four rectangular plane stress elements as shown in Fig. P9.4.

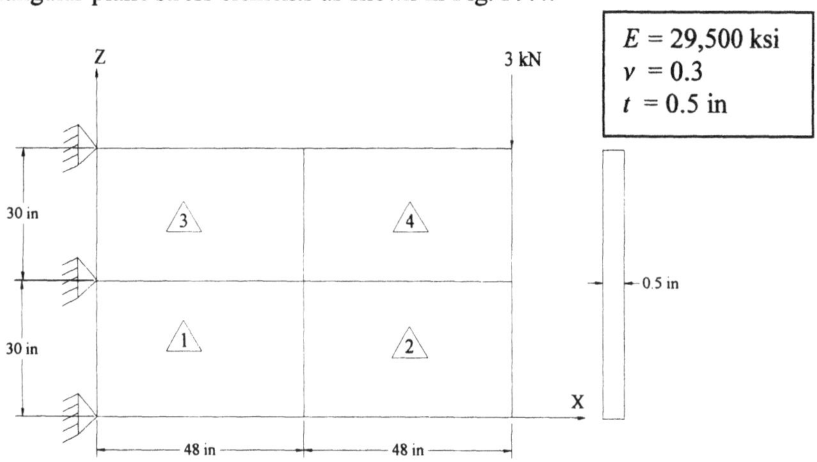

Fig. P9.4

Problem 9.5

The deep cantilever plate shown in Fig. P9.5 carries a distributed force per unit of area of magnitude $w = 0.10$ lb/in^2 applied in the opposite direction of the axis Z. Model this plate with six rectangular plane stress elements and use SAP2000 to perform the structural analysis.

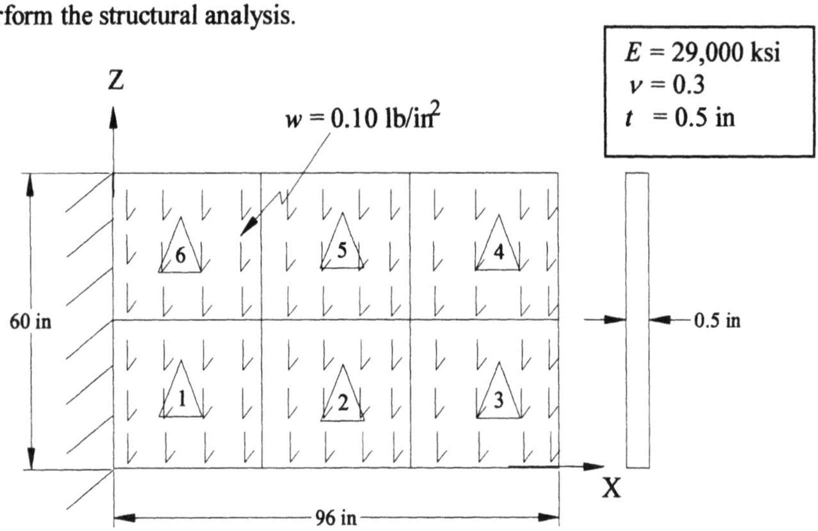

Fig. P9.5

Problem 9.6

Consider the rectangular steel plate of Illustrative Example 9.4, supporting a concentrated vertical force of magnitude $P = 8.0$ kips applied at the center of the plate downward in the vertical direction. Model this plate in bending with 16 square elements as in Illustrative Example 9.4 and use SAP2000 to determine deflections and stresses.

Problem 9.7

Solve Problem 9.6 for the distributed normal load of 0.01 kip/in^2 in addition to the concentrated force of 8 kips prescribed in Problem 9.6.

Note: The solution of Illustrative Example 9.4 (with load $w = 0.01$ kip/in^2) together with solutions of Problems 9.6 (with load $P = 8$ kips) and 9.7 (with loads $w = 0.01$ kip/in^2 and $P = 8$ kips) would allow the reader to verify the principle of superposition.

Appendix I

Equivalent Nodal Forces

Case (a): Concentrated Force

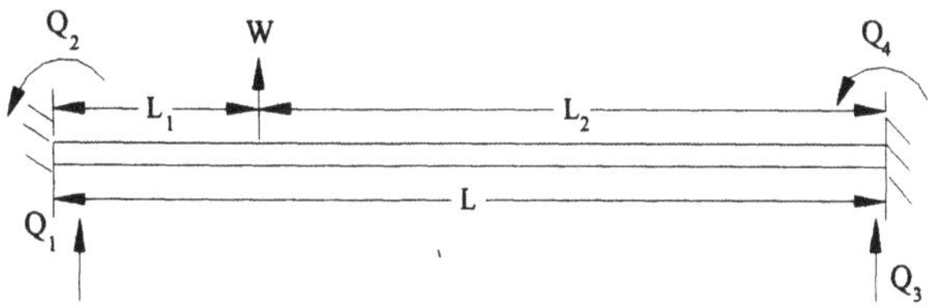

$$Q_1 = \frac{WL_2^2}{L^3}\left(3L_1 + L_2\right)$$

$$Q_2 = \frac{WL_1L_2^2}{L^2}$$

$$Q_3 = \frac{WL_1^2}{L^3}\left(L_1 + 3L_2\right)$$

$$Q_4 = -\frac{WL_1^2L_2}{L^2}$$

Case (b): Concentrated Moment

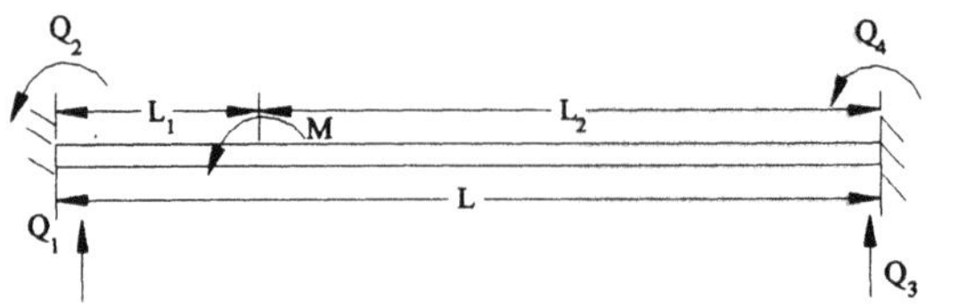

$$Q_1 = -\frac{6ML_1L_2}{L^3}$$

$$Q_2 = \frac{ML_2}{L^2}(L_2 - 2L_1)$$

$$Q_3 = \frac{6ML_1L_2}{L^3}$$

$$Q_4 = -\frac{ML_1}{L^2}(L_1 - 2L_2)$$

Case (c): Uniformly Distributed Force

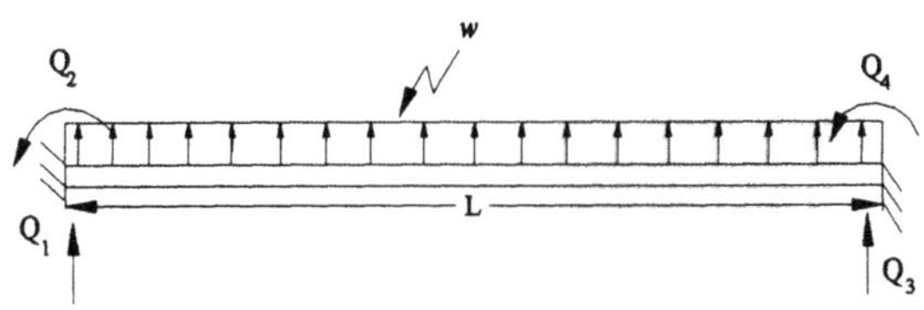

$$Q_1 = \frac{wL}{2}$$

$$Q_2 = \frac{wL^2}{12}$$

$$Q_3 = \frac{wL}{2}$$

$$Q_4 = -\frac{wL^2}{12}$$

Case (d): Trapezoidal Distributed Force

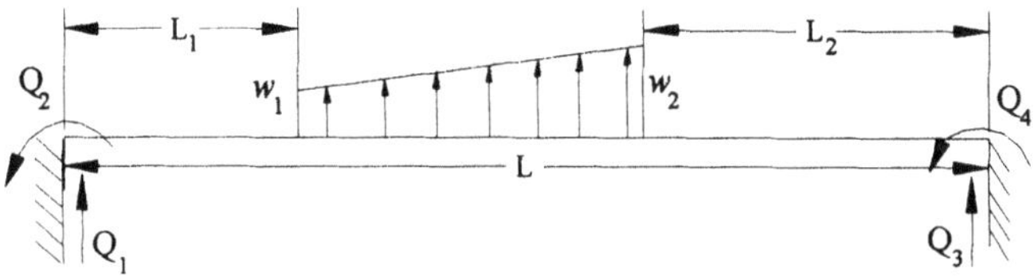

$$Q_1 = \frac{w_1(L-L_1)^3}{20L^3}\left\{(7L+8L_1) - \frac{L_2(3L+2L_1)}{L-L_1} \text{ x } \left[1+\frac{L_2}{L-L_1}+\frac{L_2^2}{(L-L_1)^2}\right] + \frac{2L_2^4}{(L-L_1)^3}\right\}$$

$$+\frac{w_2(L-L_1)^3}{20L^3}\left\{(3L+2L_1)\left[1+\frac{L_2}{L-L_1}+\frac{L_2^2}{(L-L_1)^2}\right] - \frac{L_2^3}{(L-L_1)^2}\left[2+\frac{15L-8L_2}{L-L_1}\right]\right\}$$

$$Q_2 = \frac{w_1(L-L_1)^3}{60L^2}\left\{3(L+4L_1) - \frac{L_2(2L+3L_1)}{L-L_1}\left[1+\frac{L_2}{L-L_1}+\frac{L_2^2}{(L-L_1)^2}\right] + \frac{3L_2^4}{(L-L_1)^3}\right\}$$

$$+\frac{w_2(L-L_1)^3}{60L^2}\left\{(2L+3L_1)\left[1+\frac{L_2}{L-L_1}+\frac{L_2^2}{(L-L_1)^2}\right] - \frac{3L_2^4}{(L-L_1)^3}\right\} + \frac{w_2(L-L_1)^3}{60L^2}$$

$$\left\{(2L+3L_1)\left[1+\frac{L_2}{L-L_1}+\frac{L_2^2}{(L-L_1)^2}\right] - \frac{3L_2^3}{(L-L_1)^2}\left[1+\frac{5L-4L_2}{L-L_1}\right]\right\}$$

$$Q_3 = \left(\frac{w_1+w_2}{2}\right)(L-L_1-L_2) - Q_1$$

$$Q_4 = \frac{L-L_1-L_2}{6}\left[w_1-(2L+2L_1-L_2) - w_2(L-L_1+2L_2)\right] + Q_1 L - Q_2$$

Case (e): Axial Concentrated Force

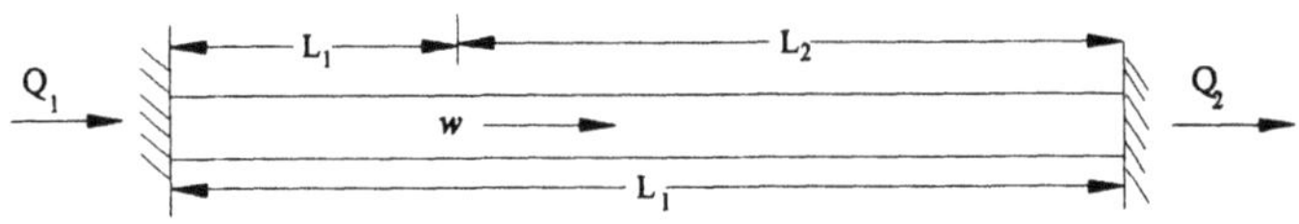

$$Q_1 = \frac{wL_2}{L}$$

$$Q_2 = \frac{wL_1}{L}$$

Case (f): Distributed Axial Force

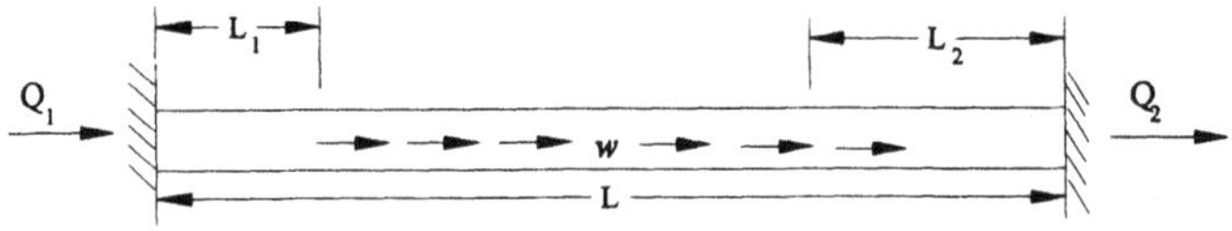

$$Q_1 = \frac{w}{2L}(L - L_1 - L_2)(L - L_1 + L_2)$$

$$Q_2 = \frac{w}{2L}(L - L_1 - L_2)(L + L_1 - L_2)$$

Case (g): Concentrated Torsional Moment:

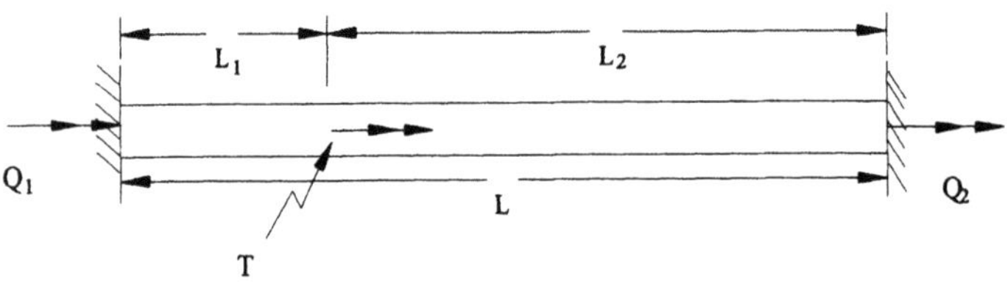

$$Q_1 = \frac{TL_2}{L}$$

$$Q_2 = \frac{TL_1}{L}$$

Appendix II

Displacement Functions for Fixed-End Beams

***Case (a): Concentrated Force**

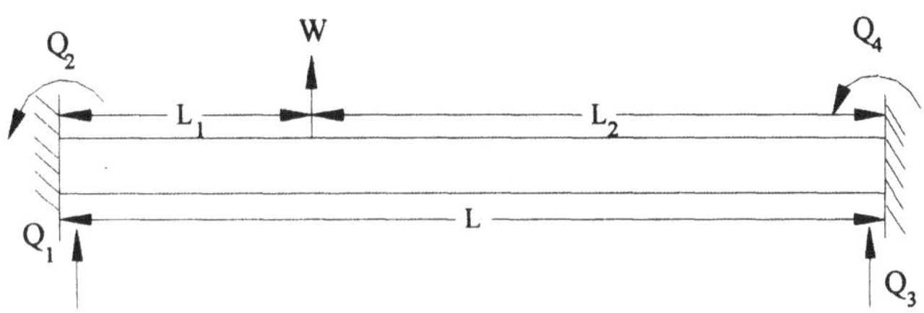

$$y_L(x) = \frac{1}{EI}[-\frac{1}{6}Q_1x^3 + \frac{1}{2}Q_2x^2] \qquad 0 \le x \le L_1$$

$$y_L(x) = \frac{1}{EI}\{\frac{1}{6}(w - Q_1)x^3 + \frac{1}{2}(Q_2 - wL)x^2 - [\frac{1}{2}(w - Q_1)L^2 + (Q_2 - wL_1)L]x + \frac{1}{3}(w - Q_1)L^3 + \frac{1}{2}(Q_2 - wL_1)L^2\} \qquad L_1 \le x \le L$$

***Case (b): Concentrated moment**

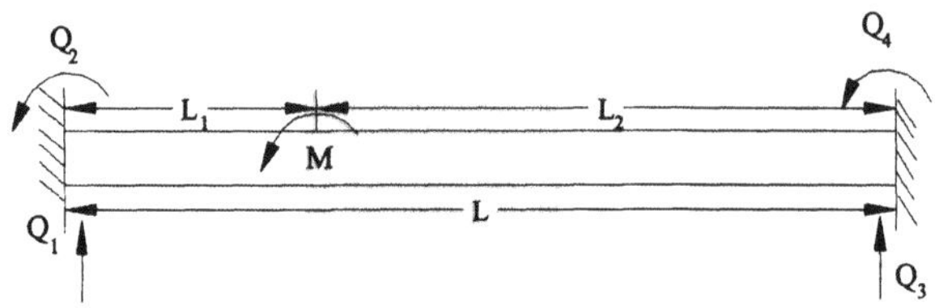

$$y_L(x) = \frac{1}{EI}\left[-\frac{1}{6}Q_1x^3 + \frac{1}{2}Q_2x^2\right] \qquad 0 \le x \le L_1$$

$$y_L(x) = \frac{1}{EI}\left[-\frac{1}{6}Q_1x^3 + \frac{1}{2}(Q_2 - M)\,x^2\right] \qquad L_1 \le x \le L$$

***Case (c): Uniformly Distributed Force**

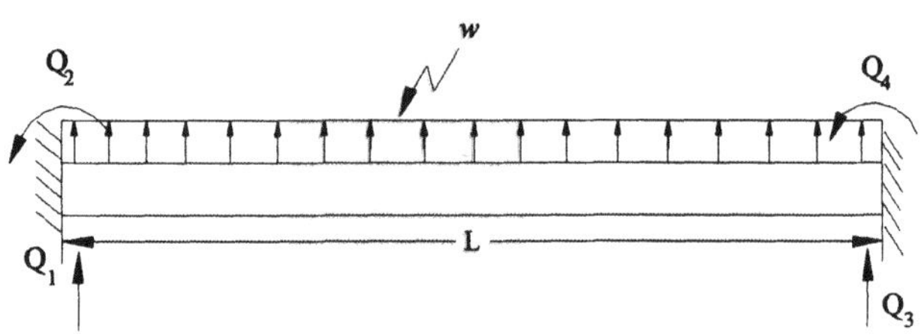

$$y_L(x) = \frac{1}{EI}\left[-\frac{1}{6}Q_1x^3 + \frac{1}{2}Q_2x^2 + \frac{1}{24}wx^4\right]$$

*Case (d): Trapezoidal Distributed Load

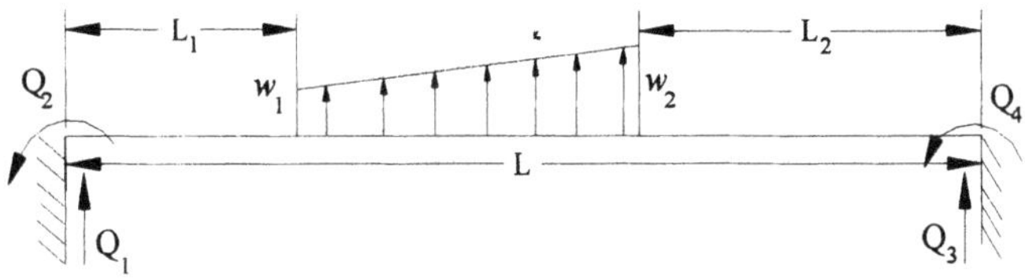

$$y_L(x)=\frac{1}{EI}\left[-\frac{1}{6}Q_1x^3+\frac{1}{2}Q_2x^2\right] \qquad 0\le x\le L_1$$

$$Y_L(x)=\frac{1}{EI}\{-\frac{1}{6}Q_1x^3+\frac{1}{2}Q_2x^2+\frac{w_1}{2}\left[\frac{x^4}{12}-\frac{L_1x^3}{3}+\frac{L_1^2x^2}{2}\right]+\frac{1}{6}\left[\frac{w_2-w_1}{L-L_1-L_2}\right]$$

$$\left[\frac{x^5}{20}-\frac{x^4L_1}{4}+\frac{x^3L_1^2}{2}-\frac{x^2L_1^3}{2}\right]-\frac{w_1L_1^3x}{6}+\left(\frac{w_2-w_1}{L-L_1-L_2}\right)\frac{L_1^4x}{24}+\frac{w_1L_1^4}{24}-\frac{l_1^5}{120}\left[\frac{w_2-w_1}{L-L_1-L_2}\right]\}$$

$$L_1\le x\le L-L_2$$

$$y_L(x)=\frac{1}{EI}\{-\frac{Q_1x^3}{6}+\frac{Q_2x^2}{2}+w_1(L-L_1)\left[\left(\frac{L-L_1-L_2}{4}\right)x^2+\frac{x^5}{20}\frac{(-L+L_2)}{2}x^2\right]$$

$$\frac{1}{2}(L-L_1-L_2)(w_2-w_1)\left[\left(\frac{L-L_1-L_2}{6}\right)x^2+\frac{x^5}{20}+\frac{(-L-L_2)}{2}x^2\right]+C_1x+C_2$$

$$L-L_2\le x\le L$$

where

$$C_1=\frac{Q_1L^2}{2}-Q_2L-w_1(L-L_1)\left[\frac{L^2-LL_1-L_{1L_2}}{2}-\frac{L^2}{2}+LL_2\right]-\frac{1}{2}(L-L_1-L_2)(w_2-w_1)$$

$$\left[\frac{L^2-LL_1-LL_2}{3}-\frac{L^2}{2}+LL_2\right]$$

$$C_2=C_1L+\frac{Q_1L^2}{2}-\frac{Q_2L^2}{2}-w_1(L-L_1)\left[\left(\frac{L-L_1-L_2}{4}\right)L^2+\frac{L^5}{5}+\frac{(L_1^3-L^2L_2)}{2}\right]$$

$$-\frac{1}{2}(L-L_1-L_2)(w_2-w_1)\left[\left(\frac{L-L_1-L_2}{6}\right)L^2+\frac{L^5}{20}+\left(\frac{-L^3-L^2L_2}{2}\right)\right]$$

Note: Expressions for the Equivalent Nodal Forces Q_1, Q_2, Q_3, *and* Q_4 for the trapezoidal load are given as Case (*d*) in Appendix I.

Glossary

Angle of Rolling --(Section 5.3) -- It is the angle by which the local axis y appears to have been rotated from the "standard" orientation. This standard orientation exists when the local plane, formed by the local axes x (centroidal axis of the beam element) and y (minor principal axis of the cross-section) is vertical, that is, parallel to the vertical global axis Z.

Assembly of the system stiffness matrix – (Section 1.5) – The process of transferring the coefficients of the element stiffness matrices to appropriate locations in the system or the global stiffness matrix. The appropriate locations is dictated by nodal coordinates in the structure assigned to the element nodal coordinates.

Beam – (Section 1.1) – A longitudinal structure supported at selected locations and carrying loads that are for the most part applied in a direction normal to the longitudinal direction of the beam.

Body force – (Section 9.2) – Distributed force per unit volume at an interior point of an element.

Differential equation of a beam – (Section 1.11) – The differential equation expressing the lateral displacement function of a loaded beam. For small deformations and considering only the effect of bending, this differential equation is given by

$$\frac{d^2y(x)}{dx^2} = \frac{M(x)}{EI} \qquad \text{(1.39) repeated}$$

in which

$y(x)$ = lateral displacement at coordinate x
$M(x)$ = bending moment at coordinate x
E = modulus of elasticity

I = cross-sectional moment of inertia

For a beam of uniform cross-sectional area , eq.(1.39) may be expressed as

$$\frac{d^4y}{dx^4} = \frac{w(x)}{EI} \qquad (1.1) \text{ repeated}$$

in which $w(x)$ is the applied external load per unit of length along the beam.

Direct Method – (Section 1.5)– A method by which the coefficients in the Element Stiffness Matrices of the structures are transferred and added appropriately to assemble the System Stiffness Matrix.

Elastic supports – (Section 1.14) – Linear elastic supports of a beam element may be considered in the analysis by simply adding the value of the spring constant (stiffness coefficient) to the corresponding coefficient in the diagonal of the system stiffness matrix.

Element displacement function – (Section 1.11) – The total lateral displacement function $y_T(x)$ for a beam element is given by

$$y_T(x) = y(x) + y_L(x) \qquad (1.38) \text{ repeated}$$

in which $y_L(x)$ is the displacement function of the loaded beam element assumed fixed at both ends (See App.II) and $y(x)$ is the lateral displacement function produced by displacements at the nodal coordinates of the beam element. The function $y(x)$ is given by

$$y(x) = N_1(x)\delta_1 + N_2(x)\delta_2 + N_3(x)\delta_3 + N_4(x)\delta_4 \qquad (1.6) \text{ repeated}$$

where $N_1(x)$, $N_2(x)$, $N_3(x)$ and $N_4(x)$ are the shape functions, respectively, corresponding to a unit displacement at the nodal coordinates δ_1, δ_2, δ_3 and δ_4 of the beam element.

Element force vector -- (Section 1.6) – A vector containing the element nodal forces either in reference to the global system of coordinates (X, Y, Z) or in reference to the local system of coordinates $(x, y, z\}$.

Element stiffness matrix – (Section 1.4) – A matrix relating the nodal displacements and the nodal forces at the nodal coordinates of an element (i.e. beam element). This relationship may be written in general as

$$\{P\} = [k]\{\delta\} \qquad (1.12) \text{ repeated}$$

in which

$\{P\}$ = element nodal forces
$\{\delta\}$ = element nodal displacements
$[k]$ = element stiffness matrix

The components of the vector $\{P\}$ and $\{\delta\}$ as well as the coefficients in the matrix $[k]$ will depend on the specific type of structure, such as Beam, Plane Frame, Space Frame, Plane Truss or Space Truss.

Equivalent nodal forces – (Section 1.3) – Forces at the nodal coordinates of a beam element producing the same displacements at the nodal coordinates as those displacements resulting from the loads applied on the element. The equivalent nodal forces are calculated as follows:

1. For loads applied on the elements: -- (Section 1.6) -- The equivalent nodal forces $\{Q\}$ may be calculated either from eq.(1.14) or by determining the element fixed-end reactions and reversing their direction. (See Appendix I)
2. For imposed displacements: (Section 1.10) The equivalent element nodal forces, $\{P\}_\Delta$ due to imposed nodal displacements, are calculated by
3.

$$\{P\}_\Delta = -[k]\{\Delta\} \qquad \text{1.37) repeated}$$

in which

$[k]$ = element stiffness matrix
$\{\Delta\}$ = imposed displacements at the element nodal coordinates.

4. For temperature changes: (Section 1.13) The equivalent vector for nodal forces $\{Q\}_T$ for a linear temperature variation on the cross-section of a beam element is given by
5.

$$\{Q\}_T = \begin{Bmatrix} 0 \\ \dfrac{-\alpha EI}{h}(T_2 - T_1) \\ 0 \\ \dfrac{\alpha EI}{h}(T_2 - T_1) \end{Bmatrix} \qquad \text{(1.51) repeated}$$

in which
T_1 = temperature at the bottom face of the beam
T_2 = temperature at the top face of the beam
α = coefficient of thermal expansion
h = height of the cross-sectional area

Finite Element Method (FEM) -- (Section 9.1) – A powerful method for the analysis of structures. The main fixtures of the FEM are:

1. Modeling the structure with elements such as beam segments, triangular or quadrilateral plates, shells and solid elements interconnected at selected points defining nodal displacements.
2. Establishing through the Element Stiffness Matrix the relationship between forces and displacements at the nodal coordinates of the elements using approximate interpolating displacement functions.
3. Transferring the coefficients in Element Stiffness Matrices to appropriate locations in the System Stiffness Matrix.
4. Solving the system stiffness equations for the unknown nodal displacements.
5. Determining the element nodal forces and stress distributions.

Fixed End Reactions *{FER}* – (Section 1.6) – These are the reactions of a loaded beam segment assumed to be fixed for translation and rotation at its two ends.

Global System of Nodal Coordinates (Barred) – (Section 1.1) – Element nodal coordinates in reference to the global system of coordinates.

Grid Frame – (Section 4.1) – It is a planar frame with the loads applied normally to the plane of the frame.

Interpolating functions – (Section 9.1) – These are functions (generally polynomial functions) used to approximate the displacements at an interior point of a finite element.

Matrix Structural Analysis – (Section 1.2) -- A method for the analysis of frame type structures (beams, frames, and trusses). The main features of this method are:

1. Modeling the structure into beam (or truss) elements interconnected at selected points defining nodal displacements.
2. Establishing through the Element Stiffness Matrix the relationship between forces and displacements at the nodal coordinate of the elements.
3. Transferring the coefficients in the Element Stiffness Matrices to appropriate locations in the System Stiffness Matrix.
4. Solving the System Stiffness Equations for the unknown nodal displacements.
5. Determining the element nodal end forces.

Member End releases – (Section 1.9) – Refers to the introduction of hinges at the ends of a beam element. The modified element stiffness matrices and modified nodal equivalent force vectors for releases (hinges) are given by the following equations:

Case 1: Beam element with a hinged node ①, eqs.(1.25) and (1.26)

Case 2: Beam element with a hinge at node ②, eqs.(1.27) and (1.28).

Case 3: Beam element with hinges at both ends, eqs.(1.29) and (1.30).

Nodal coordinates – (Section 1.2) – Designate the possible displacements at the nodes of an element or at the nodes of the structure. For example, for a plane frame, there are three nodal coordinates at each node or joint, two components of a linear displacement in the plane of the frame and a rotational displacement around an axis normal to that plane. Nodal coordinates are classified as follows :

1. Free nodal coordinates – (Section 1.2) – Nodal coordinates at which displacements are not constrained.
2. Fixed nodal coordinates – (Section 1.2) – Nodal coordinates at which displacements are constrained.

Nodes or Joints – (Section 1.2) – Points located at the ends of a beam element (member) or selected points joining elements in the structure.

Non-conforming function – (Section 9.4) – These are interpolating functions which may not satisfy fully the continuity conditions of displacements or slopes between adjacent finite elements modeling the structure.

Orthogonal Matrix -- (Section 3.5) – A matrix for which the transpose matrix is equal to its inverse.

Plane Frames – (Section 3.1) -- Structural frames in which the members as well as the loads are in the same plane.

Plane Trusses – (Section 6.1) -- Structures assembled of longitudinal members assumed to be connected at their ends by frictionless pins.

Plate bending – (Section 9.2) – A structural plate on which the loads are applied normally to the plane of the plate.

Primary Degrees of Freedom (or independent degrees of freedom) – (Section 8.2) – Those degrees of freedom selected for the analysis as independent displacements which are left after the reduction or condensation of the secondary or dependent degrees of freedom.

Principle of Virtual Work (Sections 1.15) – The Principle of Virtual Work states that for an elastic system in equilibrium, the work done by the external forces is equal to the work of the internal forces during an arbitrary virtual displacement compatible with the constraints of the structure.

Reduced system stiffness matrix – (Section 1.9) – A matrix $[K]_R$ establishing the relationship between the nodal displacements $\{u\}_R$ and the forces $\{F\}_R$ at the free nodal coordinates of the structure: $\{F\}_R = [K]_R\{u\}_R$.

SAP2000 – (Section 2.1) -- SAP2000 is an interactive, menu driven computer program for the analysis and design of structures. The student version used in this book is limited to the analysis of linear structures modeled with no more than 30 nodes or 30 elements.

Secondary Degrees of Freedom – (Section 9.2) – Those degrees of freedom selected to be condensed or reduced in order to define the system in terms of the selected primary degree of freedom.

Shape functions $N_i(x)$ – (Section 1.3) – These are functions giving the lateral displacements resulting from a unit of displacement applied to a nodal coordinate of a beam element.

Shear force and bending moment functions – (Section 1.12) – The shear force $V(x)$ and bending moment $M(x)$ functions along a loaded beam element are given, respectively, by

$$V(x) = P_1 + \int_0^x w(x)dx \qquad (1.43) \text{ repeated}$$

and

$$M(x) = -P_2 + P_1 x + \int_0^x x w(x)dx \qquad (1.44) \text{ repeated}$$

in which P_1 and P_2 are, respectively, the fixed-end force and the fixed-end moment reactions at the left node of a loaded beam element and $w(x)$ is the external load per unit of length applied along the beam element.

Skeletal structures – (Section 9.1) -- These are structures with unidirectional elements such as beams, frames or trusses.

Space Frame – (Section 5.1) – These are frames in which the members and the forces may be oriented in any direction of the three-dimensional space.

Space Trusses – (Section 7.1) -- Three dimensional structures with longitudinal members connected at their ends by hinges assumed to be frictionless.

Static Condensation (Section 8.1) – The process of reducing the number of free displacements or degrees of freedom.

Stiffness coefficients, k_{ij} – (Section 1.3) -- The forces at nodal coordinate "i" resulting from a unit displacement applied at nodal coordinates "j" with all other nodal coordinates fixed with no displacements.

Substructuring – (Section 8.1) – Is a method by which the structure is separated into component parts to be analyzed separately, thus facilitating the analysis of large or complex structures.

Support reactions – (Section 1.7) – Forces at the support of the structure. These forces may conveniently be determined from the end-member forces of those elements that are linked to a particular support.

System force vector – (Section 1.6) – A vector containing the forces at the nodal coordinates of the structure. This vector includes the forces applied directly at the nodes and the equivalent nodal forces for forces applied on the elements of the structure. It also includes equivalent nodal forces for displacements imposed at the nodes of the structure.

System of coordinates – (Section 1.1) -- Two reference systems of coordinates are used in Matrix Structural Analysis:

1. Global system of coordinates – (Section 1.1) -- A Cartesian system of coordinates with axes *X*, *Y* and *Z* fixed in the space and used to refer to any point in the structure.
2. Local system of coordinates (Section 1.1) -- A Cartesian system of coordinates with axes *x*, *y* and *z* fixed on a member or element of the structure and used to refer to any point on that element.

System stiffness matrix – (Section 1.7) – A matrix $[K]$ establishing the relationship between displacements $\{u\}$ and forces $\{F\}$ at the nodal coordinates of the structure:

$$\{F\} = [K]\{u\} \qquad (1.15) \text{ repeated}$$

Transformation Matrix -- (Section 3.5)– A matrix that transforms the element nodal displacements or forces in reference to the global system of coordinates to the local system of coordinates.

$$\{\bar{\delta}\} = [T]\{\delta\} \qquad (3.15) \text{ repeated}$$

and

$$\{\bar{P}\} = [T]\{P\} \qquad (3.17) \text{ repeated}$$

where

$\{\delta\}$ = nodal displacement vector in local coordinates
$\{\bar{\delta}\}$ = nodal displacement vector in global coordinates
$\{P\}$ = nodal force vector in local coordinates
$\{\bar{P}\}$ = nodal force vector in global coordinates
$[T]$ = transformation matrix

Selected Bibliography

- Bathe, K. J. (1982) *Finite Element Procedures in Engineering Analysis,* Prentice-hall, Englewood Cliffs, NJ.

- Guyan, R. J. (1965) *Reduction of Stiffness and Mass Matrices,* AIAA, J.13,380.

- Hibbeler, Russell C. (1995) *Structural Analysis, 3rd Edition,* Prentice-Hall, Englewood Cliffs, NJ.

- Kassimali, Aslam (1999) *Matrix Analysis of Structures,* Brooks/Cole Publishing Company, Pacific Grove, CA.

- McCormac, Jack C. and Nelson, James K., Jr. (1997) *Structural Analysis: A Classical and Matrix Approach,* Addison-Wesley, Reading, MA.

- Paz, Mario (1984) *Dynamic Condensation,* AIAA J. 22(5), 724-727.

- Paz, Mario (1997) 4th Edition, *Structural Dynamics: Theory and Computation,* Chapman and Hall, New York.

- Sennett, Robert C. (1994) *Matrix Analysis of Structures,* Prentice-Hall, Englewood Cliffs, NJ.

- Timoshenko, S. P. and Goodier, J. N. (1970) *3rd Edition, Theory of Elasticity,* McGraw-Hill, New York.

- Weaver, William, Jr. (1984) *Finite Element for Structural Analysis,* Prentice-Hall, Englewood Cliffs, NJ.

Index

ABOUT THE CD-ROM

The CD-ROM included with this book features the student version of SAP 2000, associated user manuals, and numerous sample problems and examples.

This CD-ROM will enhance student learning by the use of hands-on examples of real structural problems and solutions.

The CD requires either Windows® 95, 98, NT, or Windows® 2000.

MIX
Papier aus verantwortungsvollen Quellen
Paper from responsible sources
FSC® C105338

If you have any concerns about our products,
you can contact us on
ProductSafety@springernature.com

In case Publisher is established outside the EU,
the EU authorized representative is:
Springer Nature Customer Service Center GmbH
Europaplatz 3, 69115 Heidelberg, Germany

Printed by Libri Plureos GmbH
in Hamburg, Germany